Communications
in Computer and Information Science   **2269**

Series Editors

Gang Li⬤, *School of Information Technology, Deakin University, Burwood, VIC, Australia*
Joaquim Filipe⬤, *Polytechnic Institute of Setúbal, Setúbal, Portugal*
Ashish Ghosh⬤, *Indian Statistical Institute, Kolkata, West Bengal, India*
Zhiwei Xu, *Chinese Academy of Sciences, Beijing, China*

Rationale

The CCIS series is devoted to the publication of proceedings of computer science conferences. Its aim is to efficiently disseminate original research results in informatics in printed and electronic form. While the focus is on publication of peer-reviewed full papers presenting mature work, inclusion of reviewed short papers reporting on work in progress is welcome, too. Besides globally relevant meetings with internationally representative program committees guaranteeing a strict peer-reviewing and paper selection process, conferences run by societies or of high regional or national relevance are also considered for publication.

Topics

The topical scope of CCIS spans the entire spectrum of informatics ranging from foundational topics in the theory of computing to information and communications science and technology and a broad variety of interdisciplinary application fields.

Information for Volume Editors and Authors

Publication in CCIS is free of charge. No royalties are paid, however, we offer registered conference participants temporary free access to the online version of the conference proceedings on SpringerLink (http://link.springer.com) by means of an http referrer from the conference website and/or a number of complimentary printed copies, as specified in the official acceptance email of the event.

CCIS proceedings can be published in time for distribution at conferences or as post-proceedings, and delivered in the form of printed books and/or electronically as USBs and/or e-content licenses for accessing proceedings at SpringerLink. Furthermore, CCIS proceedings are included in the CCIS electronic book series hosted in the SpringerLink digital library at http://link.springer.com/bookseries/7899. Conferences publishing in CCIS are allowed to use Online Conference Service (OCS) for managing the whole proceedings lifecycle (from submission and reviewing to preparing for publication) free of charge.

Publication process

The language of publication is exclusively English. Authors publishing in CCIS have to sign the Springer CCIS copyright transfer form, however, they are free to use their material published in CCIS for substantially changed, more elaborate subsequent publications elsewhere. For the preparation of the camera-ready papers/files, authors have to strictly adhere to the Springer CCIS Authors' Instructions and are strongly encouraged to use the CCIS LaTeX style files or templates.

Abstracting/Indexing

CCIS is abstracted/indexed in DBLP, Google Scholar, EI-Compendex, Mathematical Reviews, SCImago, Scopus. CCIS volumes are also submitted for the inclusion in ISI Proceedings.

How to start

To start the evaluation of your proposal for inclusion in the CCIS series, please send an e-mail to ccis@springer.com.

Xijin Tang · Van Nam Huynh · Haoxiang Xia · Quan Bai

Editors

Knowledge and Systems Sciences

23rd International Symposium, KSS 2024
Hobart, TAS, Australia, November 16–17, 2024
Proceedings

 Springer

Editors
Xijin Tang 🆔
CAS Academy of Mathematics and Systems
Science
University of Chinese Academy of Sciences
Beijing, China

Haoxiang Xia 🆔
Dalian University of Technology
Dalian, China

Van Nam Huynh 🆔
Japan Advanced Institute of Science
and Technology
Nomi, Japan

Quan Bai 🆔
University of Tasmania
Hobart, TAS, Australia

ISSN 1865-0929   ISSN 1865-0937 (electronic)
Communications in Computer and Information Science
ISBN 978-981-96-0177-6   ISBN 978-981-96-0178-3 (eBook)
https://doi.org/10.1007/978-981-96-0178-3

Preface

The annual International Symposium on Knowledge and Systems Sciences aims to promote the exchange and interaction of knowledge across disciplines and borders to explore new territories and new frontiers. With over 20 years of continuous endeavors, attempts to strictly define knowledge science may be still ambitious, especially with the tremendous advances in information technologies, but a very tolerant, broad-based and open-minded approach to the discipline can be taken. Knowledge science and systems science can complement and benefit each other methodologically.

The first International Symposium on Knowledge and Systems Sciences (KSS 2000) was initiated and organized by the Japan Advanced Institute of Science and Technology (JAIST) in September of 2000. Since then our collective endeavours have resulted in KSS 2001 (Dalian), KSS 2002 (Shanghai), KSS2003 (Guangzhou), KSS 2004 (JAIST), KSS2005 (Vienna), KSS 2006 (Beijing), KSS 2007 (JAIST), KSS 2008 (Guangzhou), KSS 2009 (Hong Kong), KSS 2010 (Xi'an), KSS2011 (Hull), KSS 2012 (JAIST), KSS 2013 (Ningbo), KSS 2014 (Sapporo), KSS 2015 (Xi'an), KSS 2016 (Kobe), KSS 2017 (Bangkok), KSS 2018 (Tokyo), KSS 2019 (Da Nang), KSS 2022 (online and Beijing) and KSS2023 (Guangzhou).

Over the past two decades, KSS has become a crucial platform for researchers and scientists worldwide to share and exchange their latest findings in knowledge and systems sciences. This year, for the first time, KSS was held in Australia, on November 16–17, 2024. KSS 2024 was organized by the International Society for Knowledge and Systems Sciences and the University of Tasmania. The Systems Engineering Society of China was the co-organizer. Nowadays artificial intelligence, from deep learning to the more recent large language models, is significantly changing the way human beings produce knowledge (e.g. AI for Science) and solve problems (e.g. decision-making with decision-intelligence technologies). Correspondingly, the symbiosis and fusion of human intelligence and artificial intelligence is an extremely important topic of inquiry for the development of today's AI and its increasingly wide adoption in human societies. With the rise of Generative AI, and the challenges posed by these transformative technologies, the theme of KSS 2024 was centered on "Knowledge and Systems Sciences with Responsible AI". Three distinguished researchers were invited to deliver keynote speeches:

- **Feng Xia** (RMIT, Australia) – *"Anomaly Detection with Graph Learning"*
- **Sherry Xiwei Xu** (Data61, CSIRO, Australia) – *"Responsible Agentic Navigation of Knowledge: A Systems Perspective"*
- **Yi Zeng** (Institute of Automation, Chinese Academy of Sciences, China) – *"A Socio-Technical Approach to Building Safe and Ethical AI"*

KSS 2024 received 50 submissions from authors studying and working in Australia, China, Japan and Vietnam, and finally 23 submissions were selected for publication in the proceedings after a double-blind review process. The co-chairs of the International

Program Committee made the final decision for each submission based on the review reports from the referees, who came from Australia, China, Japan, New Zealand and Thailand. Each accepted submission received three reviews on average.

To enable KSS 2024 to happen, we received a lot of support and help from many people and organizations, including the China Association for Science and Technology and the International Federation for Systems Research. We would like to express our sincere thanks to the authors for their remarkable contributions, all the technical program committee members for their time and expertise in paper review with a very very tight schedule, and the proceedings publisher Springer for a variety of professional help. This is the 7th time the KSS proceedings are published as a CCIS volume after our successful collaboration with Springer during 2016–2019, 2022 and 2023. We greatly appreciate those three distinguished scholars for accepting our invitation to present keynote speeches at the symposium by hybrid modes. Last but not least, we are very indebted to the organizing group for their hard work.

We are happy with the thought-provoking and lively scientific exchanges in the essential fields of knowledge and systems sciences during the symposium.

November 2024

Xijin Tang
Van Nam Huynh
Haoxiang Xia
Quan Bai

Organization

General Chair

Xijin Tang   CAS Academy of Mathematics and Systems Science and University of Chinese Academy of Sciences, China

Program Committee Chairs

Van-Nam Huynh   Japan Advanced Institute of Science and Technology, Japan

Haoxiang Xia   Dalian University of Technology, China

Organizing Committee Chair

Quan Bai   University of Tasmania, Australia

Technical Program Committee

Quan Bai	University of Tasmania, Australia
Lina Cao	North China Institute of Science and Technology, China
Jindong Chen	Beijing Information Science & Technology University, China
Xuefan Dong	Beijing University of Technology, China
Yucheng Dong	Sichuan University, China
Chonghui Guo	Dalian University of Technology, China
Yuxuan Hu	University of Tasmania, Australia
Xiaohui Huang	Southern University of Science and Technology, China
Van-Nam Huynh	Japan Advanced Institute of Science and Technology, Japan
Jichao Li	National University of Defense Technology, China
Renjie Li	University of Tasmania, Australia

Weihua Li	Auckland University of Technology, New Zealand
Xiang Li	University of Tasmania, Australia
Yongjian Li	Nankai University, China
Zhenpeng Li	Taizhou University, China
Yan Lin	Dalian Maritime University, China
Peng Liu	Jiangsu University of Science and Technology, China
Shuo Liu	National University of Defense Technology, China
Xiaojun Liu	Beijing University of Technology, China
Yijun Liu	CAS Institute of Science and Development, China
Jinzhi Lu	Beihang University, China
Jing Ma	Auckland University of Technology, New Zealand
Jingli Shi	Auckland University of Technology, New Zealand
Xiaodan Wang	Yanbian University, China
Jiangning Wu	Dalian University of Technology, China
Shiqing Wu	University of Technology Sydney, Australia
Haoxiang Xia	Dalian University of Technology, China
Nuo Xu	Communication University of China, China
Xiaoying Xu	South China University of Technology, China
Zhihua Yan	Shanxi University of Finance and Economics, China
Yi Yang	Hefei University of Technology, China
Naimeng Yao	University of Tasmania, Australia
Dayong Ye	University of Technology Sydney, Australia
Soonja Yeom	University of Tasmania, Australia
Thaweesak Yingthawornsuk	King Mongkut's University of Technology Thonburi, Thailand
Wen Zhang	Beijing University of Technology, China
Yaru Zhang	China Mobile Information Technology Center, China
Zhen Zhang	Dalian University of Technology, China
Zike Zhang	Zhejiang University, China
Xiaochen Zheng	Southern University of Science and Technology, China

Keynotes

Anomaly Detection with Graph Learning

Feng Xia

School of Computing Technologies, STEM College, RMIT University, Australia
feng.xia@rmit.edu.au

Abstract. Anomaly detection is crucial because it allows us to proactively identify problems like fraud and equipment failure, but also uncovers hidden patterns that might signal exciting new discoveries or opportunities, making it a powerful tool for managing data across various fields. Traditional anomaly detection methods often struggle with the complexities of interconnected data, where anomalies can manifest as subtle deviations in node properties, edge connections, or even entire sub-graphs. The advent of graph learning techniques has revolutionized anomaly detection, empowering us to uncover subtle deviations and irregularities within complex networks. This talk explores the frontier of anomaly detection through the lens of graph learning. We will provide an overview of graph learning, highlighting its capacity to capture rich relational information inherent in diverse datasets. Drawing upon real-world applications, we will then showcase the versatility of graph learning techniques, particularly graph neural networks, in detecting both node-level and structural anomalies across diverse domains, such as knowledge graphs, social networks, and healthcare. The talk will conclude with a glimpse into the future of anomaly detection with graph learning, highlighting ongoing research and promising directions for this dynamic field.

Responsible Agentic Navigation of Knowledge: A System Perspective

Xiwei (Sherry) Xu

CSIRO Data61, Australia

Abstract. Generative AI (GenAI) have revolutionized the creation of new content, such as text, images, and music, by leveraging learned patterns from existing data. But their integration into complex software systems introduces challenges such as hallucination, outdated knowledge, and non-traceable reasoning processes. These issues raise significant concerns for responsible AI, particularly in ensuring the reliability and accountability of GenAI-generated content. Retrieval-Augmented Generation (RAG) has emerged as a promising approach to address these challenges by enabling GenAI systems to dynamically incorporate external knowledge sources.

This keynote will provide a system-level perspective on designing and implementing responsible RAG systems. It will focus on the critical decision points and trade-offs involved in building these systems, highlighting how they can be designed to meet the demands of responsible AI. Additionally, it will cover the latest advancements in agentic RAG, where GenAI agents autonomously orchestrate retrieval processes, integrating both passive knowledge bases and active domain experts. This approach enhances the factual accuracy and adaptability of GenAI systems, ensuring they operate within ethical and responsible frameworks.

A Socio-Technical Approach to Build Safe and Ethical AI

Yi Zeng

Institute of Automation, Chinese Academy of Sciences, China
yi.zeng@ia.ac.cn

Abstract. Building Safe and Ethical AI needs complementary approaches from both Sociology and Technology, both Protective thinking and Constructive thinking, both Reactive actions and Proactive actions. From the protective and reactive perspective, I will firstly introduce recent advancement on systematic AI Safety assessment, and how to quantify and control the balance between Safety and Capacity. In addition, I will discuss thinkings on AI safety redlines and the challenges to avoid catastrophic and existential risks. From the constructive and proactive perspective, I will introduce Safe and Moral AI, compared to AI Safety and Ethical AI, I will discuss Safety and Morality as first principles to develop AI, and will introduce preliminary results along this vision.

Contents

Knowledge Technologies and Systems Engineering

Knowledge Management

Complex Networks and Modeling

The Temporal Structural Pattern in Scientific Collaborative Behavior from the Perspective of Complex Network

Elina Zholdoshbaeva[1,2], Shuang Zhang[1,2], Feifan Liu[3], and Haoxiang Xia[1,2,4,5(✉)] (iD)

[1] Institute of Systems Engineering, Dalian University of Technology, No. 2 Linggong Road, Dalian 116024, Liaoning, China
hxxia@dlut.edu.cn
[2] Center for Big Data and Intelligent Decision-Making, Dalian University of Technology, Linggong Road, Dalian 116024, Liaoning, China
[3] School of Business, East China University of Science and Technology, 130 Meilong Road, Xuhui District 200237, Shanghai, China
[4] Institute for Advanced Intelligence, Dalian University of Technology, Linggong Road, Dalian 116024, Liaoning, China
[5] Key Laboratory of Social Computing and Cognitive Intelligence, Ministry of Education of China, Linggong Road, Dalian 116024, Liaoning, China

Abstract. The rapid advancement of science has underscored the importance of collaboration among scientists. While many studies have explored the structure of collaboration networks, there is a lack of comprehensive quantitative research on the temporal patterns of these collaborations. This study addresses this gap by analyzing temporal scientific collaboration networks in physics, spanning from 1960 to 2021, using complex network theories and methods. By constructing five-year temporal collaborative networks and employing metrics such as network diameter, shortest path, and clustering coefficient, this research reveals an increase in connectivity and small-world characteristics over time. Additionally, community detection highlights a shift towards specialized collaboration patterns and increased cross-community interactions. The study also uncovers growing centrality and social stratification within the research system, with the "rich club" phenomenon indicating strong and dynamic collaborations among highly connected scientists. Temporal analysis shows that while ultra-high connectivity scientists engage in fluctuating relationships, medium-high connectivity scientists maintain more stable collaborations. These findings provide insights into the evolution of collaborative behaviors and inform science policy.

Keywords: Temporal rich-club phenomenon · Scientific collaboration network · Network analysis · Co-authorship patterns · Temporal network dynamics

© The Author(s), under exclusive license to Springer Nature Singapore Pte Ltd. 2025
X. Tang et al. (Eds.): KSS 2024, CCIS 2269, pp. 3–14, 2025.
https://doi.org/10.1007/978-981-96-0178-3_1

1   Introduction

In contemporary science, collaboration among researchers is crucial for promoting progress and innovation [1,2]. The intersection of disciplines has made these relationships more complex and diverse, necessitating a deeper exploration of scientific collaboration patterns and dynamics to enhance cooperation and formulate science policies. The interdisciplinary field of the science of science combines insights from sociology, economics, and information science to investigate how collaboration influences knowledge development and dissemination [3].

Science can be viewed as a complex, self-organizing network of scholars, projects, papers, and ideas [4,5]. This perspective reveals patterns that characterize the emergence of new scientific fields and the path of impactful discoveries. Collaborative research networks, where scientists are nodes and co-authorship relationships are edges, quantitatively reveal overall collaboration patterns. These networks highlight the importance of collective effort in driving research breakthroughs, accelerating idea exchange, and fostering innovation [6,7].

A critical element within these networks is the "rich-club" phenomenon, where high-profile nodes form tightly-knit groups [8]. Prominent scientists tend to collaborate closely, concentrating core resources and shaping the scientific agenda [9,10]. Understanding this hierarchical organization is key to grasping the influential role of top researchers [11].

As scientific fields evolve, collaboration patterns change, resulting in distinct structures of collaboration networks [12]. Recent advances in large-scale academic data and complex network analysis provide opportunities to investigate the temporal patterns of collaboration, particularly the evolution of "rich club" structure. Measures of network heterogeneity reveal centralization and potential fragility, accentuating the "rich-club" phenomenon where elite scientists form densely interconnected clusters with far-reaching implications for research dissemination and innovation [13].

However, a critical gap remains in the systematic, quantitative exploration of the temporal dynamics of collaboration networks, particularly regarding the "rich club" phenomenon. Exploring temporal collaborative patterns is essential for understanding the dynamic interactions among scientists and the mechanisms driving scientific collaboration evolution and impact [14]. It highlights how diverse scientific collaborations address complex challenges and pave the way for revolutionary discoveries and technological progress [15].

Addressing this gap necessitates comprehensive, large-scale analysis focusing on the temporal patterns of collaboration networks, with an emphasis on unraveling the intricacies of the "rich club" structure. Such an endeavor would bridge the divide between individual-level behaviors and macro-level network properties, enriching our understanding of the dynamics that drive scientific innovation and knowledge dissemination across disciplines. Additionally, it underscores strategic collaboration's significance in enhancing research effectiveness and advancing collective knowledge.

2 Material and Methods

2.1 Dataset

We utilized a dataset from the American Physical Society (APS) journals, sourced from the Microsoft Academic Graph (MAG), and disambiguated author names, covering publications from 1960 to 2021 [16]. This dataset focuses on physics, offering extensive historical and disciplinary coverage. Physics was chosen for its robust data on collaborative patterns and its influence on fundamental science and other domains. The APS journals, renowned globally for their quality and openness to diverse contributions, include leading titles like Physical Review Letters and Reviews of Modern Physics. This comprehensive dataset comprises 692,839 papers authored by 775,990 distinct authors, providing a rich resource for analyzing the evolution of scientific collaboration over several decades. The dataset's scale and longevity enable us to map the historical landscape of physical science, understand future trends, and contribute insights into the dynamics of scientific impact and collaborative networks within physics.

2.2 Methods

Network Construction of Physicist's Collaboration. In this study, we employed a comprehensive data preprocessing methodology aimed at constructing an enriched dataset for analyzing collaboration networks within the domain of physics. This methodology involves the usage of academic paper records from distinct source:APS dataset and the MAG. Initially, we performed data cleansing operations, including the normalization of identifier fields, specifically by converting the 'id' attributes of APS records to lowercase, to ensure consistency across datasets. Subsequently, we conducted a systematic matching process, wherein APS papers from the year 2021 were identified within the MAG dataset, followed by the extraction of authorship information. This integrative approach not only facilitates the comprehensive mapping of physics collaboration networks but also enhances the reliability of subsequent network analyses by leveraging the strengths and breadth of data.

In the context of our extensive study on the physicists' collaboration networks, the data spanning from 1960 to 2021 was meticulously constructed from lists of papers, inclusive of authors' names, paper titles, abstracts, dates, and references to journal publications. Given the evolution of scientific collaborations over time, the dataset was segmented into different periods for a more granular analysis. The dataset was collated in five-year intervals, reflecting a more dense and rich set of information.

Complex Network Analysis. Collaboration networks in physics offer insights into interconnectedness and interdependence across research groups. Network analysis provides tools to uncover structural and dynamic aspects crucial for scientific advancement: Centrality measures like Closeness Centrality [17] identify pivotal researchers facilitating efficient knowledge dissemination.

$$C_C(\vartheta) = \frac{N-1}{\sum_{u \in V} d(\vartheta, u)}$$

quantifies how quickly information spreads through the network, crucial for cohesive scientific communities.

Betweenness Centrality [18] highlights researchers bridging disparate groups, essential for integrating subdomains. Algorithmic advancements like Brandes' algorithm

$$C_B(\vartheta) = \sum_{s,t \in V} \frac{\sigma(s,t|\vartheta)}{\sigma(s,t)}$$

enable scalable computation, revealing key figures in information flow.

The modularity metric via the Louvain method [19,20] detects cohesive groups, illuminating network dynamics over time.

Examining clustering coefficient C and average path length L reveals efficient information flow within densely connected physics networks. The small-world [21] coefficient

$$\sigma = \frac{(C/C_{\text{random}})}{(L/L_{\text{random}})}$$

indicates balanced clustering and global reach, fostering collaborative environments.

Network diameter and radius [22] assess overall connectivity and reachability. These metrics provide insights into network robustness and potential vulnerabilities to targeted disruptions.

Degree distribution [23,24] analysis identifies hubs and network resilience. Comparing with theoretical distributions elucidates network formation principles and growth dynamics.

Network density

$$D = \frac{2E}{N(N-1)}$$

and average degree [25]

$$\bar{k} = \frac{1}{N} \sum_{i=1}^{N} k_i$$

metrics quantify interconnectivity and collaboration intensity within physics research. They reflect network efficiency and innovation potential.

Network analysis offers a comprehensive view of physics collaboration networks, highlighting structural patterns and dynamics crucial for fostering scientific innovation and community cohesion.

Temporal Rich-Club Coefficient of the Physicist's Collaboration Network. A recent advancement in the science of science area is the introduction of the temporal rich club phenomenon [26]. This concept involves quantitatively analyzing the tendency of well-connected nodes to form simultaneous and stable structures in temporal networks [27] . This measure is crucial in understanding the dynamics of various natural, technological, and social systems, including communication and transportation infrastructures, biological and ecological systems, and social interactions. In studying the temporal dynamics of scientific collaborations, we construct a temporal network, conceptualized as a sequence of distinct network snapshots spanning the time interval [1, T]. Each snapshot captures the collaborative interactions, or temporal edges, between pairs of researchers at a given timestamp. By aggregating these temporal interactions over the entire period, we form a static network G = (V, E) [28] In this network, an edge is established between two researchers if they have collaborated at least once. The degree k of a node in this network represents the count of unique collaborators for a researcher. Our objective is to examine whether scientists with a high number of collaborations (high-degree nodes) exhibit a propensity to collaborate more frequently and for longer durations than random chance would suggest [29]. This phenomenon is referred to as the temporal rich-club (TRC) effect.

To quantify the TRC effect, we introduce the TRC coefficient $M(k, \Delta)$, which measures the highest density of collaborations that persist over a duration Δ among scientists with a degree greater than k [30]. This coefficient helps determine if observed collaboration patterns among highly connected scientists are genuinely interlinked over time or merely coincidental.

For validation, the TRC coefficient is compared with that derived from a randomized model of the network, denoted $M_{\mathrm{ran}}(k, \Delta)$. The ratio $\mu(k, \Delta) = \frac{M(k,\Delta)}{M_{\mathrm{ran}}(k,\Delta)}$ indicates the extent to which scientists of higher degrees are interconnected over a time interval Δ compared to random expectation. Our randomization approach maintains the original network's temporal edge count, achieved by shuffling the timestamps of collaborations.

Furthermore, we investigate the temporal evolution of the network's cohesion, $\epsilon_{>k}(t, \Delta)$, to identify periods of peak connectivity among highly collaborative scientists and evaluate the stability of these connections [31]. This enables us to discern between enduring and ephemeral rich-club phenomena in scientific collaborations.

In essence, this methodology integrates temporal network analysis with static network measures to reveal the underlying structure and temporal characteristics of scientific collaboration networks. By doing so, it offers insights into the patterns of collaboration among highly connected researchers, contributing to our understanding of the dynamics within scientific communities.

Null Model. We analyze scientific collaboration networks in the physics community using randomization techniques to distinguish real structures from ran-

dom ones. Using the Erdös-Rényi model [32] and Gauvin et al.'s temporal shuffling [33], we create null models to validate network properties.

Null models help establish the significance of network features like diameter,
degree distributions, and clustering coefficients. Gauvin et al.'s method incorporates timing, allowing us to explore how collaboration timing impacts network
dynamics. By comparing real data to temporally randomized models, we determine if observed patterns are deliberate or random. This framework uncovers
organized, non-random features driving scientific collaboration, aiding strategies
for fostering innovative scientific ecosystems.

3 Results

3.1 Basic Network Characteristics of the Physics Collaboration

The analysis of the Physics Collaboration Network, as presented in Table 1,
reveals a significant evolution in its structure and properties over time. The
number of nodes and edges grows markedly from 1,497 nodes and 4,938 edges
at t0 to 173,250 nodes and 1,118,646 edges at t11. This increase indicates a substantial expansion in the number of participating physicists and collaborations.
The average degree also rises from 6.597 to 12.913, showing that on average,
each physicist collaborates with more peers as time progresses.

Despite the network's growth, its density decreases from 4.4×10^{-3} to
7.5×10^{-5}, suggesting that while more connections are formed, the proportion of
possible connections that are realized diminishes, a common trait in large-scale
networks. The network's diameter and radius also exhibit interesting changes:
the diameter initially fluctuates but decreases from 35 to 19, while the radius
reduces from 18 to 11. These changes imply that the network becomes more
interconnected and efficient over time.

The modularity and number of communities provide insights into the network's mesoscopic structure. The high modularity, peaking at 0.940 and later
decreasing to 0.774, reflects strong community structures that become more
consolidated over time. The number of communities increases from 27 to 108,
indicating the formation of more distinct research groups. The average clustering coefficient remains high, ranging from 0.655 to 0.735, highlighting a strong
propensity for local clustering and group formation.

Moreover, the average path length decreases significantly from 12.224 to
6.193, demonstrating enhanced efficiency in information flow within the network.
The small-world coefficient's dramatic rise from 47.966 to 7725.96 underscores
the network's evolution into a highly interconnected small-world structure, facilitating efficient communication pathways.

However, the random network with the same size exhibits smaller diameters
and radii, indicating a more compact structure. Its modularity is significantly
lower, revealing weaker community structures compared to the physics collaboration network. The number of communities is also slightly lower, reflecting
less distinct clustering. The average clustering coefficient is much lower, ranging
from 0.367 to 0.197, indicating reduced local clustering and group formation.

Table 1. Properties of the real-world network of the physics collaboration

	Nodes	Edges	Avg. degree	Density	Diameter	Radius	Modularity	No. comm.	Avg. clustering coef.	Avg. path length	Small-world coef.
t_0	1497	4938	6.597	$4.4 * 10^{-3}$	35	18	0.901	27	0.655	12.224	47.966
t_1	4615	15274	6.619	$1.4 * 10^{-3}$	33	18	0.932	49	0.675	12.447	159.876
t_2	8399	28858	6.872	$8.2 * 10^{-4}$	30	18	0.940	69	0.691	12.590	273.688
t_3	13520	54466	8.057	$6.0 * 10^{-4}$	32	17	0.932	88	0.727	10.517	662.384
t_4	18745	75434	8.048	$4.3 * 10^{-4}$	27	15	0.908	94	0.725	9.378	1084.94
t_5	33411	136285	8.158	$2.4 * 10^{-4}$	30	16	0.898	114	0.726	8.402	1756.78
t_6	50403	211403	8.389	$1.7 * 10^{-4}$	26	14	0.895	103	0.735	8.188	3610.85
t_7	69865	309988	8.874	$1.3 * 10^{-4}$	25	14	0.871	120	0.728	7.672	4150.40
t_8	83825	395645	9.440	$1.1 * 10^{-4}$	21	12	0.847	96	0.723	7.252	4173.16
t_9	102541	534733	10.430	$1.0 * 10^{-4}$	22	12	0.827	101	0.723	6.884	5323.73
t_{10}	113888	658938	11.572	$1.0 * 10^{-4}$	24	14	0.803	108	0.729	6.624	5588.55
t_{11}	173250	1118646	12.913	$7.5 * 10^{-5}$	19	11	0.774	108	0.735	6.193	7725.96

The average path length in the random network is shorter, underscoring its more compact nature. These comparative results highlight the physics collaboration network's higher modularity, clustering coefficients, and average path lengths, indicating a more structured and complex network. The real-world network demonstrates strong community structures and efficient information flow, characteristic of collaboration networks. The trend towards increased interconnectedness and a small-world structure over time, as evidenced by the small-world coefficient and decreasing average path length, further distinguishes the physics collaboration network from its random counterpart.

This analysis underscores the differences between real-world collaboration networks and random networks, with the former exhibiting characteristics that support robust collaboration and efficient knowledge dissemination.

3.2   Temporal Rich-Club Coefficient(TRC) of the Physicist's Collaboration Network

Detecting the Temporal Rich Club. Our investigation of the physics collaboration network reveals significant insights into the temporal rich club phenomenon. Figure 1 shows that the degree distributions follow a heavy-tailed, scale-free pattern, indicating a few highly connected hubs within the network. These hubs are crucial for network functionality and resilience. The $E(k)$ and $N(k)$ metrics highlight a power-law distribution, pointing to heterogeneity in node connectivity. Steeper distributions suggest networks with a few highly influential nodes.

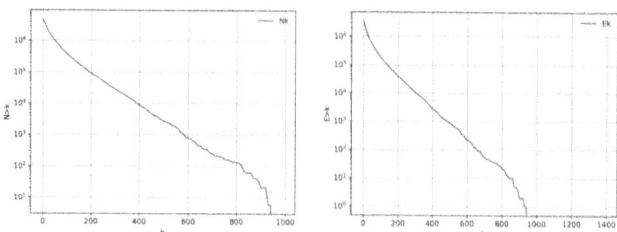

Fig. 1. Number of connections between nodes and edges with an aggregate degree $> k$

Figure 2 tracks the rich-club coefficient over time, illustrating dynamic collaboration patterns. At lower degree thresholds, the rich-club coefficient remains modest, indicating limited cohesion among less connected physicists. Higher thresholds exhibit fluctuations, with peaks suggesting periods where highly connected researchers form tightly knit communities.

Figure 3's heatmap analysis uncovers different collaboration dynamics. Mid-range degree thresholds (700–800) display stable and persistent collaborations, characteristic of long-term research partnerships. In contrast, high-degree thresholds (900) show intense but transient collaborations, likely reflecting short-term projects or specific scientific endeavors.

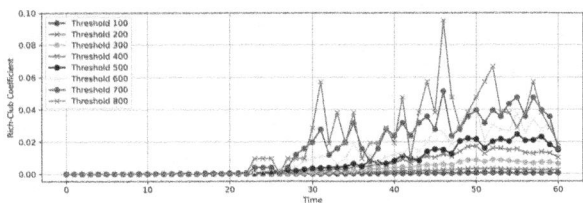

Fig. 2. Line graph of the rich-club coefficient over time, with different degree threshold

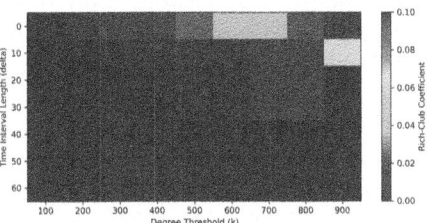

Fig. 3. Rich-club coefficient across different degree thresholds and delta

These findings highlight the network's reliance on hubs for stability and efficiency. Figure 4 further demonstrates the rich-club ratio across various degree thresholds and time intervals. Lower degree thresholds show high rich-club ratios, indicating significant temporal cohesion. Higher thresholds reveal a more complex interplay, with some intervals approaching the randomness of a baseline network.

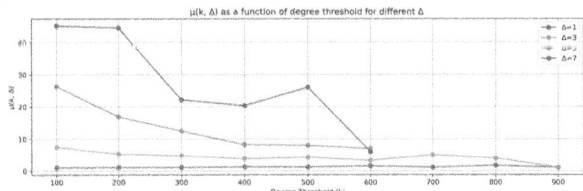

Fig. 4. The ratio $\mu(k, \Delta)$ between $M(k, \Delta)$, computed for the data, and $M_{\mathrm{ran}}(k, \Delta)$, for different values of Δ

Overall, the analysis underscores the importance of both stable and dynamic clusters within the network, crucial for understanding and maintaining its structural integrity. The network's adaptive nature supports evolving scientific collaborations, driven by internal and external factors.

4   Conclusion

Analyzing temporal behavioral patterns in scientific collaboration networks is crucial for understanding the dynamic nature and intricate structures of these networks. Initially mapping the evolution of these networks reveals broad trends in growth, diversification, and the impact of external factors on collaboration. This is followed by a focus on the rich club phenomenon, highlighting influential clusters of highly connected individuals who dominate information flow and significantly contribute to network resilience and efficiency.

In this study, a comprehensive exploration into the temporal behavioral patterns of physicists' collaboration networks from 1960 to 2021, using data from over 692,839 papers and 775,990 unique contributors, reveals significant growth and increased interconnectedness. The network evolved from loose associations to a structured web, with visualizations depicting the consolidation of research communities. On a microscopic level, the network exhibits a small-world nature, characterized by efficient information flow and tight-knit group formation, with dynamic core stability across different node thresholds.

The following temporal rich club analysis highlights the rich club phenomenon, where highly connected nodes form tightly-knit groups exerting substantial influence over the network's function and evolution. The analysis reveals a persistent rich-club phenomenon with a dynamic rich-club coefficient over time.

The study of rich club ordering in physicist collaboration networks provides insights into enhancing collaboration, innovation, and efficiency. Understanding and applying the principles of these complex and temporal collaboration patterns can help various disciplines develop strategies to optimize their networks for better outcomes. However, all the results are limited to the field of physics. Future research should expand the analysis to other disciplines to explore the generality of the obtained patterns. Additionally, the future plan could delve deeper into the underlying mechanisms of the phenomena and their specific implications for science policy. Nevertheless, this study provides a comprehensive understanding of both macroscopic trends and microscopic intricacies of influence concentration, guiding policymakers, institutions, and the scientific community in fostering robust and innovative collaborations.

Acknowledgements. This work is supported by the National Natural Science Foundation of China (Grant No. 71871042 and 72371052).

Disclosure of Interests. The authors have no competing interests to declare that are relevant to the content of this article.

References

1. Wuchty, S., Jones, B.F., Uzzi, B.: The increasing dominance of teams in the production of knowledge. Science **316**(5827), 1036–1039 (2007)
2. Scarazzati, S., Wang, L.: The effect of collaborations on scientific research output: the case of nanoscience in chinese regions. Scientometrics **121**(4), 839–868 (2019)
3. Milojević, S.: Principles of scientific research team formation and evolution. Proc. Natl. Acad. Sci. **111**(11), 3984–3989 (2014)
4. Zeng, A., Shen, Z., Zhou, J., et al.: The science of science: from the perspective of complex systems. Phys. Rep. **714**(2), 1–73 (2017)

5. Fortunato, S., et al.: Science of science. Science **359**(6379), eaao185 (2018)
6. Jones, B.F., Wuchty, S., Uzzi, B.: Multi-university research teams: shifting impact, geography, and stratification. Science **322**(5905), 1259–1262 (2008)
7. Barabási, A.L., Jeong, H., Néda, Z., et al.: Evolution of the social network of scientific collaborations. Phys. A **311**(3–4), 590–614 (2002)
8. Colizza, V., Flammini, A., Serrano, M.Á., Vespignani, A.: Detecting rich-club ordering in complex networks. Nat. Phys. **2**(2), 110–115 (2006)
9. Szell, M., Sinatra, R.: Research funding goes to rich clubs. Proc. Natl. Acad. Sci. U.S.A. **112**(48), 14749–14750 (2015)
10. Ma, A., Mondragón, R.J., Latora, V.: Anatomy of funded research in science. Proc. Natl. Acad. Sci. U.S.A. **112**(48), 14760–14765 (2015)
11. Yan, E., Ding, Y.: Scholarly network similarities: how bibliographic coupling networks, citation networks, co-citation networks, topical networks, coauthorship networks, and co-word networks relate to each other. J. Am. Soc. Inform. Sci. Technol. **63**(7), 1313–1326 (2012)
12. Ajiferuke, I., Grácio, M., Yang, S.: Editorial: research collaboration and networks: characteristics, evolution and trends. Front. Res. Metrics Anal. **6**, 690986 (2021)
13. Nakajima, K., Shudo, K., Masuda, N.: Higher-order rich-club phenomenon in collaborative research grant networks. Scientometrics **128**(2), 2429–2446 (2023)
14. Petersen, A.M., Pavlidis, I., Semendeferi, I.: Together we stand. Nat. Phys. **10**(11), 700–702 (2014)
15. Shi, F., Foster, J.G., Evans, J.A.: Weaving the fabric of science: dynamic network models of science's unfolding structure. Soc. Networks **43**, 73–85 (2015)
16. Wang, K., et al.: A Review of Microsoft academic services for science of science studies. Front. Big Data **2**, 45 (2019). https://doi.org/10.3389/fdata.2019.00045
17. Bavelas, A.: Communication patterns in task-oriented groups. J. Acoust. Soc. Am. **22**(6), 725–730 (1950)
18. Brandes, U.: A faster algorithm for betweenness centrality. J. Math. Sociol. **25**(2), 163–177 (2001)
19. Newman, M.E.: Modularity and community structure in networks. Proc. Natl. Acad. Sci. U.S.A. **103**(23), 8577–8582 (2006)
20. Blondel, V.D., Guillaume, J.L., Lambiotte, R., Lefebvre, E.: Fast unfolding of communities in large networks. J. Stat. Mech: Theory Exp. **2008**(10), P10008 (2008)
21. Watts, D.J., Strogatz, S.H.: Collective dynamics of 'small-world' networks. Nature **393**(6684), 440–442 (1998)
22. Hage, P., Harary, F.: Eccentricity and centrality in networks. Soc. Networks **17**(1), 57–63 (1995)
23. Newman, M.E.: The structure and function of complex networks. SIAM Rev. **45**(2), 167–256 (2003)
24. Boccaletti, S., Latora, V., Moreno, Y., Chavez, M., Hwang, D.J.: Complex networks: structure and dynamics. Phys. Rep. **424**(4–5), 175–308 (2006)
25. Caldarelli, G., Capocci, A., De Los Rios, P., Muñoz, M.A.: Scale-free networks from varying vertex intrinsic fitness. Phys. Rev. Lett. **89**(25), 258702 (2002)
26. Pedreschi, N., Battaglia, D., Barrat, A.: The temporal rich club phenomenon. Nat. Phys. **18**(8), 931–938 (2022)
27. Holme, P.: Modern temporal network theory: a colloquium. Eur. Phys. J. B **88**(9), 234 (2015)
28. Pan, R., Saramáki, J.: Path lengths, correlations, and centrality in temporal networks. Phys. Rev. E **84**(1), 016105 (2011)

29. Williams, O.E., Lengyel, I., Di Clemente, R.: Measuring the temporal stability of inter-firm collaboration networks. Front. Big Data **2**, 1–11 (2019)
30. Vestergaard, C.L., Génois, M., Barrat, A.: How memory generates heterogeneous dynamics in temporal networks. Phys. Rev. E **90**(4), 042805 (2014)
31. Gallos, L.K., Makse, H.A., Sigman, M.: A small world of weak ties provides optimal global integration of self-similar modules in functional brain networks. Proc. Natl. Acad. Sci. **109**(8), 2825–2830 (2012)
32. Erdős, P., Rényi, A.: On the evolution of random graphs. Trans. Am. Math. Soc. **286**, 257–257 (1984)
33. Gauvin, L., Panisson, A., Cattuto, C.: Detecting the community structure and activity patterns of temporal networks: a non-negative tensor factorization approach. PLoS ONE **8**(1), e86028 (2013)

The Analysis of Innovation Network in China's Hydrogen Energy Industry from the Perspective of Patents

Qi Zhong and Yuxuan Jin[✉]

Dongbei University of Finance and Economics, Dalian 116025, China
jyxuan10@163.com

Abstract. As an essential component of China's future national energy system, analyzing the technological innovation of hydrogen energy is important in promoting the green and low-carbon transformation of energy production and consumption. This paper combines the technology life cycle theory and the social network method to construct the IPC co-occurrence network of upstream, midstream and downstream in China's hydrogen energy industry based on the patent data from 1985 to 2021, and analyzes the innovation entities, hot and core innovation fields. The results show that: (1) China's hydrogen energy industry innovation is in a period of rapid growth, and is expected to enter the maturity period in 2028. (2) With the development evolution, the hydrogen energy industry has formed a technological innovation pattern dominated by enterprises and supplemented by universities and research institutions. (3) At different stages of development, there are big differences in the evolution of the hydrogen energy industry's upstream, midstream, and downstream technical fields.

Keywords: Network analysis · Visual analytics · Hydrogen energy industry · IPC co-occurrence · Technology life cycle

1 Introduction

Since the 20th century, fossil energy—primarily coal, oil, and natural gas, have been depleted. Under the global trend of green and low-carbon technology transformation, Hydrogen energy has sparked a global development boom due to its abundant resources, renewable and pollution-free nature, and diverse application scenarios. China's hydrogen energy industry started late but is now developing rapidly and has formed a complete hydrogen energy industry chain. The hydrogen energy industry serve as the key development target for strategic emerging industries in China, its development is significant in adjusting the energy structure, promoting energy production, and building an ecological civilization.

Innovation plays a crucial role in driving economic development. Understanding an industry's capacity for innovation can help in promoting the strategy for development driven by innovation. However, quantifying innovation is challenging due to its complexity and uncertainty. Common indicators used to measure innovation include patents,

X. Tang et al. (Eds.): KSS 2024, CCIS 2269, pp. 15–30, 2025.
https://doi.org/10.1007/978-981-96-0178-3_2

research papers, and R&D funding, etc., with patents often serving as the sole metric in many studies [1, 2]. According to the World Intellectual Property Organization (WIPO), patent information is one of the largest sources of publicly available technical data. The strong theoretical connection between patents and innovation, combined with the abundance of accessible patent data and a stable patent system, has led to the widespread use of patent data in innovation research, where its reliability and robustness have been verified in studies [3, 4]. Therefore, analyzing patents can reveal the developmental stage of industrial technological innovation and predict the development trend, which holds strategic significance for enhancing industrial innovation capacity.

2 Literature Review

The hydrogen energy industry is a typical strategic emerging industry with high complexity, integration, and uncertainty [5]. The development of strategic emerging industries is of great significance in promoting high-quality economic growth, and enhancing national competitiveness. In addition, due to the general applicability of patent data in industrial innovation research, the research on innovation network of strategic emerging industries based on patent perspective has gradually received more attention from scholars.

Existing research on the innovation network of strategic emerging industries based on the patent perspective is mainly categorized into the following three perspectives. From the perspective of cooperation networks, Sun et al. [6] constructed a cooperation network of the equipment manufacturing industry based on cooperative patent data, analyzing and predicting industrial innovation through the aspects of innovation entities and collaborative technologies. Wu et al. [7] utilized patent data from the 2016 pharmaceutical manufacturing industry-university-research (IUR) cooperation to construct a cooperation network, analyze the influencing factors of innovation performance, and provide recommendations for industrial development planning. From the perspective of citation networks, Zhang et al. [8] constructed a citation network and extracted the main path for the field of fuel cells, revealing that proton exchange membrane fuel cell is currently a research hotspot. From the perspective of co-occurrence networks, Xu et al. [9] constructed a co-occurrence network related to new energy vehicle technology, analyzing the hotspots and directions of industrial research.

However, while research on the innovation networks of strategic emerging industries has received much attention, few scholars have examined the development and evolution of the innovation network specific to the hydrogen energy industry. The hydrogen energy industry is different from other strategic emerging industries because of its technology-intensive and wide range of applications. Therefore, it is necessary to conduct focused research on the innovation network of the hydrogen energy industry. In light of this, this paper focuses on China's hydrogen energy industry, constructing upstream, midstream, and downstream IPC co-occurrence networks based on patent data, analyzing the technological trend and innovation hotspots to enhance the innovation capability and competitiveness of China's hydrogen energy industry.

3 Methodology and Data

3.1 International Patent Classification

International Patent Classification (IPC), as an internationally recognized system for classifying patent technologies, is widely utilized for the classification and search of patent documents across various countries and regions. The IPC code categorizes patents based on their technological content, with each patent assigned at least one IPC code. When multiple IPC codes are present in the same patent, this situation is called the IPC co-occurrence phenomenon [10]. In this paper, IPC codes are characterized as technology fields, and IPC co-occurrence reflects the integration between different technology fields. This integration reveals the cross-technology applications of patents and is significant for researching technological hotspots characterized by a high frequency of co-occurrence, as well as for exploring new directions in technological innovation [11]. This paper use complex network to represent the co-occurrence of 6-digit IPC codes and analyze the correlation and strength of relationships between the technology fields represented by each IPC code. Specifically, the IPC co-occurrence network is constructed according to whether the IPC code is co-occurring in the same patent as shown in Fig. 1.

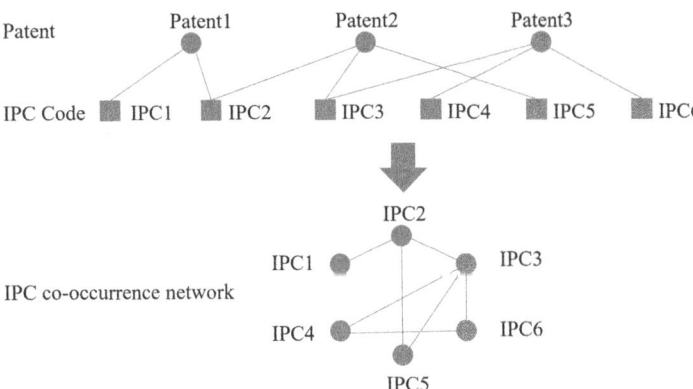

Fig. 1. Example of IPC co-occurrence network construction

3.2 Social Network Analysis

Social Network Analysis (SNA) is a quantitative graphical method for analyzing relational data, with a mature theoretical framework and extensive applications. This study employs SNA to construct IPC co-occurrence networks for the upstream, midstream, and downstream links of China's hydrogen energy industry to analyze the co-occurrence relationships between technical fields. The IPC co-occurrence network is constructed with IPC codes as nodes, the co-occurrence relationships between IPC codes as edges, and the frequency of co-occurrence as edge weights. The network structure is analyzed through the indices of degree centrality, closeness centrality, and betweenness centrality.

A higher degree centrality indicates a stronger influence of the node within the network. A higher closeness centrality reflects an enhanced capacity for resource transmission by the node. Similarly, a higher betweenness centrality signifies greater control over the network by the node. In addition, this paper utilizes Gephi to generate visual representations of technology co-occurrence diagrams, thereby exploring the technical structures, hotspots, and development trends of various links of the hydrogen energy industry.

3.3 Technology Life Cycle Theory

The development of technology has significant life cycle characteristics. In 1984, Harvey [12] embarked on a study of technological life cycles, dividing them into six stages: Development, Argumentation, Technological Adoption, Expansion, Maturity, and Decline. In 1986, Foster [13] introduced the S-curve model, which further refined the theory of technological life cycles by dividing it into four stages: Germination, Growth, Maturity, and Decline. The S-curve encompasses both the Logistic curve and the Gompertz curve, with the Logistic curve being particularly suitable for scenarios where the development of a subject is influenced by both its existing quantity and the potential for future growth [14]. This paper employs the Logistic curve to divide the technological lifecycle and evolution stages of China's hydrogen energy industry, explores the IPC co-occurrence network structure and the evolution characteristics of upstream, midstream, and downstream of the industry at different stages, and evaluates the innovation level and evolution path of China's hydrogen energy industry.

3.4 Data Acquisition and Processing

According to the industrial chain, the hydrogen energy industry can be divided into three links: upstream production of hydrogen, midstream storage, transportation, and refueling of hydrogen, and downstream utilization of hydrogen. Patent data related to the hydrogen energy industry was collected from the China National Intellectual Property Administration (CNIPA). Considering that it usually takes around 18 months for a patent application to be granted, data after 2022 is excluded from the analysis, with data retrieved as of December 31, 2021.

The main countries for the development of the hydrogen energy industry include the United States, Japan, South Korea, China, and Germany. Due to the limited availability of patent data before 1970, Fig. 2 shows the trends of hydrogen energy industry patent applications in five major countries from 1970 to 2021. By examining trends in patent applications across countries, it can be discerned that patent applications have evolved from being primarily dominated by advanced economies, such as Japan and the United States, in the early 21st century to developing nations such as China catching up in recent years. The United States, Japan, South Korea, and Germany reached the peak of patent applications around 2005. In contrast, China's hydrogen energy industry began its development relatively late but has since entered a high-growth phase as patent filings from other countries have gradually declined. Over the past five years, patent applications in China showed an explosive growth trend, with an average annual increase of 27.8%, ranking first in the world.

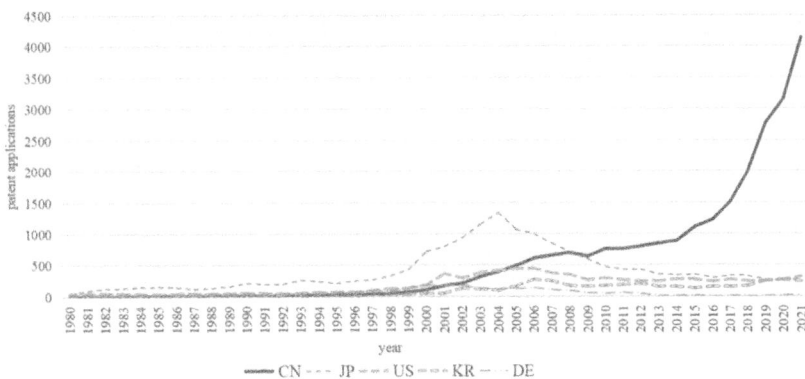

Fig. 2. Patent application trends of the hydrogen energy industry in major countries

Figure 3 shows the trend of patent application across each link of China's hydrogen energy industry from 1985 to 2021. A comparison of these link reveals that the development of China's hydrogen energy industry innovation is uneven. Fuel cell technology in the downstream link is the core field of the hydrogen energy industry innovation. Consequently, the patent applications in the downstream link is significantly higher than in other links, and it entered the rapid development stage earlier. The midstream link represents the technical bottleneck in the hydrogen energy system, yet they exhibit the least innovation and receive inadequate attention. Given the uneven development trends, it is not accurate enough to assess the overall level of the industry chain. Therefore, this paper considers the heterogeneity of innovation across different links, analyzing the innovation networks within each link of China's hydrogen energy industry chain separately.

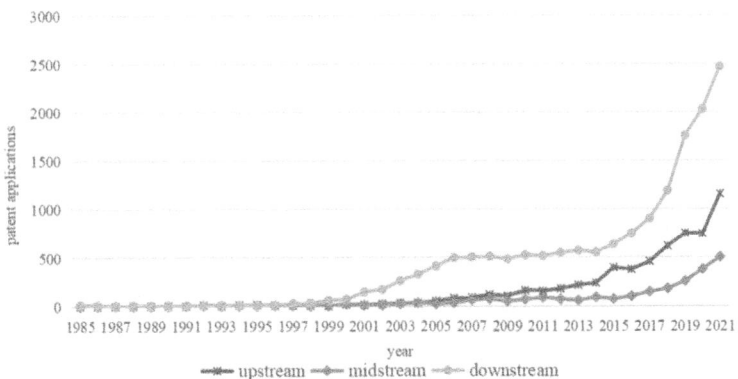

Fig. 3. Patent application trends in China's hydrogen energy industry and various links

4   Development Status of China's Hydrogen Energy Industry Innovation Network

Based on the patent data from the upstream, midstream, and downstream links of the hydrogen energy industry, this paper constructs IPC co-occurrence networks for each link (see Fig. 4). IPC co-occurrence network takes IPC code as the node, node Color darkness and Label size indicate the node degree; that is, nodes with darker colors and larger labels have co-occurrence relationship with a greater number of technical fields. A line between two nodes indicates that the corresponding IPC codes are present in the same patent. The thickness of the line represents the frequency of co-occurrence between the two technical fields; a thicker line indicates a higher number of occurrences. Table 1 shows the IPC codes and their descriptions involved in this study.

Upstream   Midstream   Downstream

Fig. 4. IPC co-occurrence network in various links of China's hydrogen energy industry

Table 1. Detailed description of recommended technologies

IPC code	Description
B01J	Chemical or physical processes
B01J20	Solid adsorbent composition or filter aid composition
B01J23	Catalysts comprising metals or metal oxides or hydroxides
B01J27	Catalysts comprising the elements or compounds of halogens, sulfur, selenium, tellurium, phosphorus or nitrogen; Catalysts comprising carbon compounds
C01B3	Hydrogen; Gaseous mixtures containing hydrogen; Separation of hydrogen from mixtures containing it
C07C29	Preparation of compounds having hydroxy or O-metal groups bound to a carbon atom

(continued)

Table 1. (*continued*)

IPC code	Description
C07C31	Saturated compounds having hydroxy or O-metal groups bound to acyclic carbon atoms
C22C1	Making non-ferrous alloys
C22C1	Making non-ferrous alloys
C22C19	Alloys based on nickel or cobalt
C25B	Electrolytic or electrophoretic processes for the production of compounds or non-metals
C25B1	Electrolytic production of inorganic compounds or non-metals
C25B11	Electrodes; Manufacture thereof not otherwise provided for
C25B15	Operating or servicing cells
C25B9	Cells or assemblies of cells
F17C	Vessels for containing or storing compressed, liquefied, or solidified gases
F17C1	Pressure vessels
F17C13	Details of vessels or the filling or discharging of vessels
F17C5	Methods or apparatus for filling pressure vessels with liquefied, solidified, or compressed gases
F17D	Pipe-line systems
F17D1	Pipe-line systems
F17D3	Arrangements for supervising or controlling working operations
H01M2	Structural parts or manufacturing method of inactive parts
H01M4	Electrodes
H01M8	Fuel cells; Manufacture thereof

4.1 Analysis of Innovation Entities

Innovative entities can be categorized into four groups: enterprises, universities, research institutions, and individuals. Figure 5 shows the statistical results of the attributes of innovation entities in each link of China's hydrogen energy industry. Enterprises serve as the primary innovative entities in all links of China's hydrogen energy industry, followed by universities and research institutions, while individuals innovation demonstrates the least advantage. Notably, enterprises' innovation predominantly targets the midstream and downstream, while universities' innovation is concentrated in the upstream, and research institutions' innovation is centered around the midstream.

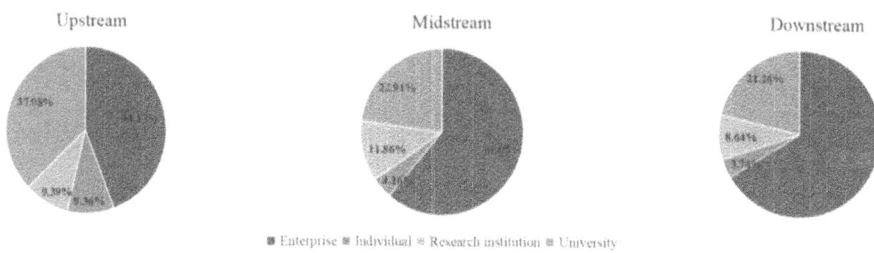

Fig. 5. The attribute of innovation entities in various links of China's hydrogen energy industry

The statistics on collaborative innovation across various links of China's hydrogen energy industry are shown in Table 2. Although the midstream link has the fewest patents, the number of cooperation patents significantly exceeds those in both the upstream and downstream. Hydrogen transportation, storage, and hydrogenation technology in the midstream form the core part of industrial innovation, which has always been a crucial focus area for industrial collaborative innovation. By establishing an industrial alliance among innovation entities to enhance technological cooperation, the capacity to innovate core technologies can be significantly strengthened. Currently, China's hydrogen energy storage and transportation standard system is inadequate, with hydrogen energy primarily stored and transported in high-pressure gaseous form. This limitation further restricts the technical options of hydrogen refueling stations. Therefore, improving the efficiency of hydrogen energy storage and transportation while reducing costs—without compromising safety—will significantly influence the pace of development in China's hydrogen energy industry. Addressing this issue is crucial, as it represents one of the key technological bottlenecks that must be overcome for the advancement of the current hydrogen energy sector.

Table 2. The proportion of collaborative innovation patents in various links of China's hydrogen energy industry

Echelon	Number of collaborative patents	Percentage of collaborative patents
Upstream	531	8.72%
Midstream	343	14.30%
Downstream	1518	9.45%

4.2 Analysis of Hot Innovation Fields

Hot innovation fields refer to technical fields that attract significant attention and substantial resources [15]. The top ten IPC codes of patent applications in various links of China's hydrogen energy industry are shown in Table 3. Upstream hot innovation fields are distributed in C25B and B01J, indicating that the hotspot of hydrogen production is green hydrogen production, with electrolytic hydrogen production technology at its core. Currently, the industrialization of electrolytic hydrogen production technology in China includes alkaline electrolysis water and proton exchange membrane electrolysis water. The hot innovation fields in midstream mainly include F17C13, F17C5, F17C1, and C22C1, indicating that hydrogen storage tanks, bottles, and alloys are the hot fields in midstream. Furthermore, F17D1 ranks sixth, suggesting that the hot research direction for hydrogen delivery in China is pipeline delivery. Pipeline hydrogen transportation has received increasing attention as a vital method for achieving large-scale, long-distance, and low-cost transportation of hydrogen energy. Downstream link can be broadly categorized into hydrogen energy transportation, industrial hydrogen, and hydrogen energy storage. The proportion of H01M8 is 79.35%, indicating that hydrogen energy transportation is the hot innovation direction of hydrogen energy utilization, particularly concentrated in hydrogen fuel cell technology.

Table 3. Top 10 hot innovation fields in various links of China's hydrogen energy industry

Rank	Upstream		Midstream		Downstream	
	IPC	Percentage	IPC	Percentage	IPC	Percentage
1	C01B3	40.51%	F17C13	31.14%	H01M8	79.35%
2	C25B1	39.90%	C01B3	21.63%	H01M4	12.84%
3	C25B9	17.86%	F17C5	19.67%	C01B3	7.45%
4	B01J27	14.29%	F17C1	14.01%	B01J23	4.76%
5	C25B11	12.30%	C22C1	12.26%	C07C29	3.76%
6	B01J23	11.43%	F17D1	11.00%	H01M2	3.51%
7	C25B15	10.94%	F17D3	10.46%	C07C31	3.25%
8	H01M8	9.05%	H01M4	9.42%	C08J5	3.23%
9	B01J35	7.67%	C22C19	8.13%	C25B1	2.60%
10	B01J37	6.23%	F17C7	7.29%	C01C1	2.24%

4.3   Analysis of Core Innovation Fields

The core innovation fields refer to the technical fields that occupy a crucial position in the IPC co-occurrence network, significantly contributing to the development of industrial technology and exerting a substantial influence on other technical fields [15]. In this paper, the degree centrality, closeness centrality, and betweenness centrality of network nodes are used to measure the IPC co-occurrence network nodes at each link of China's hydrogen energy industry.

Using Gephi to calculate the node centrality of the IPC co-occurrence network, Table 4 presents the technical fields with degree centrality, closeness centrality, and betweenness centrality in the top ten as the core innovation fields. Combining Fig. 2 and Table 4, the upstream core innovation fields are primarily found in electrolysis production (C25B1) and hydrogen separation (C01B3). The midstream is mainly distributed in gas storage tanks (F17C) and pipeline systems (F17D), while the downstream focuses mainly on electric energy conversion devices (H01M), catalytic action (B01J), and pentacyclic or carbocyclic compounds (C07C). These core innovation fields have propelled the advancement of hydrogen industry technology through their integration with other technology fields. It is noteworthy that in the downstream, the fuel cell technology (H01M8) occupies a central position within the network, ranking first in all centrality indicators. It is the most central and technologically advanced field in the downstream link, characterized by abundant resources and significant influence on other technology fields. China's fuel cell patent applications concentrate on the electric reactor system and control system. Core components of the electric reactor system technology include catalysts, proton exchange membranes, bipolar plates, membrane electrodes, diffusion layers and seals, while the control system technology focuses on DC converters, air compressors, electronic control valves, energy management systems, circulating pumps, hydrogen tanks.

Table 4. Core innovation fields in various links of China's hydrogen energy industry

Echelon	Core innovation fields
Upstream	C25B1, C01B3
Midstream	F17C13, F17C5, C01B3, F17D1, F17D3, F17C1
Downstream	H01M8, H01M4, C01B3, B01J23, C07C29, H01M2, C07C31, C25B1

5 Evolution of China's Hydrogen Energy Industry Innovation Network

5.1 Technical Life Cycle Analysis of China's Hydrogen Energy Industry

In this paper, a logistic model is used to fit an S-curve that measure the technology life cycle of China's hydrogen energy industry by using the patent data of China's hydrogen energy industry from 1985 to 2021 as the basis for prediction. The fitting formula is as follows.

$$y = \frac{K}{1 + e^{-r(t-t_m)}} \tag{1}$$

$$r = \frac{\ln(81)}{T_{0.1-0.9}} \tag{2}$$

where y is the number of patent accumulation applications, t is the time. K is the maximum value of the S-curve, $T_{0.1-0.9}$ is the time required for the patent accumulation applications to reach 90%K from 10%K, t_m is the time required for the patent accumulation applications to reach 50%K [16].

Logistic curve fitting was conducted using Loglet Lab4 to obtain the fitted values of the parameters of the technology life cycle of China's hydrogen energy industry. Table 5 shows that the model fitting goodness-of-fit value is 0.976, indicating a fairly satisfactory fit. Based on the parameter fitting values, the technology life cycle of China's hydrogen energy industry is divided into four stages: (1) The budding stage (2000–2014), characterized by few patent applications and slow growth; (2) The growth stage (2015–2028), during which the rate of patent applications continues to rise; (3) The maturity stage (2029–2042), in which the growth rate of patents gradually decreases; and (4) The decline stage (2043–2056), during which the patent accumulation applications is expected to reach saturation. Figure 6 shows the predicted technological development trend of China's hydrogen energy industry. The solid line represents the actual cumulative patent applications, while the dotted line represents the predicted trend of cumulative patent applications.

Based on the above research, this paper divides the evolution of China's hydrogen energy industry into three stages: the first stage is 1985–2000, during which China's hydrogen energy industry was still in its exploratory startup stage. Hydrogen energy technology research was concentrated among a few patent applicants; the second stage is 2001–2014, the domestic market demand for hydrogen energy continued to expand, leading to an increase in patent applications; the third stage is 2015–2021, when Toyota offered 5680 patents related to hydrogen fuel cell technology for free in 2015, which led to a large number of innovation entities investing in the hydrogen industry, triggering a domestic hydrogen boom and leading to a surge in hydrogen patent applications. Based on the patent data corresponding to various evolutionary stages of each link within the hydrogen energy industry, the evolution maps of IPC co-occurrence networks of three links were constructed respectively.

Table 5. Fitting results of technical life cycle parameters

Logistic	Statistics	Logistic	Statistics
R2	0.976	1%	2000
K	94479	10%	2014
a	26.5	50%	2028
tm	2028	90%	2042
r	0.166	99%	2056

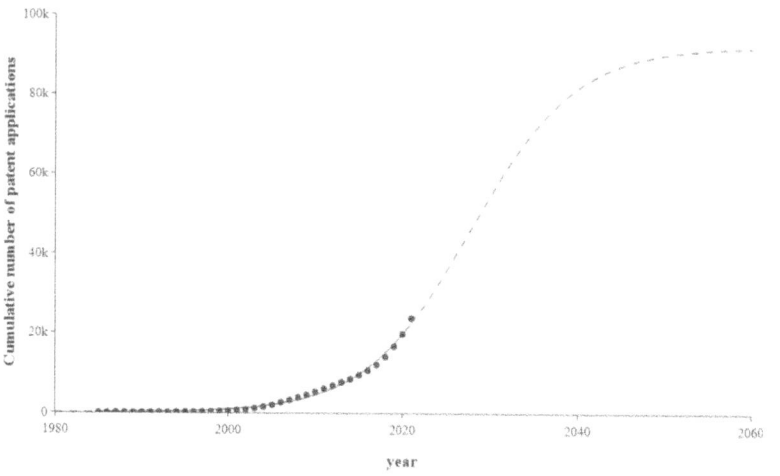

Fig. 6. S-curve diagram of the technological life cycle of China's hydrogen energy industry

5.2   Evolution Analysis of Innovation Entities

Table 6 gives the types of innovation entities and collaborative innovation in each link of China's hydrogen energy industry at each evolutionary stage. Comparing the collaborative innovation situation in each stage, it can be found that the proportion of collaborative innovation in each link is the highest in the first stage, decreases in the second stage, and then increases again in the third stage. In the first stage, the patent development of the domestic hydrogen energy industry is almost blank, with individual innovation entities occupying a larger proportion. Thus, collaborative innovation was easier to achieve. In the second stage, as demand for hydrogen energy in the domestic market continued to rise, most innovation entities engaged in independent technological research and development activities to secure competitive advantages, resulting in a decline in the proportion of collaborative innovation to its lowest point. In the third stage, as knowledge within the hydrogen energy sector matured, there was a notable increase in technical collaborations among enterprises, and resources were exchanged and integrated with educational institutions and research units, leading to an increase in the proportion of collaborative innovation once again.

Table 6. The types of innovation entities and the proportion of collaborative innovation

Type	Upstream			Midstream			Downstream		
	Stage 1	Stage 2	Stage 3	Stage 1	Stage 2	Stage 3	Stage 1	Stage 2	Stage 3
Enterprises	38.00%	38.26%	48.80%	57.14%	44.99%	74.48%	71.92%	68.87%	70.33%
Universities	1.00%	36.31%	40.27%	23.81%	38.68%	18.56%	5.82%	20.77%	23.84%
institutions	20.00%	13.45%	8.36%	12.70%	14.47%	11.97%	14.38%	10.91%	3.13%
Individuals	46.00%	16.01%	6.89%	19.05%	6.30%	3.11%	12.67%	4.81%	7.80%
cooperation	16.00%	7.80%	8.87%	15.87%	5.59%	17.95%	14.73%	8.36%	16.43%

5.3   Evolution Analysis of Hot Innovation Fields

Table 7 shows the evolution of hot innovation fields in each link of the hydrogen energy industry. In the upstream link, C25B1, C01B3, and C25B9 consistently ranked among the top five hot innovation fields in the three periods, while the ranking of B01J27 gradually increased. This trend indicates that water electrolysis for hydrogen production has garnered significant attention since the early stages of development, and the importance of hydrogen production catalysts has increasingly been recognized by researchers as evolved. Water electrolysis for hydrogen production is known for its high conversion efficiency and lack of pollutant emissions, making it a promising technology in hydrogen production. Additionally, hydrogen production catalysts play a crucial role in accelerating reactions, which is vital for enhancing hydrogen production efficiency. In the midstream link, H01M4, C22C19, and C22C1 have seen a gradual decline in their rankings over time, while F17C and F17D are coming up later, occupying the top ten positions in the hot innovation fields. Currently, China's hydrogen transportation technology is insufficient to meet the anticipated demand for large-scale hydrogen transmission in the future. Consequently, the hot technology fields of hydrogen storage and transportation has changed from hydrogen storage alloy to hydrogen storage cylinder and pipeline hydrogen transmission. In the downstream link, H01M8 has maintained its position as the leading hot innovation field across all periods, while B01J23 and C25B1 have improved their ranking as the evolution progressed. It indicates that fuel cell is the hot research field of Hydrogen application throughout all periods, with resource investment is far more than other technology fields. As hydrogen energy technology advances, industrial hydrogen and hydrogen energy storage are also beginning to attract more attention, although resource investment in these areas remains relatively modest.

Table 7. Evolution of hot innovation fields in each link of China's hydrogen energy industry

Rank	Upstream			Midstream			Downstream		
	Stage 1	Stage 2	Stage 3	Stage 1	Stage 2	Stage 3	Stage 1	Stage 2	Stage 3
1	C25B1	C01B3	C25B1	H01M4	C01B3	F17C13	H01M8	H01M8	H01M8
2	C01B3	C25B1	C01B3	C22C19	C22C1	F17C5	C01B3	H01M4	H01M4
3	C25B9	B01J23	C25B9	C22C1	H01M4	C01B3	C07C29	C01B3	C01B3
4	C25B15	C25B9	B01J27	C01B3	C22C19	F17D1	H01M4	H01M2	B01J23
5	B01D53	B01J27	C25B11	F17C1	C01B6	F17C1	C07C31	B01J23	C25B1
6	B23K5	H01M8	C25B15	C22C23	C01B3	F17D3	B01J23	C08J5	C07C29
7	C25B11	C12P3	B01J23	B01J20	C01B31	F17C7	C01C1	C07C29	B60L50
8	C01B31	C10J3	H01M8	C01B31	C22C30	C22C1	H01M	C07C31	C07C31
9	C10J3	C25B15	B01J35	H01M10	H01M8	F17D5	H01M2	C01C1	B60L58
10	C25B13	B01J21	B01J37	B22F9	C22C23	H01M4	C10G45	C25B1	H02J3

5.4 Evolution Analysis of Core Innovation Fields

Table 8 shows the evolution of core innovation fields in each link of the hydrogen energy industry. In the first stage of the upstream, C25B1 and C01B3 dominate the network, with water electrolysis for hydrogen production garnering significant attention as a fundamental technical field. In the second stage, water electrolysis for hydrogen production still occupies the center of the network. However, the increase in patents related to B01J23, along with a growing association with other technology fields, suggests a breakthrough in the technical field of hydrogen catalyst. In the third stage, C25B9, C25B11, C25B9, C25B11, and C25B15 are once again receiving greater attention, indicating that electrolyzers—key components in hydrogen production from electrolyzed water—are experiencing rapid advancements in technological innovation and warrant future focus. The core innovation fields in the first stage of the midstream mainly focuses on H01M4. The technology fields most closely integrated with it are C22C1 and B01J20, indicating that the core innovation fields of the midstream are concentrated on solid hydrogen storage. In contrast, the core innovation fields identified in the third stage are predominantly found in F17C and F17D. This shift suggests that hydrogen storage and transportation have transitioned from solid to gaseous forms. Consequently, the development of hydrogen storage bottle, hydrogen storage tanks and pipeline construction—though initiated later—has progressed rapidly, establishing them as emerging core fields of innovation that are likely to remain significant in the future. In the downstream link, alongside fuel cells and catalysts, which are the core innovation during each period, C07C29 and C07C31 were incorporated in the second stage, and C25B1 was introduced in the third stage as the core innovation field. It shows that research on hydrogen energy applications in China primarily focuses on fuel cells, with the application scenarios of hydrogen energy expanding into the domains of chemical and energy storage.

Table 8. Evolution of core innovation fields in each link of China's hydrogen energy industry

Echelon	Stage	Core innovation fields
Upstream	1	C25B1, C01B3, C25B11, C25B9, B01D53, C25B13, C01B31, B23K5
	2	C25B1, C01B3, B01J23, H01M8, B01D53, C10J3, C02F1
	3	C25B1, C01B3, C25B9, C25B11, C25B15, C02F1, B01J23, B01J35
Midstream	1	H01M4, B01J20, C01B3, C22C1, C01B31, H01M8
	2	C01B3, H01M4, H01M8
	3	F17C13, F17C5, F17D1, F17D3, C01B3, F17C1, F17D5
Downstream	1	H01M8, C01B3, B01J23, H01M4, C01B31
	2	H01M8, H01M4, C01B3, B01J23, C07C29, C07C31
	3	H01M8, C01B3, B01J23, H01M4, C25B1, C07C29, C07C31

6 Conclusion

This study constructed the IPC co-occurrence network of upstream, midstream, and downstream in China's hydrogen energy industry, analyzed the evolution process of innovation entities, hot innovation fields, and core innovation fields. The main conclusions and recommendations are as follows.

Firstly, there is an imbalance in its development of China's hydrogen energy industry. Research of innovative technologies for upstream hydrogen production, midstream hydrogen storage and transport should be strengthened. Secondly, China's hydrogen energy industry has formed a innovation pattern dominated by enterprises, supplemented by universities and research institutions. The participation of research institutes, compared to enterprises and universities, gradually decreases with the development of the industry, so the patent construction of research institutes should be promoted. Thirdly, there are significant differences in the technical fields of each link in the hydrogen energy industry at different stages of development. For the upstream link, the mainstream technology field is electrolytic hydrogen production, and it is urgent to promote lower carbon technology such as biomass hydrogen production and photolytic hydrogen production; for midstream link, high-pressure gaseous hydrogen storage is the main technical field. Research on storage and transportation technology of low-temperature liquid and solid materials should be enhanced to achieve a low-cost and large-scale supply of hydrogen; for downstream link, patents are concentrated in hydrogen fuel cells, and should continue to expand the application of hydrogen energy in industry and electric power.

Limited by the length of the article, there are some limitations in this paper. Firstly, this paper analyzes the innovation and development of China's hydrogen energy industry using only patent data. Future research can be enhanced by integrating indicators such as papers, research funding, researchers, and others. Secondly, this paper analyses hydrogen energy industry innovation only from the perspective of the IPC co-occurrence network. Future research could incorporate additional analytical methods, such as keyword co-occurrence analysis and patent citation analysis, to provide a more comprehensive examination of innovation in the hydrogen energy industry, thereby providing more valuable insights for policymakers, industry stakeholders, and researchers in the hydrogen energy industry.

Acknowledgments. This work is supported by the National Social Science Fund of China(23BGL041), and the National Natural Science Foundation of China(72293563).

References

1. Carree, M., et al.: Factors favoring innovation from a regional perspective: a comparison of patents and trademarks. Int. Entrep. Manage. J. **11**(4), 793–810 (2014)
2. Albert, M., et al.: Direct validation of citation counts as indicators of industrially important patents. Res. Policy **20**(3), 251–259 (1991)
3. Ji, Y., Li, C.: Innovation driving and industrial upgrading: the spatial econometric analysis based on China's provincial panel data. Stud. Sci. Sci. **33**(11), 1654–1659 (2015)
4. Hagedoorn, J., Cloodt, M.: Measuring innovative performance: is there an advantage in using multiple indicators? Res. Policy **32**(8), 1365–1379 (2003)

5. Zhang, L., Xue, L., Zhou, Y., Zhang, X.: Evolution mechanism of the strategic emerging industries' innovation network from—based on the new energy automobile empirical data from 2000 to 2015. Stud. Sci. Sci. **36**(06), 1027–1035 (2018)

6. Sun, X., Zhuang, W., Li, B., Chen, N.: Research on characteristics and cooperation prediction of industry-university-research institute collaboration based on patents in regional equipment manufacturing industry. Sci. Manage. **40**(01), 31–40 (2020)

7. Wu, H., Gu, X.: The social network analysis of cooperative innovation performance of industry-university-research institution. Stud. Sci. Sci. **35**(10), 1578–1586 (2017)

8. Zhang, F., Liu, L., Liao, X.: Fuel cell hot research areas and key technologies: an empirical research based on patent data. J. Intell. **36**(01), 54–58 (2017)

9. Xu, X., Huang, T., Zhang, X.: Analysis of the current status of china's new energy vehicle technology based on co-word network. China Sci. Technol. Inf. (05), 68–70+72 (2017)

10. Xu, K., Chen, X.: Patent value study based on patent breadth measures. Stud. Sci. Sci. **28**(02), 202–210 (2010)

11. Li, D., Zhai, D., Feng, X., Zhang, J.: Study of the evolution of technological innovation network in TCM technology based on dynamic network. J. China Soc. Sci. Tech. Inf. **34**(11), 1164–1172 (2015)

12. Harvey, M.G.: Application of technology life cycles to technology transfers. J. Bus. Strategy **5**(2), 51–58 (1984)

13. Foster, R.N.: Working the S-curve: assessing technological threats. Res. Manage. **29**(4), 17–20 (1986)

14. Yang, X., Yu, X.: A study of the technological innovation model of the graphene industry based on life cycle. Sci. Res. Manage. **41**(09), 12–21 (2020)

15. Yuan, Y., Li, Y., Wu, C., Wu, J.: An analysis on the international competition situation and frontier of digital logistics technology based on patent mining. Innov. Sci. Technol. **24**(01), 76–91 (2024)

16. Zhang, B.: Comparison and enlightenment of life cycle characteristics of high-speed rail technology at home and abroad: the patent perspective. J. Intell. **39**(01), 83–90 (2020)

Construction and Evaluation of a Subjects Synergistic Network for Cross-Regional Major Infectious Disease Emergency Response

Cong Ming[✉], Ning Wang, and Lili Rong

Dalian University of Technology, Dalian 116024, China
CM@mail.dlut.edu.cn

Abstract. The emergency response process for major infectious diseases is very complex and requires synergy among multiple emergency subjects across regions, levels, and sectors to deal with the emergency response process. This study aimed to investigate the use of a complex network to construct a subject synergistic network and evaluate synergy to improve the efficiency of emergency response in cross-regional major infectious diseases. Emergency response subject synergy was identified using a "missions-resources-subjects" approach. Subject synergy was extracted from the text about cross-regional major infectious disease emergency response to construct subjects synergistic network. A synergy evaluation model was constructed and synergy was calculated by UCINET. The measurements were shown to lead to improving the efficiency of emergency response by emergency subjects synergy. This indicates that there is a possibility of improving emergency response efficiency by building a subjects synergistic mechanism and suggests that the synergy of emergency subjects needs to be transformed.

Keywords: Cross-regional Major Infectious Disease · Subjects Synergy · Complex Network · Synergy Evaluation

1 Introduction

In recent years, the rapid spread and unpredictability of major infectious diseases have drawn considerable attention due to their significant economic impact. These diseases represent a critical public health challenge, affecting the health and safety of humans. The management of these emergencies is a complex process that encompasses a variety of intricate and unpredictable tasks, which necessitate the coordinated efforts of various emergency response subjects [1]. However, a lack of synergy among emergency response subjects, due to poor documentation, hinders effective action. Enhancing response efficiency requires shifting from a traditional hierarchical command to a synergy model. Research by Kettl [2] and Piraino and Trucco underscores the importance of strategic partnerships and cross-sectoral collaboration in emergency management. However, qualitative studies on emergency response synergy are plentiful [3–5], and quantitative research is scarce. Therefore, this paper presents a model for building subjects synergistic networks for infectious disease emergencies and evaluates them using social network

X. Tang et al. (Eds.): KSS 2024, CCIS 2269, pp. 31–45, 2025.
https://doi.org/10.1007/978-981-96-0178-3_3

analysis and synergy evaluation methods. The results found that emergency response subjects synergy improved the efficiency of emergency response, yet grassroots-level synergy remained a challenge, indicating a need for further improvement in emergency synergy strategies.

2 Literature Review

Scholars have highlighted the significance of subject synergy in emergency response as a solution to complex public crises. Lyu [6] explored key issues in synergy across levels, experts, and sectors through a thorough analysis of collaborative behaviors. Fan [7] emphasized the utility of social network analysis (SNA) in uncovering the structure and efficacy of cross-departmental emergency networks. Comfort and Kapucu [1] highlighted the importance of resource-based organizational synergies in emergency contexts, showing their impact on response effectiveness.

Research indicates that emergency response benefits from multi-level, cross-organizational collaboration, forming intricate network structures. Tang [8] applied SNA to map and quantify the roles of key actors in emergency networks, enhancing disaster relief synergy. The COVID-19[9] response's synergy network topology exposed its scale-free nature and showed that centralized leadership of key decision-makers was vital for network evolution. Minyoung [10] analyzed the MERS outbreak response in South Korea, constructing an inter-organizational collaboration network. Through social network analysis, they revealed how a state-controlled, multisectoral network evolved, with a few government agencies becoming central to structural changes.

SNA has been widely employed to assess network synergy through various metrics. Wang [11] gave a method to measure the performance of an emergency cooperation network, using network metrics to measure the network collaboration performance comprehensively. Hu [12] noted that assessing the formation and performance of emergency management networks was a research priority, emphasizing the use of holistic structural features and specific, indicators for evaluation.

Previous studies have recognized the advantages of emergency subject synergy in improving response effectiveness but have been deficient in methods to identify emergency subjects and their synergy relationships. This paper introduces a "missions-resources-subjects" framework to identify subjects and their relationships, creating a synergy network. While current research often employs social network analysis (SNA) for overall network metrics, it often overlooks individual responder assessments. This study bridges this gap by evaluating the network and building an assessment model for individual emergency responder synergy, potentially enriching evaluation methodologies.

3 Cross-Regional Major Infectious Disease Emergency Response Subjects Synergistic Network Construction

3.1 Subjects Synergistic Network Construction Ideas

This paper initiates from the emergency mission and identifies emergency subjects based on the mission's resource needs, preventing redundancy and facilitating rapid action planning. In practice, subjects need to address diverse missions throughout the emergency response lifecycle. This paper considers that there are multiple relationships between emergency missions, such as correlation, process, or hierarchical relationships. The completion of an emergency mission roughly required knowledge of what was needed and by whom, in other words, an emergency mission could be thought of as consisting of resources as well as performers. Therefore, emergency subjects synergy was defined as a behavioral relationship arising from two or more emergency subjects, to accomplish a certain emergency mission, while having a common need for a certain resource and needing to participate in or jointly carry out an emergency response mission using resource transfer or sharing. In cross-regional major infectious disease emergency response missions, resources most referred to emergency response resources, such as masks, blood, population flow information, etc.; And performers most referred to emergency response subjects, such as governments, public interest organizations, and enterprises, etc. According to the definition, emergency missions could be divided into three levels, which are mission level, resource level, and subject level, and they could be related and form a hierarchical network. This framework is used to construct a model for the emergency response subject synergistic network, as illustrated in Fig. 1.

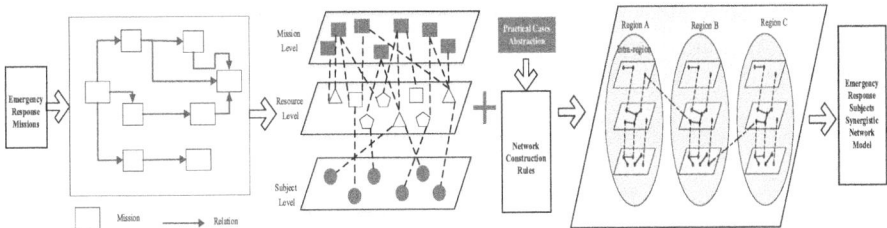

Fig. 1. The idea of constructing a synergistic network model for emergency response subjects

3.2 Response Mission Analysis for Major Infectious Disease

Data Source. In this research, we utilized a comprehensive data collection strategy, amassing over 300,000 words from various sources, including:

(1) Government documents and legal regulations were published on the official websites of governments and relevant departments as the primary data sources, supplemented by news reports from unofficial websites such as Weibo, public accounts, and the Red Cross.
(2) Actual case studies on the emergency response processes to major infectious diseases in Hubei, Shanghai, Nanjing, Yangzhou, Jilin, Dalian, Harbin, and other locations.

The initial textual data was meticulously reviewed, and ultimately, approximately 50,000 words of valid information were extracted. This information encompassed the relevant details of the emergency response process for major infectious disease events, including the associated emergency response subjects, their responsibilities, and emergency guarantees. Utilizing this data, a model was developed to represent the synergy network among emergency response subjects.

Macro-emergency Missions Identification. This study has systematically gathered 43 government documents encompassing emergency plans and regulations at various administrative levels across different regions in China. Based on the analysis of the structure of China's emergency response plans at all levels, this paper carries out a preliminary extraction of the secondary headings in the textual information and extracts the headings consisting of "noun + verb" or those that can be expressed as general actions as macro emergency missions. At last, a total of 13 macro-emergency missions necessary in the process of emergency response to major infectious diseases were compiled, including professional teams, material support, publicity and education, medical rescue, public security maintenance, financial support, transport, communication support, basic life support, scientific and technological support, legal protection, personnel protection, and emergency command.

Macro-emergency Mission Relationships. In this section, Interpretative Structural Modeling (ISM) was employed to elucidate the interconnections among macro-emergency missions. Introduced by Warfield in 1973, ISM is a pivotal technique within the realm of modern systems engineering. This approach facilitates the examination of binary relationships among system components, enabling the construction of a simplified, hierarchical topology that preserves system functionality. Ultimately, a multi-tiered hierarchical structural model is generated with the aid of computer software. Using the content of Macro-emergency missions as a reference, the binary relationships among

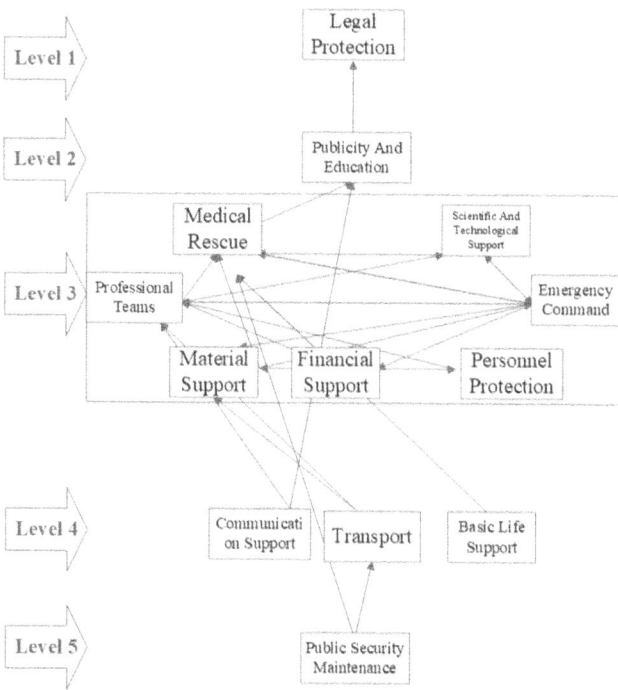

Fig. 2. ISM of Macro-emergency Missions

these elements were established. Subsequently, a multi-level interpretative structural model was crafted using MATLAB. This model illustrated the intricate relationships between Macro-emergency missions, as depicted in Fig. 2.

3.3 Response Subjects Analysis for Major Infectious Disease

In this study, a keyword search methodology was utilized to identify emergency subjects from the previously mentioned textual data. The principles of the extract are as follows:

(1) Extracting nouns from textual information about ministries, commissions, bureaus, offices, agencies, departments, etc.
(2) Merging the same functional departments and similar social group organizations at different levels.

In reality, departments in different administrative regions and at different levels have slightly different names, and there are affiliations between departments at different levels. Thus, departments with the same name or similar functions were merged, and only the name prefixes were retained for departments with affiliations, but nodes were established at different administrative levels. For example, only the government was listed, but was divided into the provincial government, municipal government, and district government in the emergency response subjects synergistic network. Finally, 54 emergency subjects were extracted: Party Central Committee, Healthcare Institutions, Public Security, Government, and so on.

3.4 Association Rules of Subjects Synergistic Network

The emergency response subjects synergistic network was structured with nodes representing the subjects and edges representing the relationships between them. These relationships, both horizontal and vertical, were derived from real-case scenarios to form a comprehensive network model.

The Horizontal Association Rules. In blood transport missions, blood was transferred from blood stations to hospitals via the transportation department. This transfer involved direct interactions between the blood station, transportation department, and hospital, as they all handled the same blood resources. Such direct interaction indicates a synergistic relationship. Consequently, rule 1 was established: network nodes where a synergistic relationship exists were directly associated.

In medical rescue operations, a clear sequence of events defined the process: Patients first underwent a nucleic acid test, and if the result was positive, they were notified by designated personnel, transported to the hospital by ambulance, and subsequently received treatment and monitoring by medical staff. These steps constituted a logical progression that could not be rearranged, reflecting a process relationship among the associated emergency subjects, as depicted in Fig. 3. Given the fixed order of these operations, the associated emergency subjects had to maintain this sequence in the network model. For instance, there should have been no direct association between testing organizations and medical institutions that bypassed the necessary intermediate steps. Thus, rule 2 was established: When there were multiple emergency subjects with process relationships between emergency missions, the network nodes they represented needed to be associated according to the mission process relationships.

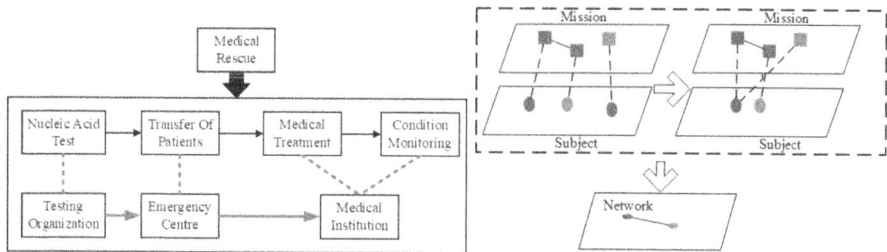

Fig. 3. Breakdown of Medical Treatment Mission

Fig. 4. Horizontal Association Rule 3

For instance, epidemiological investigation missions required timely information synergy between the health department, the big data department, and the public security department, after which the public security department would carry out specific investigations by telephone. Another example was that public area disinfection missions needed to be completed by village committees, and similarly, household screening missions were also required to be completed by village committees; two completely different missions necessitated the same emergency response subjects. The former involved three different emergency subjects for the same emergency mission, while the latter involved the same emergency subjects for different emergency missions. It was evident that there was not an exact one-to-one correspondence between emergency missions and emergency subjects. Consequently, rule 3 was established: In case of duplication of emergency subjects under different emergency missions, the relevant network nodes would be merged directly, as shown in Fig. 4.

In mask production missions, numerous enterprises participated, bearing different names but serving the same function within the mission. Separating these into distinct network nodes was deemed unnecessary. Thus, unifying such nodes under the category of "enterprises" would streamline the network by reducing the number of nodes and eliminating redundancy. Consequently, rule 4 was established: When network nodes were linked horizontally, network nodes with similar functions or similar names at the same level were consolidated.

The Vertical Association Rules. The present paper reviews typical cases and classifies interactions between subjects at different levels of emergency response into three categories. In Category 1, higher levels of government offered support to lower levels without direct intervention. Consequently, rule 1 was established that the upper network nodes were directly associated with the corresponding lower network nodes without re-association through the upper and lower government nodes. Category 2 involved higher levels of government establishing local command structures that participated directly in emergency response, with these local structures integrated into the higher-level command framework. Then, rule 2 was established that the relevant network nodes of the upper level directly replaced the relevant network nodes of the lower level, and the network nodes associated with the original lower level were directly associated with the relevant network nodes of the upper level. For Category 3, where multiple command levels functioned concurrently in synergy to manage emergency command missions, rule 3 was established: when multi-level network nodes were associated across levels, the network nodes associated with them were unnecessarily associated across levels.

3.5 Cross-Regional Major Infectious Disease Emergency Response Subjects Synergistic Network in X Village and Z City

Emergency Response Subjects Synergistic Network. For X Village. An outbreak of a major infectious disease in Village X was of a minor scale, and the response was managed by the jurisdiction to which Village X belongs. On January 15, 2020, Village X received notification from its superiors to initiate the emergency plan and implement suitable emergency measures. Throughout the response process, Village X engaged in several macro-emergency missions, such as professional teams, material support, medical rescue, transport, financial support, publicity and education, and public security maintenance. The present paper divides the macro-emergency missions into sub-missions based on the actual emergency response measures and identifies the emergency subjects through the "mission-resource-subject" approach. The result of the mission breakdown is shown in Fig. 5, where the right column is the sub-tasks of the left column.

The major infectious disease emergency response subjects synergistic network model was utilized to construct a network for X village's response to a major infectious disease outbreak. This network was visualized using UCINET, with different colors used to distinguish the attributes and hierarchies of the emergency response subjects, as illustrated in Fig. 6.

Macro-Emergency Missions	Missions Breakdown		Subjects
Professional Teams	Daily Sanitisation	Disinfection Of Public Areas	Village Council
		Disinfection Of Business Premises	Enterprise
		Residential Disinfection	Individuals
	Population Screening	Household Surveys	Village Officials
		Big Data Verification	Big Data Centre
		Self-Reporting	Individuals
		Data Screening	Cmc
		Checkpoint Screening	Village Officials
	Surveillance	Body Temperature Monitoring	Individuals
			Village Government
			Enterprise
			County Hospital
			Village Council
			Checkpoint
		Health Code Monitoring	Individuals
		Nucleic Acid Monitoring — Organisation	County CDC
		Nucleic Acid Monitoring — Nucleic Acid	County Hospital
		Vaccination Monitoring	Village Health
	Epidemiological Disposal	Household Population Census	Village Officials
		Information Reporting	Village Officials

Macro-Emergency Missions	Missions Breakdown		Subjects
Medical Rescue	Vaccination Screening	Organisation	County CDC
			Village Health
		Get Vaccinated	Village Health
		Vaccinations — Vaccine Financial Support	County Finance
		Oversight Advancement	Enterprise
			Village Government
			Village Health
		Release Of Vaccination News	Village Government
		Follow-Up On Site	Village Council
		Arranging The Vaccination Process	Village Health
			Village Health
		Supervision	County CDC
	Professional Rescue	Rescue Service	Village Health
		Medical Material Security	Village Health Centre
Financial Support	Quarantine	Centralised Quarantine	County Hospital
		Home Quarantine	Village Health
	Basic Financial Support	Financial Funds	Village Finance
		Donation Management	Village Finance
	Donation		Enterprise
			Individuals

Macro-Emergency Missions	Missions Breakdown			Subjects
Transport	Traffic Control	Checkpoint Guarding	Checkpoint Stands	Village Council
			Checkpoint Guard	Volunteers
			Checkpoint Life	Village Council
			Checkpoint Medical Supplies	Village Government
		Quarantine And Disinfestation Of Transport		Village Council
				Volunteers
	Transport Of Goods	Distribution Of Goods		Village Council
				Volunteers
Publicity And Education	Publicity And Education	Information Release	News Report	Propaganda
		Public Opinion Management	Public Opinion Leadership	
			Public Opinion Monitoring	
Public Security Maintenance	Public Security			Village Police
				Relevant Persons From
		Crime Investigation And Prosecution		Law Enforcement Teams
Material Support	Material Support	Issuance Of Protective Materials	Provision Of Material	Village Health
			Receiving Materials	Village Health

Fig. 5. Breakdown of tasks and emergency response subjects in Village X

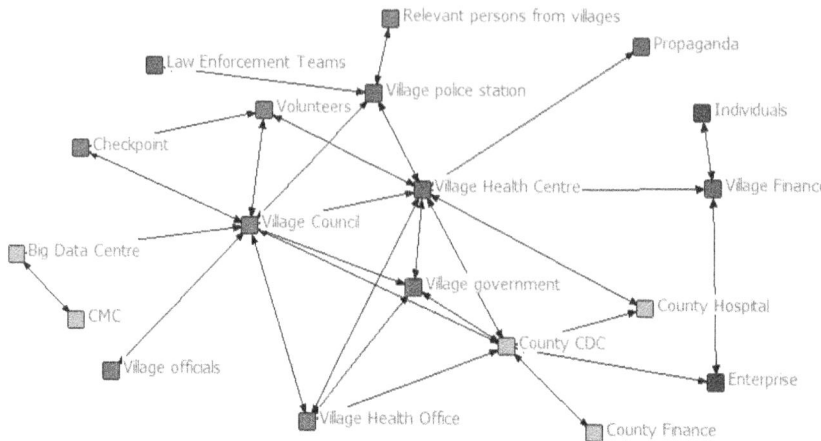

Fig. 6. Synergistic Network of Emergency Response Subjects in Village X

Emergency Response Subjects Synergistic Network for Z City. City Z, situated in Guangdong Province, China, confronted a significant disease outbreak linked to an individual in Zhuhai. Upon confirming the first case on January 13, 2022, the city swiftly initiated its emergency response plan. Measures outlined in Documents 1 and 2 included school closures, restrictions on non-essential businesses, and comprehensive nucleic acid testing, with normalcy restored by January 27. The surge in cases exceeded City Z's testing capabilities, leading to an urgent deployment of over 3,000 medical personnel from surrounding areas and the engagement of 10 external labs to aid in testing efforts. In the same way, macro-emergency missions were broken down into sub-missions. The result of the mission breakdown is shown in Fig. 7, where the right column is the sub-tasks

Macro-Emergency Missions	Missions Breakdown	Subjects
Professional Teams	Information Collection	Information Collation → Health Department
		Information dissemination → Publicity
		Public Opinion Leadership → Publicity
	Epidemiological Surveys	Field Epidemiological Investigation → Public Security
		Municipal CDC
		Community
		Data Epidemiological Investigation → Health Department
		Big Data Center
		Public Security
	Nucleic Acid Testing	Domain Detection → Health Department
		Public Security
		Municipal CDC
		Township Government
		Volunteer
		Community
		Extraterritorial Dispatch → Provincial Healthcare Departments
		Extra-Area Testing Organisations
		Extra-Area Hospitals
Financial Support	Emergency Fund	Finance Department
		Medical Insurance
		Medical Institutions

Macro-Emergency Missions	Missions Breakdown		Subjects
Emergency Command	Multilevel Linkage	Internal Linkage	Committee
			Government
			Discipline Inspection
			Publicity
			Big Data Center
			Development Reform
			Culture And Tourism
			Health Department
			Township Government
		Cross-level Linkages	Community
		Extraterritorial Dispatch	Committee
			Government
			Provincial Government
			Provincial Healthcare Departments
Material Support	Prevention And Control of Material Security		Industry And Information Technology
			Development Reform
			Market Supervision
			Health Department
	Transport of Materials		Transportation
	Material Donation		Township Government

Macro-Emergency Missions	Missions Breakdown	Subjects
Public Security Maintenance	Social Stability	Politics And Law
		Public Security
		Judicial
		Transportation
		Township Government
Medical Rescue	Linkage Guarantee	Committee
		Government
	Medical Services	Medical Institutions
		Township/Medical Institution
	Transport of Personnel	Medical Institutions
Basic Life Support	Daily Life Security	Civil Affairs
		Industry And Information Technology
		Commerce
		Township Government
	Lead Management	Committee

Fig. 7. Breakdown of tasks and emergency response subjects in City Z

of the left column. In addition, the cross-regional major infectious disease emergency subjects synergistic network in City Z is shown in Fig. 8.

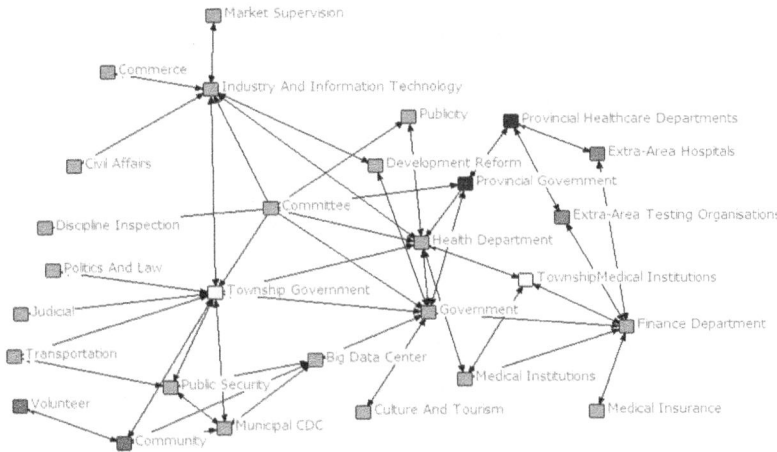

Fig. 8. Cross-regional Synergy Network of Emergency Response Subjects in City

4 Assessment of Synergy for Subjects Synergistic Network and Emergency Response Subjects

4.1 Assessment of Synergy for Subjects Synergistic Network

Assessment of Metrics. Research indicates that the majority of scholars have utilized social network analysis (SNA) metrics to evaluate collaborative networks, as reviewed in the literature [9, 10, 13, 14]. Therefore, the present paper employs similar network metrics to assess the synergy within the emergency response subjects synergistic network. Six network metrics of the subjects synergistic network were calculated, which were network density, average distance, network centralization, clustering coefficients, network cohesion, and network efficiency. The first five metrics were used to analyze the network's structural properties, while network efficiency was applied to evaluate the overall synergy of the network.

Impact of the Absence of Emergency Subjects on Subjects Synergistic Network. For the emergency response in X village, the synergistic network was analyzed, showing a network density of 0.170 and an average distance of 2.38. The network centralization stood at 36.93%, which was notably higher than the 15.03% average for general public health events, suggesting a more obvious core within the emergency subjects' network for this outbreak.

The clustering coefficients reached 0.472, significantly higher than the 0.199 observed in Wuhan's normalized emergency cooperation network, indicating a modular clustering of emergency subjects within the network, with a tendency for subjects to cluster together. The cohesion was 0.502, indicating a high level of network cohesion, low reliance on core nodes, and good robustness. This implied that the network was stable and relatively unaffected by external factors, suggesting that the event's subjects synergistic network were stable. Lastly, the network efficiency was 0.928, indicating a very high level of interaction efficiency between nodes. This high efficiency points to an effective synergy among the emergency response subjects in the event. Although the network density was not high and the interaction cost of the network nodes was high, the interaction efficiency was also high, and the network structure was stable, which ultimately achieved a good synergy effect.

The centrality (Degree, Closeness, and Between) of each node of this network was calculated and the results are shown in Table 1. In practical emergency response situations, the participation of certain emergency subjects might be hindered due to factors such as a lack of awareness of their responsibilities or a significant number of staff infections that render them unable to contribute. To examine the impact of such absences on the network's overall synergy, this study simulated the removal of the nodes with the highest and lowest centrality values. The results are shown in Table 2.

Table 1. The centrality of network nodes of emergency response subjects in the domain

Emergency Response Subjects	Degree	Closeness	Betweenness
Village Council ★★★	9	28	68.50
Village Health Centre	9	28	62.83
Prefectural Disease Activity Prevention	7	33	32.00
Village government	4	36	0.00
Village Health Office	4	36	0.00
Village police station	4	36	33.00
Volunteers	3	39	2.33
Village Finance	3	41	19.33
County Hospital	2	42	0.00
Big Data Centre	2	43	17.00
Checkpoint	2	44	0.00
Village officials	1	45	0.00
Propaganda	1	45	0.00
Enterprise	2	46	2.00
County Finance	1	50	0.00
Relevant persons from villages	1	53	0.00
Law Enforcement Teams	1	53	0.00
Individuals	1	58	0.00
CMC ★	1	60	0.00

Table 2. Network indicator results of networks with deleted nodes

Metrics	Village Council	CMC	The Lowest Degree Centrality	Village Health Centre	The Lowest Betweenness Centrality
Centralization	38.20%	39%	39%	38.20%	39%
Density	0.131	0.183	0.183	0.131	0.183
Avg Distance	2.368	2.275	2.373	2.735	2.735
Compactness	0.354	0.521	0.509	0.408	2.288
Clustering Coefficient	0.379	0.515	0.478	0.434	0.472
Efficiency	0.9451	0.9191	0.9191	0.9667	0.9191

The results indicated that the impact of the absence of emergency subjects on the synergy within the major infectious disease emergency response network was not linearly correlated. Analyzing the effects from the perspectives of degree centrality, betweenness centrality, and closeness centrality, it was found that removing nodes with low closeness centrality had the most significant impact on the network's synergistic degree. This suggested that in actual emergency scenarios where the subjects remained constant, the primary consideration should be the proximity between subjects to minimize unnecessary interaction and enhance the efficiency of the emergency response synergy.

Impact of the Absence of Edges on Subjects Synergistic Network. The absence of synergy among emergency subjects may also occur in the actual emergency response process. For example, emergency subjects did not know with whom they should synergize. Or, strict traffic control could cut off the channels for transferring materials between emergency subjects, etc. To address this, the present paper explores the impact of missing edges on the network by deleting network edges. To this end, the study simulated the deletion of 10%, 20%, 30%, 40%, and 50% of the edges in the subjects synergistic network to assess the effects of such omissions on network performance and the results are shown in Table 3.

Table 3. Remove the Randomized Edge in the Synergy Network Indicator

Missing Edges	Efficiency	Difference
0%	0.928	0%
5%	0.9083	1.97%
10%	0.8333	7.50%
20%	0.7500	8.33%
30%	0.5333	21.67%
40%	0.3333	20.00%

The findings indicated that minor disruptions to the synergy among emergency response subjects caused only slight damage to the overall network. However, as the level of damage increased, the network synergy decreased progressively. When the damage reached 30%, the network synergy dropped significantly, nearly paralyzing the emergency response subject synergistic network. This also demonstrated that during the actual emergency response to major infectious diseases, strict control measures such as city-wide lockdowns and stringent traffic controls, if implemented, could severely impact the synergy among emergency subjects. It was shown that emergency command departments should have carefully considered the serious impacts of such strict control measures on the network's collaborative effectiveness.

4.2 Assessment of Synergy for Cross-Regional Emergency Response Subjects

This present paper constructed the emergency response subjects synergy assessment model. The formulas are as follows.

$$ce_i = \frac{a_i}{n-1} \tag{1}$$

$$Hc_i = -ce_i \log ce_i \tag{2}$$

$$W_i = \frac{1 - Hc_i}{n - \sum_{i=1}^{n} Hc_i} \tag{3}$$

$$C_i = W_i a_i \tag{4}$$

In the formulas, a_i was synergy factor, n was the number of emergency subjects, ce_i was the probability of synergy among emergency subjects, Hc_i was the synergetic entropy, W_i was the weights of emergency subjects, and C_i was the synergy of emergency subjects. The present paper constructs an emergency response network for City Z based on the epidemic in the city, using the emergency subjects as the nodes and the verbs between the subjects as the connecting edges. Then, the synergy of emergency subjects in the two networks in City Z was calculated separately and compared to each other. The results are shown in Table 4.

The results showed that nine emergency subjects need to improve their synergy in a major cross-regional infectious disease event in City Z. Among them, the municipal finance department, the township government, and the municipal CDC need to significantly increase their participation. There was also potential to improve the participation of the provincial government, township healthcare organizations, provincial healthcare departments, communities, extra-area hospitals, and extra-area testing organizations. In addition, the results also indicated that greater involvement of extra-aera emergency subjects in the intra-area emergency response process would be conducive to improving the effectiveness of the emergency response. Analyzed from another perspective, the epidemic involved three grass-roots emergency subjects, namely the township government, township medical institutions, and the community. They all had the potential to improve their emergency response synergy. They were at the forefront of the emergency response. This demonstrated that there was an urgent need for China's grassroots emergency subjects to strengthen their synergistic capacity and response capability to deal with major infectious disease epidemics.

Table 4. Synergy of Response Subjects and in Synergy Subjects City Z.

Response Subjects	a_i	c_i	Synergy Subjects	a_i	c_i
Big Data Center	9	0.300	Township Government	10	0.333
Government	8	0.267	Government	8	0.267
Health Department	8	0.267	Health Department	8	0.267
Committee	8	0.267	Industry And Information Technology	7	0.235
Development Reform	8	0.267	Committee	7	0.235
Industry And Information Technology	7	0.235	Finance Department	6	0.203
Township Government	6	0.203	Municipal CDC	4	0.139
Publicity	6	0.203	Provincial Healthcare Departments	4	0.139
Discipline Inspection	6	0.203	Public Security	4	0.139
Culture And Tourism	6	0.203	Community	4	0.139
Public Security	6	0.203	Big Data Center	4	0.139
Transportation	4	0.139	Provincial Government	3	0.106
Politics And Law	3	0.106	Township Medical Institutions	3	0.106
Medical Institutions	3	0.106	Medical Institutions	3	0.106
Judicial	3	0.106	Extra-Area Hospitals	2	0.073
Medical Insurance	2	0.073	Publicity	2	0.073
Provincial Healthcare Departments	2	0.073	Transportation	2	0.073
Finance Department	2	0.073	Extra-Area Testing Organisations	2	0.073
Community	2	0.073	Development Reform	2	0.073
Extra-Area Hospitals	1	0.038	Discipline Inspection	1	0.038
Municipal CDC	1	0.038	Politics And Law	1	0.038
Volunteer	1	0.038	Volunteer	1	0.038
Township Medical Institutions	1	0.038	Culture And Tourism	1	0.038
Market Supervision	1	0.038	Market Supervision	1	0.038
Extra-Area Testing Organisations	1	0.038	Medical Insurance	1	0.038
Provincial Government	1	0.038	Judicial	1	0.038
Civil Affairs	1	0.038	Civil Affairs	1	0.038
Commerce	1	0.038	Commerce	1	0.038

5   Conclusion

This paper defines emergency response subject synergy from the "missions-resources-subjects" perspective and proposes methods for determining synergy relationships. It constructs models for the synergy network of emergency response subjects and their evaluation, which are then validated through case studies. The results demonstrate that a synergy approach can significantly enhance emergency response efficiency compared to traditional command styles. In emergency management, this method assists governments in developing emergency plans and regulations, especially during the response phase, where it aids command centers in swiftly devising and assessing action plans. Emergency response subjects were proposed from the perspective of 'task-resource-subject'.

However, the study has limitations, such as the use of traditional methods for collecting textual data. Future research could employ web crawlers and Natural Language Processing (NLP) to improve efficiency, reduce the risk of incomplete data, and more accurately reflect the complexity of real emergencies. Moreover, the synergy network models presented in this paper are static and do not capture the dynamics of the emergency response process.

Acknowledgment. This work was supported by the National Natural Science Foundation of China (72271041; 72434001).

References

1. Comfort, L.K., Kapucu, N.: Inter-organizational coordinationin extreme events: the world trade center attacks, September 11, 2001. Nat. Hazards **39**(2), 309–327 (2006)
2. Kettl, D.F.: The job of government: interweaving public functions and private hands. Public Adm. Rev. **75**(2), 219–229 (2015)
3. Piraina, M., Trucco, P.: Emergency management capabilities of interdependent systems: framework for analysis. Environ. Syst. Decis. **42**(2), 149–176 (2022)
4. Lu, Y., Li, Y.: Cross-sector collaboration in times of crisis: findings from a study of the funing tornado in China. Local Gov. Stud. **46**(3), 459–482 (2020)
5. Myomin, T., Lim, S.: The emergence of multiplex dynamics between information provision ties and rescue collaboration ties: a longitudinal network analytic approach to flooding cases in Myanmar. Nat. Hazards **114**(1), 645–663 (2022)
6. Xyu, X., Fu, S., Zhu, X., Xue, L.: Joint crisis sensemaking: a review and research agenda. Bull. Natl. Nat. Sci. Found. China **34**(06), 693–702 (2020)
7. Fan, B., Zhang, Y.: Digital collaboration in emergency management: a case study on a megacity in China. China Soft Sci. (10), 76–87 (2023)
8. Tang, M., Liu, B., Li, S., Li, P.: Network analysis of sudden disaster emergency response partner relationship based on social network perspective—taking the "6·24" diexi landslide as an example. Oper. Res. Manage. Sci. **30**(04), 103–108 (2021)
9. Zhang, G., Lei, Y., Zhou, F.: Research on collaborative governance network of government emergency organization from the perspective of social network: taking the central-lejoint document policy as an example. Jinan J. **43**(11), 90–104 (2021)
10. Ku, M.: The dynamics of cross-sector collaboration in centralized disaster governance: a network study of interorganizational collaborations during the MERS epidemic in South Korea. Int. J. Environ. Res. Public Health 19–18 (2022)
11. Xie Y., Wang, H.: Air pollution control of "synergy-performance" system in Beijing-Tianjin-Hebei from perspective the dynamic space. Res. Socialism Chin. Characteristics (04), 57–66 (2021)
12. Hu, Q.: A review of recent research on emergency management networks. Public Adm. Policy Rev. **9**(01), 36–43 (2020)

13. Wang, H., Shi, Y., Guo, X.: Research on the evolution and performance of the emergency cooperation network for major public health emergencies: the case of Wuhan's response to COVID-19 response to COVID-19. Manage. Rev. **1** (2023)
14. Hang, Q., Wu, Z., et al.: Identifying dominant factors of waterlogging events in metropolitan coastal cities: the case study of Guangzhou, China. J. Environ. Manage. **271**, 11095 (2020)

Evolution of Cumulative Reciprocity in Structured Populations

Shuangling Luo[1], Zhenjia Tian[1], Juan Li[2,3,4(✉)] ⓘ, and Haoxiang Xia[2,3,4] ⓘ

[1] School of Maritime Economics and Management, Dalian Maritime University, Dalian 116026, China
[2] Institute of Systems Engineering, Dalian University of Technology, Dalian 116024, China
juan@dlut.edu.cn
[3] Center for Big Data and Intelligent Decision-Making, Dalian University of Technology, Dalian 116024, China
[4] Key Laboratory of Social Computing and Cognitive Intelligence (Dalian University of Technology), Ministry of Education, Dalian 116024, China

Abstract. Direct and network reciprocity are key mechanisms that promote cooperation in social dilemma games. Direct reciprocity focuses on exploring successful strategies in repeated interaction, while network reciprocity often simplifying players' strategies to pure strategies or a limited set of classic strategies. Consequently, understanding the evolutionary dynamics of large-scale and diversity strategies in structured populations remains a challenge. Recent study reveals the cumulative reciprocity strategy's extraordinary evolutionary advantages in well-mixed populations, but its performance in structured populations remains unclear. This study addresses this research gap by investigating the impact of the cumulative reciprocity strategy on the evolution of large-scale reactive strategy populations across different networks including lattice, small-world, and random networks. Through ecological evolutionary simulations, the results reveal that introducing cumulative reciprocity leads to noticeable changes in the evolutionary dynamics of populations employing reactive strategies. Cumulative reciprocity replaces the traditional generous tit-for-tat strategy as an evolutionarily dominant strategy, thereby fostering stable cooperation within the population. This conclusion holds true across various network densities, population size, mutation mechanisms, and network structures.

Keywords: Network Reciprocity · Complex Networks · Evolutionary Dynamics · Prisoner's Dilemma Games · Reactive Strategies

1 Introduction

The question of how cooperative behavior evolves is considered one of the most important scientific issues [1]. The social dilemma game effectively captures the theoretical conflict inherent in this question: individuals often choose to defect to maximize their own benefit, while cooperation can maximize the group's overall benefit [2]. Over the past few decades, researchers from various fields, including sociology [3, 4], biology [5–7], computer science [8–10], physics [11–14], and mathematics [15–17], have investigated the evolution of cooperation. Evolutionary game theory provides a valuable framework for studying this issue. Within this framework, researchers have proposed five main mechanisms that promote cooperation: direct reciprocity, indirect reciprocity, network reciprocity, kin selection and group selection [18]. Among these, direct reciprocity is the most fundamental, as it seeks to identify effective strategies in repeated games by focusing on the complexity of the strategies themselves. In contrast, network reciprocity aims to reveal the mechanisms underlying cooperation by exploring the complexity of group structures. Due to their different focuses, these two mechanisms are often studied independently. The prisoner's dilemma game is the most commonly used models for examining these mechanisms among self-interested individuals.

In repeated prisoner's dilemma games (PDG), players use historical interaction information to decide whether to cooperate or defect subsequent rounds. Previous research has primarily focused on memory-one strategies, proposing several well-known strategies such as tit-for-tat (TFT) [19], generous tit-for-tat (GTFT) [20, 21], win-stay lose-shift (WSLS) [22], and zero-deterministic (ZD) [23] strategies. Long-memory strategies, which capture more complex decision-making behaviors, have also been explored. Researchers have proposed several effective memory-2, memory-3 and even full-memory strategies, including CAPRI [24], TFT-ATFT [25], and All-Or-None (AON) [26] strategies. Despite these advancements, the role of memory length in promoting cooperation remains debated. Recently, Li et al. introduced a cumulative reciprocity (CURE) [27] strategy, which demonstrates that while long-memory strategies may not always yield higher payoffs against specific opponents, they exhibit strong adaptability in population evolution. This suggests that winning is not everything and adaptability is crucial. However, while the CURE strategy has shown excellent performance in well-mixed populations, its evolutionary effectiveness in structured populations remains unknown.

Network reciprocity examines how the complexity of group structures influences the evolution of cooperation. Nowak and May first introduced lattice networks into evolutionary game models [28]. Building on this foundation work, researchers have explored how different network types affect cooperation, including static topologies such as lattice, small-world, and scale-free networks [29–31], dynamic structures involving edge rewiring and migration mechanisms [32], and higher-order network structures [33]. Additionally, studies have combined spatial reciprocity with other mechanisms, such as rewarding and punishment [34, 35], reputation mechanisms [36–39], multilayered populations with asymmetric social interactions [40, 41], and group selection [42]. Network structures can either promote or hinder the evolution of cooperation, depending on the specific context [43]. Despite extensive research on network reciprocity, most studies have focused on the evolution of pure strategies, or a limited number of three or four strategies within structured populations. The effects of network structures on evolutionary dynamics of large-scale strategy populations still requires further investigation.

In summary, while the evolutionary advantages of the CURE strategy in large-scale well-mixed populations have been demonstrated, its effectiveness and evolutionary dynamics within the context of spatial reciprocity remain insufficiently understood.

Previous research [27] has highlighted the CURE strategy's excellent performance in prisoner's dilemma games, public goods games, stochastic games, and behavioral experiments. Further investigation into how network structures affect the evolution of the CURE strategy is crucial for expanding its applicability. Additionally, studying the evolution of large-scale strategies within structured populations holds important theoretical implications for broadening our understanding of spatial reciprocity mechanisms.

This study addresses this gap by integrating both direct and spatial reciprocity to investigate whether the CURE strategy retains its evolutionary advantage when applied to structured populations. It also examines the evolutionary dynamics of large-scale strategy groups across different network structures, providing new insights into the robustness and adaptability of CURE under varying conditions. Specifically, this study focuses on reactive strategies alongside the CURE strategy, conducting simulation experiments on classic static network structures such as lattice, small-world, and random networks, which represent a range of connectivity patterns and clustering properties. By doing so, this research not only deepens our understanding of CURE's performance in more complex and realistic settings but also contributes to the broader field of evolutionary game theory by exploring the interplay between different types of reciprocity in diverse network configurations.

2   The Model

2.1   Repeated Prisoner's Dilemma Games with Cumulative Reciprocity and Reactive Strategies

Our model is based on the repeated prisoner's dilemma game (PDG). In each round, players choose to cooperate or defect according to their strategy. If both players cooperate, they each receive a payoff of R. If both defect, they each receive a payoff of P. If one player cooperates while the other defects, the cooperator receives a payoff of S and the defector receives a payoff of T. The PDG requires the payoff to satisfy $T > R > P > S$ and $2R > T + S$. We use Axelrod's standard payoff matrix [44, 45], specifically $T = 5, R = 3, P = 1, and S = 0$, unless otherwise stated.

For the strategy set, we consider 121 reactive strategies and one cumulative reciprocity strategy. Following the methodologies outlined by Nowak and Sigmund [20] and Grim [46, 47], the reactive strategies are represented by the form (p, q), where both p and q range between 0 and 1. Here, p denotes an individual's probability of cooperating if the opponent cooperated in the previous round, while q denotes the probability of cooperating if the opponent defected in the previous round. This formulation allows each individual to adapt their behavior based on one step direct historical interactions, embodying direct reciprocity. Notably, p and q are not restricted to absolute values of 0 or 1. In an environment with a noise level of 0.1, the unconditional cooperation (ALLC) strategy approximates (0.99, 0.99), the unconditional detection (ALLD) strategy is (0.01, 0.01), and the TFT is expressed as (0.99, 0.01). The p and q values are incremented by 0.1 across 11 levels from 0.01 to 0.99, resulting in 121 distinct two-dimensional reactive strategies ranging from (0.01, 0.01) to (0.99, 0.99). This approach ensures a comprehensive exploration of reactive strategy dynamics within two-dimensional strategy space.

For the cumulative reciprocity strategy, we adhere to its original definition as described in reference [27]. This strategy relies on full historical interaction memory and is formalized using two counter variables, $n_A(k)$ and $n_B(k)$, which record the frequency of defections by the two players before round k. The difference $d(k) = n_B(k) - n_A(k)$ is referred to as the defection difference statistic. The cumulative reciprocator cooperates if and only if this statistic is below a predefined threshold, $d(k) \leq \Delta_A$. The threshold $\Delta_A \geq 0$ represents the player's tolerance level. A zero threshold means the opponent must be at least as cooperative as the cumulative reciprocator, while larger values indicate greater tolerance. In our simulations, we set the tolerance threshold to $\Delta_A = 1$, unless otherwise stated.

To increase the efficiency of the evolutionary experiment, we calculated the theoretical payoffs between various strategies before the game. For the payoffs between reactive strategies, the payoff $\pi(x, y)$ obtained by an individual x using strategy (p_x, q_x) and another individual y using strategy (p_y, q_y) in the game is defined as the average payoff that individual x can achieve over infinite repetition. According to Nowak and Sigmund [20], the formula for calculating $\pi(x, y)$ is as follows:

$$\pi(x, y) = c_x c_y R + c_x(1 - c_y)S + (1 - c_x)c_y T + (1 - c_x)(1 - c_y)P \qquad (1)$$

$$c_x = \frac{q_x + (p_x - q_x)q_y}{1 - (p_x - q_x)(p_y - q_y)} \qquad (2)$$

$$c_y = \frac{q_y + (p_y - q_y)q_x}{1 - (p_x - q_x)(p_y - q_y)} \qquad (3)$$

Here, c_x and c_y represent the limiting probability of individuals x and y adopting cooperative behavior, respectively. For the payoffs between reactive strategies and CURE, we conducted 1,000,000 rounds of repeated games between CURE and each reactive strategy. Each game consisted of 1000 repetitions, and the average payoff over these rounds was computed. For the payoffs between two cumulative reciprocators, For the payoffs between two accumulative reciprocators, we use their theoretical calculations as outlined in reference [27]. Specifically, for a CURE player with a tolerance level of 1, the payoff in a noise environment of 0.1 is approximately 2.986557. These average payoffs are used as fixed values for subsequent simulation experiments.

2.2 Evolutionary Dynamics in Structured Populations

Evolutionary games on complex networks can be examined through three main aspects: the game rules, the network model and the strategy updating mechanism. We have already described the game rules. Next, we focus on the network structure and evolutionary mechanisms.

We investigate evolutionary dynamics across various network structures, including lattice networks, Watts-Strogatz (WS) small-world networks, and random networks. These structures help explore spatial reciprocity by defining how local clusters of cooperative or defecting individuals interact and propagate strategies across the network. We primarily use lattice networks with $N = L \times L$, incorporating a periodic boundary condition where an individual on one edge of the lattice is a neighbor of an individual on the opposite edge. Each square in the lattice is occupied by an individual who interact only with its immediate neighbors. We adopt the classic Von Neumann neighborhood, where individuals interact with four immediate neighbors, up, down, left, and right. In addition, we use WS small-world networks with $N = 10,000$ nodes and a reconnection probability $p = 0.03$, resulting in an initial average node degree $k = 4$. To explore the influence of network density, we also vary k to 2 and 10 in different experiments. The results include experiments conducted on a fully random network with $N = 10,000$ nodes and an average node degree $k = 4$. By adjusting network density and node connectivity, we analyze the impact of network topology on strategy evolution, thereby linking the spatial arrangement of agents to their evolutionary trajectories.

In the evolution of populations, we employ Monte Carlo simulations. Each individual in the network undergoes two sequential steps: playing games and updating strategies. During each time step t, after all individuals have completed their games with their neighbors, the entire population undergoes the strategy updating process. The strategy update rule used in all experiments is the Fermi update rule, proposed by Szabó and Tőke [48]. The total payoff of individual i at time step t is denoted as $P_i(t)$. Individual i randomly selects one of its neighbor j and considers updating its strategy s_i based on the following function:

$$W\left(s_j(t) \rightarrow s_i(t+1)\right) = \frac{1}{1 + exp\left[\frac{P_i(t) - P_j(t)}{K}\right]} \tag{4}$$

where P_i and P_j represent the total payoffs of individual i and individual j, respectively, in the last generation of evolution. The parameter K represents the degree of rationality of an individual. When K approaches 0, individuals exhibit complete rationality. As K increases, rationality decreases, meaning the probability of an individual imitating a less successful strategy increase. In our simulations, we take $K = 0.1$. This mechanism simulates the nuanced decision-making processes occurring in both spatial and direct reciprocity contexts.

In addition to strategy updating, we also incorporate a mutation mechanism. After each round of the game, an individual updates its strategy by the Fermi function with probability $1 - \mu$ or changes its strategy through mutation with probability μ. When a mutation occurs, the individual's strategy randomly changes to one of the strategies in the set.

By combining these mechanisms, our model effectively integrates spatial and direct reciprocity, offering a comprehensive framework for examining the evolutionary dynamics of cooperation within large-scale populations across different network structures. This dual focus enables a deeper understanding of how strategies evolve in environments that reflect both direct interaction information and spatial interaction information.

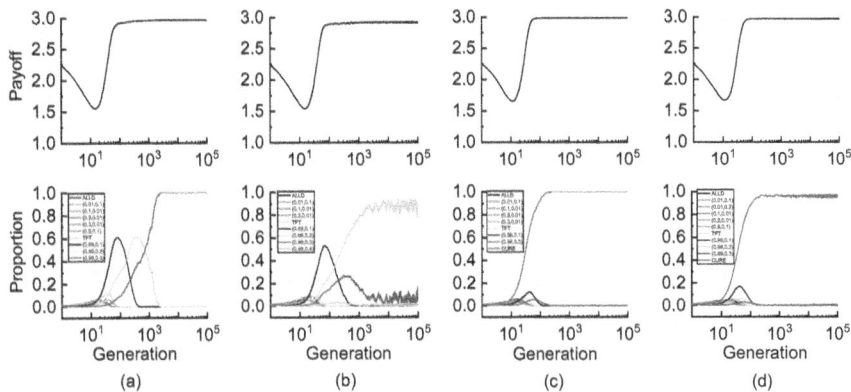

Fig. 1. Evolutionary dynamic of reactive strategy populations with and without the cumulative reciprocity strategy on lattice networks. Upper panels show population average payoffs, whereas lower panels display the strategies' proportions across the evolutionary process. In order to clarify the results, only typical strategies (like ALLD, TFT) and strategies reaching a certain proportion are shown here. We consider for scenarios: (a) evolution of populations including 121 reactive strategies without mutations; (b) evolution of populations including 121 reactive strategies with mutations; (c) evolution of populations including 121 reactive strategies and the cumulative reciprocity strategy without mutation; (d) evolution of populations including 121 reactive strategies and the cumulative reciprocity strategy with mutation. The scale of the lattice network is L = 100.

3   Results

3.1   Evolutionary Dynamics on the Lattice Network

First, we focus on a strategy set containing only reactive strategies and explore the evolutionary dynamics without and with mutation mechanisms, as shown in Fig. 1 (a) and (b). The results show that in the environment without mutation, the (0.99, 0.1) strategy initially dominates the evolution. As evolution progresses, the (0.99, 0.2) strategy replaces the (0.99, 0.1) strategy, ultimately leading to the gradual replacement of the (0.99, 0.2) strategy by the (0.99, 0.3) strategy, which subsequently occupies the entire evolutionary space. In a mutated environment, the (0.99, 0.1) strategy initially retains its dominance. However, the introduction of a mutation mechanism results in the (0.99, 0.3) strategy failing to displace the (0.99, 0.2) strategy. Consequently, the (0.99, 0.2) strategy ultimately occupies the entire evolutionary space.

In a system composed of multiple strategies, those that are excessively tolerant are more vulnerable to invasion and suppression by other strategies. As GTFT strategies become more tolerant, they begin to exhibit characteristics that are increasingly similar to those of ALLC strategies. In our experiments, ALLC did not perform well across various environments. This effect is especially pronounced when a mutation mechanism is introduced, increasing the complexity of the strategy composition within the population. In such scenarios, GTFT strategies exhibiting 10% and 20% tolerance demonstrate enhanced resilience, maintaining stability and competitiveness even when confronted with adversarial strategies.

Next, we add CURE strategies to the strategy set, resulting a total of 122 strategies, and the evolution of these strategies is simulated under conditions without mutation and with mutation, as shown in Fig. 1 (c) and (d).. The results show that in the absence of mutations, the proportion of CURE strategy rapidly increases and occupies the entire space, successfully suppressing the originally advantageous strategies. In an environment with mutations, although the proportion of CURE strategies fluctuated, the overall proportion remains above 90%, indicating that CURE strategies achieve evolutionary stability.

In order to elucidate this result, we initially attempt to provide an explanation from a macro-level perspective. The CURE strategy differs from the GTFT strategy by not relying on a fixed level of leniency when facing betrayal. Instead, it determines whether to retaliate based on the actual number of betrayals. This more precise punishment mechanism enables it to respond to treachery more swiftly and effectively, thereby reducing the risk of exploitation by defective strategies. By adjusting its responses according to the specific behaviors of its opponents, the CURE strategy can make more targeted decisions and is better suited to different game environments and a wide range of opponents.

To understand the formation of cooperation from the micro level, we use snapshots to observe the evolutionary dynamics. We explore the evolutionary dynamics of populations with 122 strategies on lattice networks by varying the population size and mutation mechanism. The results show that changing the network size do not affect the rapid expansion of CURE strategies in the population, and the mutation also do not alter the overall evolution trend. It only influences whether a single strategy remain after the population evolution stabilized, as shown in Fig. 2.

Specifically, in the initial stage of evolution, each individual on the lattice network randomly selects one of the 122 strategies, resulting in an initial proportion of each strategy being approximately equal at 1/122. In the 1st to 10th generations of evolution, defecting strategies absorbed neighboring strategies, while cooperative strategies began forming clusters to resist invasion. As the evolutionary process continued, the CURE strategy rapidly spread through space, quickly occupying most of it. This led to more than 85% of the population adopting CURE strategies, with a small number of GTFT strategies exhibiting varying degrees of tolerance. In the absence of mutation, the entire space is eventually occupied by the CURE strategy. It can be seen that the 'catalytic effect' of the CURE strategy on cooperation is relatively stable in the game between CURE strategy and reactive strategies.

In the next, we explore the effects of different R values on the evolution of 122 strategies in the lattice network. We find that as R increases, the advantage of CURE strategies in spatial evolution decreases, eventually coexisting with GTFT strategies of varying tolerance levels. The population payoff and ratios of strategies are shown in Fig. 3, and the snapshots of strategy evolution are shown in Fig. 4. An R value of 2.6 theoretically favors the expansion of defective strategies, but the CURE strategy effectively changes this situation. As the R value increases, the payoff of inter-strategy cooperation becomes higher, benefiting cooperative strategies. Therefore, from the strategy evolution ratio graph and snapshot, it can be seen that the proportion of GTFT strategies with different tolerance continues to rise with increasing R values, but the CURE strategy is not completely defeated. This shows that CURE is a strategy that can effectively counter

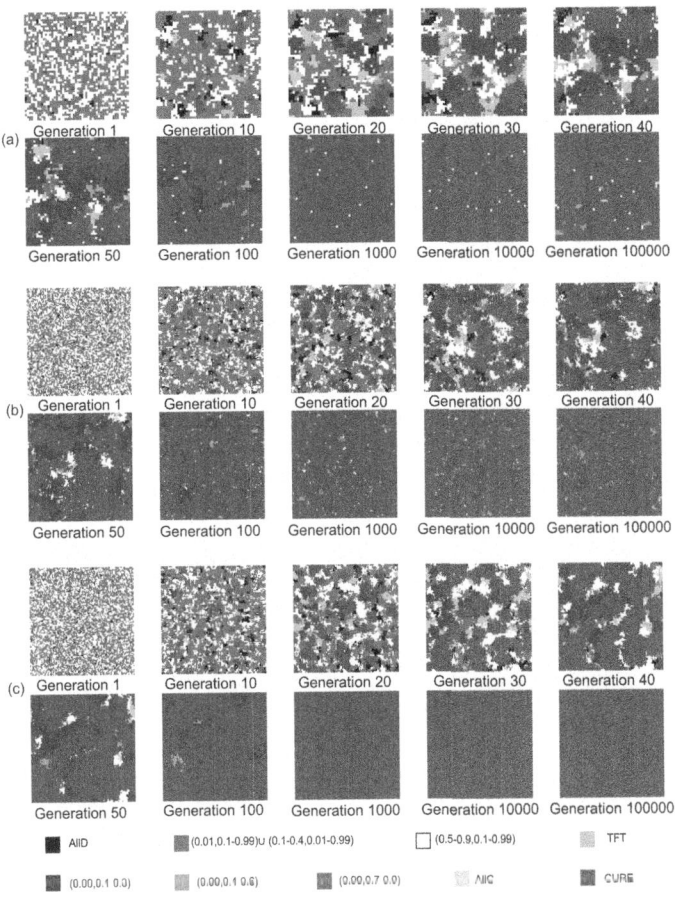

Fig. 2. Snapshots of strategies on lattice networks. We show the distribution of strategies under different evolutionary generations, respectively. (a) Strategy evolution snapshot in an L = 50 lattice network (with mutation); (b) Strategy evolution snapshot in an L = 100 lattice network (with mutation); (c) Strategy evolution snapshot in an L = 100 lattice network (without mutation). To facilitate observation, 122 strategies are divided into nine categories: black represents ALLD strategies; blue represents strategies with a 1%–40% probability of cooperating when the opponent cooperated in the last round (excluding ALLD); white represents strategies with a 50%–90% probability of cooperating when the opponent cooperated in the last round; light green represents TFT strategies; dark green represents GTFT strategies with 10%–30% tolerance; orange represents GTFT strategies with 40%–60% tolerance; magenta represents GTFT strategies with 70%–90% tolerance; yellow indicates ALLC policies; red represents the CURE strategy. (Color figure online)

defection and prevail over cooperative strategies at low or normal R values, and at least maintain its presence at high R values.

As the R-value increases, the reward for both parties choosing to cooperate becomes more appealing compared to betrayal or other non-cooperative behaviors, making the

R = 2.6
R = 3
R = 3.5
R = 4
R = 4.5
Payoff
Proportion
Generation

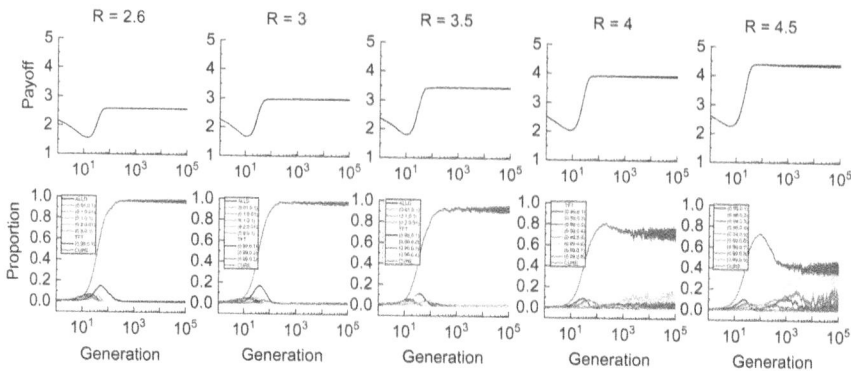

Fig. 3. Average payoff of populations and strategies' proportion for varying R values on lattice networks. We vary the value of R in payoff matrix used in Axelrod's classical computer experiment to test the evolution trend of the CURE strategy. We consider five conditions: R = 2.6, 3, 3.5, 4, and 4.5, while keeping S, T, and P values unchanged.

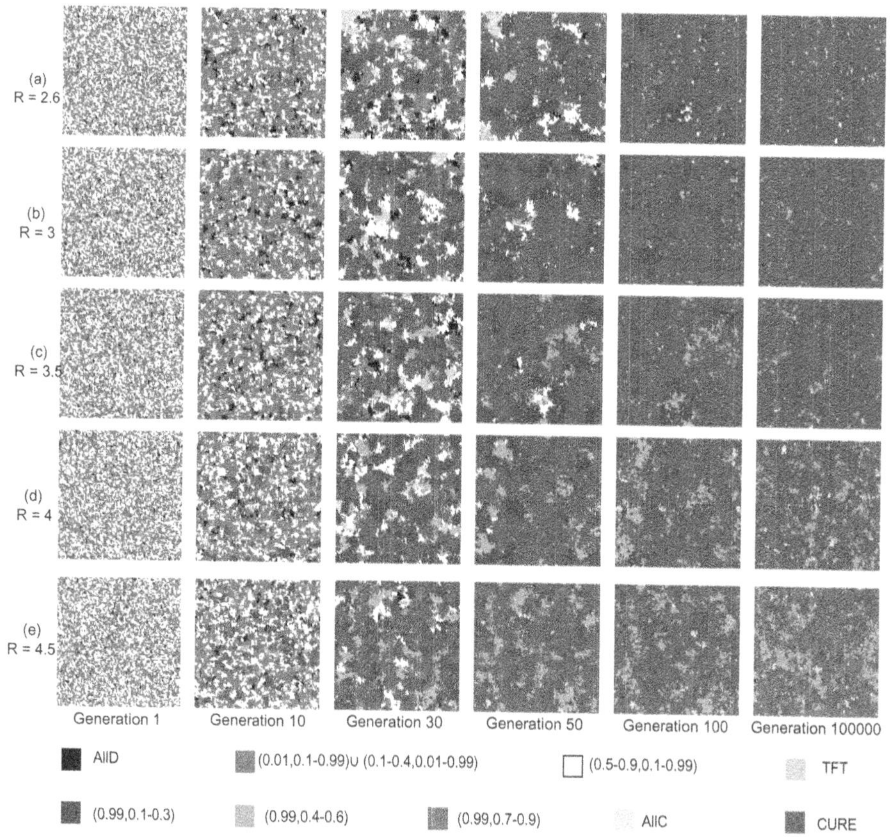

(a) R = 2.6
(b) R = 3
(c) R = 3.5
(d) R = 4
(e) R = 4.5
Generation 1
Generation 10
Generation 30
Generation 50
Generation 100
Generation 100000
AllD
(0.01,0.1-0.99)∪(0.1-0.4,0.01-0.99)
(0.5-0.9,0.1-0.99)
TFT
(0.99,0.1-0.3)
(0.99,0.4-0.6)
(0.99,0.7-0.9)
AllC
CURE

Fig. 4. Snapshots of strategies on lattice networks for varying R. (a) R = 2.6; (b) R = 3; (c) R = 3.5; (d) R = 4; (e) R = 4.5. Strategies classification and color are consistent with Fig. 2.

group more inclined toward a cooperative equilibrium state. This is evident in the snap-shot from the 30th generation, where the black ALLD strategy clusters struggle to survive in a cooperative environment with R-values of 4 and 4.5. In such an environment, the tolerant nature of the GTFT strategies allows it to better establish and maintain cooper-ative networks, resulting in stable and high returns. Furthermore, repeated one-on-one games between the CURE and ALLC strategies demonstrate that the CURE strategy struggles to win against the ALLC strategy. Thus, in an evolutionary space with high R-values that favor cooperation, GTFT strategies with high tolerance and near-complete cooperation form clusters that are difficult for the CURE strategy to invade.

3.2 Evolutionary Dynamics on Various Other Network Structures

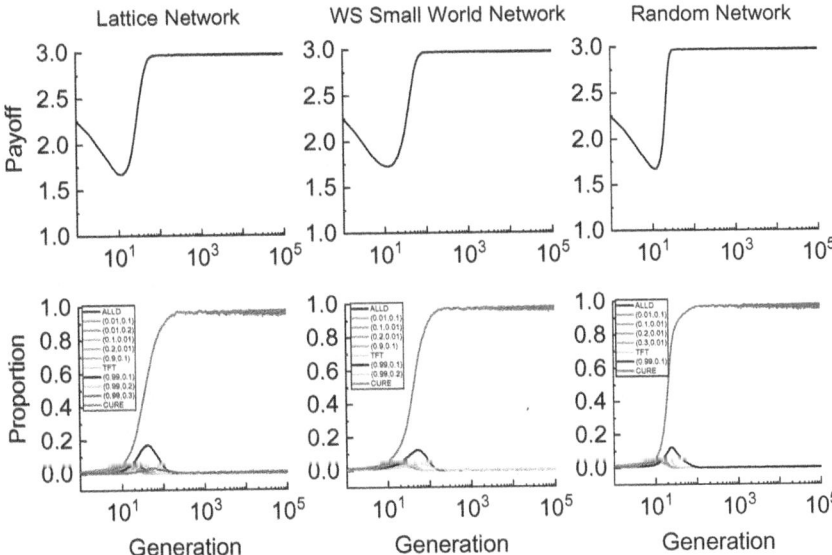

Fig. 5. Evolutionary dynamics of populations on different network structures. We investigate the average payoff of populations and proportions of strategies including 121 reactive strategies and CURE strategies on lattice network, WS small-world network, and random network, respectively. The results are observed in the present of mutations.

In addition to exploring the evolutionary dynamics on lattice networks, we also examined the evolutionary dynamics of populations including reactive strategies and the cumulative reciprocity strategy, on WS small-world networks and random networks, as shown in Fig. 5. The results show that in the three different network structures, the CURE strategy steadily maintains an evolutionary advantage. Given that the networks in question possess an identical network scale, adhere to a four-neighbor structure, and employ the same strategy update rules, it can be reasonably inferred that their evolutionary speeds are approximately equivalent.

The CURE strategy relies on full-history memory to track opponents' actions, allow-ing it to consistently make sound decisions across different network structures. Whether in a lattice network, a WS small-world network with a mix of short- and long-range connections, or a highly connected yet random network, the CURE strategy maintains high adaptability. In WS small-world and random networks, despite the presence of long-range or random connections, the evolution of strategies is still significantly influenced by local group behavior. The CURE strategy is effective in forming stable clusters in these network structures due to its local defence capabilities and ability to sustain cooperation. Consequently, CURE appears to have an evolutionary advantage across various network structures.

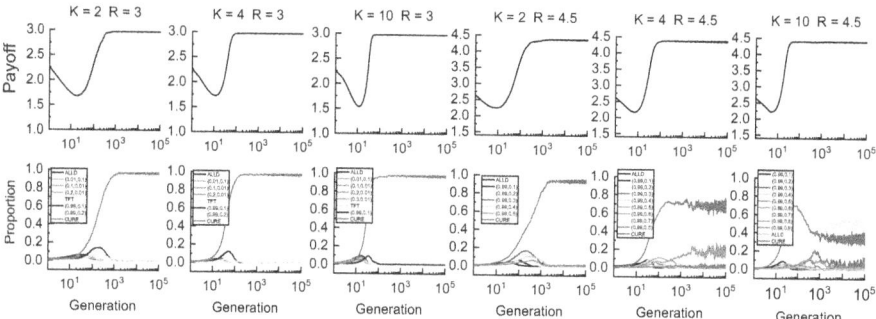

Fig. 6. The influence of network density k on the evolutionary dynamics of WS small-world network for varying R values. The upper panels show the evolution of averaged population payoffs, and the lower panels show the evolution of strategies. The k values are 2, 4, and 10 respectively and the R values are 3 and 4.5.

Next, we explore the evolutionary dynamics of 122 strategies with different k and R values in the WS small-world network, as shown in Fig. 6. In network theory, the k value refers to the number of neighbors of each node in the network. An increase in node degree k means that each node is connected to more neighbors. A higher k allows individuals to interact with more neighbors in each round of the game, leading to faster transmission of strategies and information. As strategy changes spread more rapidly across the network, the evolutionary process accelerates, as shown in Fig. 6.

Additionally, when both the R value and k value increased simultaneously, the evolutionary advantage of the CURE strategy gradually weakened, while the proportion of GTFT strategies with varying levels of tolerance steadily increased. We believe that this phenomenon occurs because the simultaneous increase of both values enhances the tendency for cooperation and the speed of propagation within the network. This allows the GTFT strategy, which already has some evolutionary advantage, to further accelerate its spread and suppress the CURE strategy group.

4   Conclusions

Currently, there is a significant gap in understanding the dynamic mechanisms of cooperation evolution that integrate spatial and direct reciprocity. This study aims to address this gap by examining the evolutionary advantages of cumulative reciprocity strategies (CURE) within large scale populations on complex networks. Our findings indicate that when CURE strategies are introduced into populations initially dominated by reactive strategies, CURE becomes the dominant strategy, promoting sustained, near-complete cooperation. These results are robust across various network structures, densities, population sizes, and conditions with or without mutations.

Specifically, in the lattice network, our results consistently demonstrate that the introduction of CURE strategy confers evolutionary advantages regardless of the presence of mutation, effectively overcoming evolutionary strategies predominant in groups composed solely of reactive strategies. Furthermore, by varying the network size of the lattice network, and combined with the snapshot analysis, we observed that the CURE strategy can maintain stable evolutionary advantages across different network sizes, and facilitating the formation of cooperative clusters that rapidly proliferate within the population. Additionally, analysis of the game return matrix (R value) revealed that as the R value increases, the evolutionary advantage of CURE strategies diminishes, yielding to strategies akin to Generous Tit-for-Tat (GTFT) that exhibit higher tolerance. Our exploration extended to WS small-world networks and random networks, where we consistently observed the evolutionary superiority of CURE strategies across diverse complex networks. Adjusting network density and cooperation income R-values within WS small-world networks further affirmed that while changes in network density do not impact the evolutionary advantage of CURE strategies, higher R-values in denser networks weaken the efficacy of CURE strategies similar to findings in lattice networks.

In the future, this paper needs to carry out further research. On the one hand, the co-evolution mechanism of strategy and network structure is explored in spatial reciprocity, and on the other hand, the influence of network structure on the evolution of CURE strategy is discussed from a microscopic perspective.

Acknowledgments. This work was supported by the National Natural Science Foundation of China under grant 72371052 (to H.X.) and China Postdoctoral Science Foundation 2023TQ0044 (to J.L.).

Disclosure of Interests.   The authors have no competing interests to declare that are relevant to the content of this article.

References

1. Pennisi, E.: How did cooperative behavior evolve? Science **309**, 93 (2005)
2. Dawes, R.M.: Social dilemmas. Annu. Rev. Psychol. **31**, 169–193 (1980)
3. Perc, M., Gomez-Gardenes, J., Szolnoki, A., Floria, L.M., Moreno, Y.: Evolutionary dynamics of group interactions on structured populations: a review. J. R. Soc. Interface. **10**, 20120997 (2013). https://doi.org/10.1098/rsif.2012.0997

4. Hauert, F.F.C.: Reputation-based partner choice promotes cooperation in social networks. Phys. Rev. E **78** (2008)
5. Szolnoki, A., Perc, M.: Antisocial pool rewarding does not deter public cooperation. Proc. Biol. Sci. **282**, 20151975 (2015). https://doi.org/10.1098/rspb.2015.1975
6. Su, Q., Zhou, L., Wang, L.: Evolutionary multiplayer games on graphs with edge diversity. PLoS Comput. Biol. **15**, e1006947 (2019). https://doi.org/10.1371/journal.pcbi.1006947
7. Boyd R, Richerson, P.J.: Group beneficial norms can spread rapidly in a structured population. J. Theor. Biol. **215** (2002)
8. Bachrach, Y., Porat, E., Rosenschein, J.S.: Sharing rewards in cooperative connectivity games. J. Artif. Intell. Res. (2013)
9. Ma, J., Zheng, Y., Wang, L.: Nash equilibrium topology of multi-agent systems with competitive groups. IEEE Trans. Industr. Electron. **64**, 10 (2017)
10. Zhang, Y., Wang, J., Ding, C., Xia, C.: Impact of individual difference and investment heterogeneity on the collective cooperation in the spatial public goods game. Knowl.-Based Syst. **136**, 150–158 (2017)
11. Szolnoki, A., Perc, M.: Reward and cooperation in the spatial public goods game. Europhys. Lett. **92**, 38003 (2010)
12. Xia, C.-Y., Meloni, S., Perc, M., Moreno, Y.: Dynamic instability of cooperation due to diverse activity patterns in evolutionary social dilemmas. Europhys. Lett. **109**, 58002 (2015)
13. Perc, M.: Double resonance in cooperation induced by noise and network variation for an evolutionary prisoner's dilemma. New J. Phys. **8**, 183 (2006)
14. Szolnoki, A., Perc, M.: Evolutionary advantages of adaptive rewarding. New J. Phys. **14**, 093016 (2012)
15. Li, C., Xu, H., Fan, S.: Synergistic effects of self-optimization and imitation rules on the evolution of cooperation in the investor sharing game. Appl. Math. Comput. **370**, 124922 (2020)
16. Nagashima, K., Tanimoto, J.: A stochastic Pairwise Fermi rule modified by utilizing the average in payoff differences of neighbors leads to increased network reciprocity in spatial prisoner's dilemma games. Appl. Math. Comput. **361**, 661–669 (2019)
17. Pan, Q., Wang, L., He, M.: Social dilemma based on reputation and successive behavior. Appl. Math. Comput. **384**, 125358 (2020)
18. Nowak, M.A.: Five rules for the evolution of cooperation. Science **314**, 1560–1563 (2006)
19. Axelrod, R., Hamilton, W.D.: The evolution of cooperation. Science **211**, 1390–1396 (1981)
20. Nowak, M.A., Sigmund, K.: Tit for tat in heterogeneous populations. Nature **355**, 250–253 (1992)
21. Molander, P.: The optimal level of generosity in a selfish, uncertain environment. J. Conflict Resolut. **29**, 611–618 (1985)
22. Nowak, M., Sigmund, K.: A strategy of win-stay, lose-shift that outperforms tit-for-tat in the Prisoner's Dilemma game. Nature **364**, 56–58 (1993)
23. Press, W.H., Dyson, F.J.: Iterated Prisoner's Dilemma contains strategies that dominate any evolutionary opponent. Proc. Natl. Acad. Sci. **109**, 10409–10413 (2012)
24. Murase, Y., Baek, S.K.: Five rules for friendly rivalry in direct reciprocity. Sci. Rep. **10**, 1–9 (2020)
25. Do Yi, S., Baek, S.K., Choi, J.-K.: Combination with anti-tit-for-tat remedies problems of tit-for-tat. J. Theor. Biol. **412**, 1–7 (2017)
26. Hilbe, C., Martinez-Vaquero, L.A., Chatterjee, K., Nowak, M.A.: Memory-n strategies of direct reciprocity. Proc. Natl. Acad. Sci. **114**, 4715–4720 (2017)
27. Li, J., et al.: Evolution of cooperation through cumulative reciprocity. Nat. Comput. Sci. (2022). https://doi.org/10.1038/s43588-022-00334-w
28. Nowak, M.A., May, R.M.: Evolutionary games and spatial chaos. Nature **359**, 826–829 (1992)

29. Szabó, G., Vukov, J., Szolnoki, A.: Phase diagrams for an evolutionary prisoner's dilemma game on two-dimensional lattices. Phys. Rev. E **72**, 047107 (2005)
30. Hauert, C., Szabó, G.: Game theory and physics. Am. J. Phys. **73**, 405–414 (2005)
31. Santos, F.C., Pacheco, J.M.: Scale-free networks provide a unifying framework for the emergence of cooperation. Phys. Rev. Lett. **95**, 098104 (2005)
32. Su, Q., McAvoy, A., Plotkin, J.B.: Strategy evolution on dynamic networks. Nat. Comput. Sci. **3**, 763–776 (2023)
33. Sheng, A., Su, Q., Wang, L., Plotkin, J.B.: Strategy evolution on higher-order networks. Nat. Comput. Sci. 1–11 (2024)
34. Fehr, E., Gächter, S.: Altruistic punishment in humans. Nature **415**, 137–140 (2002)
35. Wang, Z., Xia, C.-Y., Meloni, S., Zhou, C.-S., Moreno, Y.: Impact of social punishment on cooperative behavior in complex networks. Sci. Rep. **3**, 3055 (2013)
36. Chen, M.-H., Wang, L., Sun, S.-W., Wang, J., Xia, C.-Y.: Evolution of cooperation in the spatial public goods game with adaptive reputation assortment. Phys. Lett. A **380**, 40–47 (2016)
37. Quan, J., Tang, C., Zhou, Y., Wang, X., Yang, J.-B.: Reputation evaluation with tolerance and reputation-dependent imitation on cooperation in spatial public goods game. Chaos Solitons Fractals **131**, 109517 (2020)
38. Yang, H.-X., Yang, J.: Reputation-based investment strategy promotes cooperation in public goods games. Phys. A **523**, 886–893 (2019)
39. Tanimoto, J.: Does information of how good or bad your neighbors are enhance cooperation in spatial Prisoner's games? Chaos Solitons Fractals **103**, 184–193 (2017)
40. Su, Q., McAvoy, A., Plotkin, J.B.: Evolution of cooperation with contextualized behavior. Sci. Adv. **8**, eabm6066 (2022)
41. Su, Q., Allen, B., Plotkin, J.B.: Evolution of cooperation with asymmetric social interactions. Proc. Natl. Acad. Sci. **119**, e2113468118 (2022)
42. Shi, Z., Wei, W., Perc, M., Li, B., Zheng, Z.: Coupling group selection and network reciprocity in social dilemmas through multilayer networks. Appl. Math. Comput. **418**, 126835 (2022)
43. Qi, S., Shilin, X., Qionglin, D., Haihong, L., Junzhong, Y.: Spatial structure might impede cooperation in evolutionary games with reinforcement learning. Int. J. Mod. Phys. C, Phys. Comput. (2022)
44. Axelrod, R.: Effective choice in the prisoner's dilemma. J. Conflict Resolut. **24**, 3–25 (1980)
45. Axelrod, R.: More effective choice in the prisoner's dilemma. J. Conflict Resolut. **24**, 379–403 (1980)
46. Grim, P.: The greater generosity of the spatialized prisoner's dilemma. J. Theor. Biol. **173**, 353–359 (1995)
47. Grim, P.: Spatialization and greater generosity in the stochastic Prisoner's Dilemma. BioSystems **37**, 3–17 (1996)
48. Szabó, G., Tőke, C.: Evolutionary prisoner's dilemma game on a square lattice. Phys. Rev. E **58**, 69 (1998)

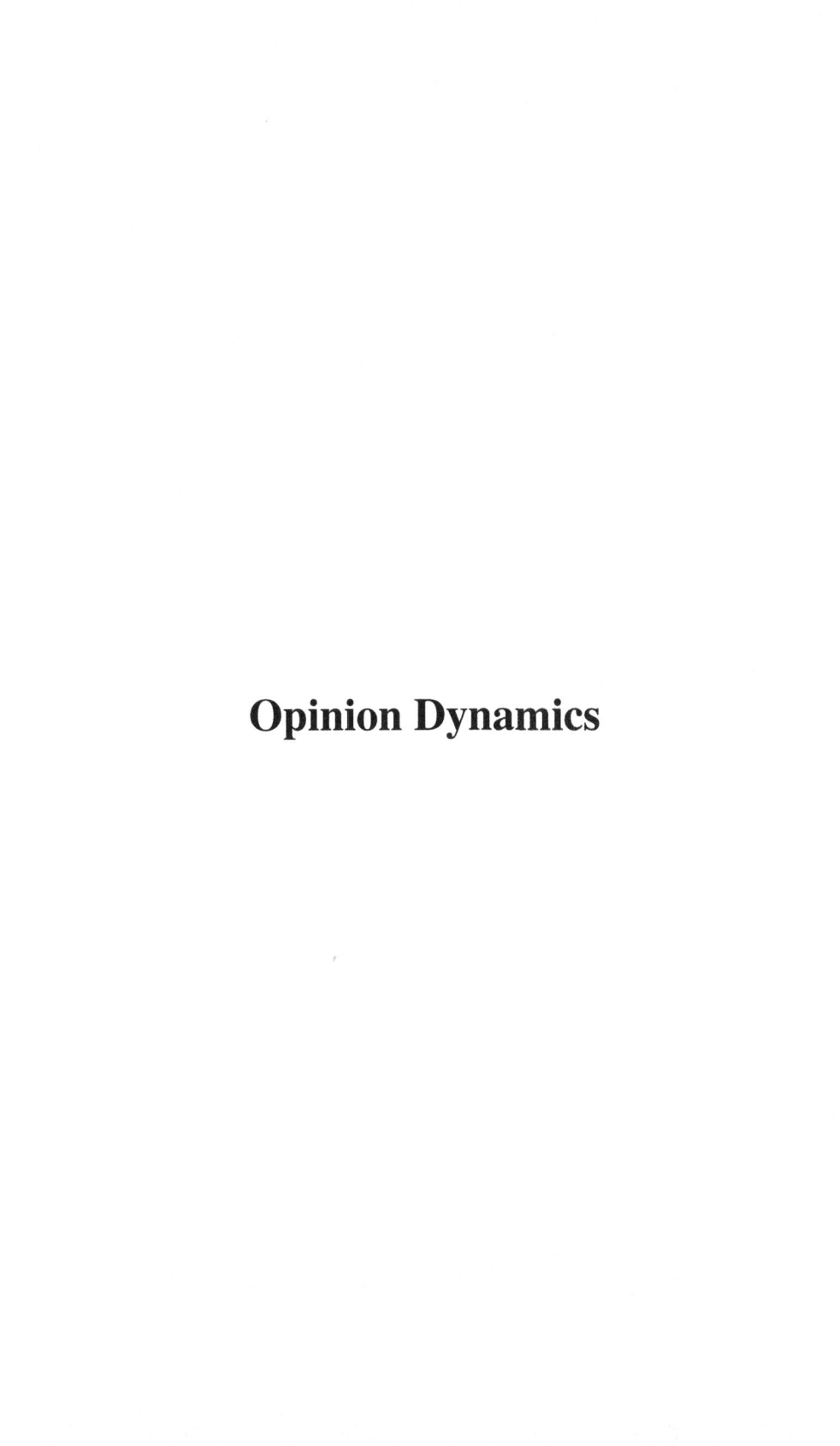

Opinion Dynamics

Simulating Social Network with LLM Agents: An Analysis of Information Propagation and Echo Chambers

Wenzhen Zheng[1,2] and Xijin Tang[1,2(✉)]

[1] Academy of Mathematics and Systems Science, Chinese Academy of Sciences, Beijing 100190, China
xjtang@iss.ac.cn
[2] University of Chinese Academy of Sciences, Beijing 100139, China

Abstract. In this paper, we leverage the capabilities of GPT-4o, which is integrated with cognitive modules such as memory, reflection, and chain of thought in simulating social network dynamics. Focusing on the Roe v. Wade case, which significantly influenced abortion laws and women's rights in the United States, we model public opinion propagation on Twitter through likes, retweets, comments, and posts. Multiple simulation experiments are conducted to observe agents over different time intervals. By employing small-world networks and scale-free networks, we analyze the impact of various network structures. Our comparative study reveals variations in interaction patterns and potential echo chamber effects.

Keywords: Social Network Simulation · LLM Agent · Public Opinion Propagation · Echo Chamber Effect · Agent-Based Modeling · Information Dynamics

1 Introduction

The dynamics of social networks have become a significant area of study, particularly with the rise of social media platforms such as Twitter, which significantly influence public opinion and behavior [1,2]. Recent advancements in natural language processing (NLP) and large language models (LLMs) like GPT-4 offer new possibilities for simulating and analyzing these complex interactions [3,4]. By integrating cognitive modules such as memory, reflection, and chain of thought, these models can more accurately mimic human decision-making processes [5]. This paper explores the use of these advanced models to simulate public opinion propagation, focusing on the Roe v. Wade debate as a case study.

In this paper, we model the mechanisms of public opinion propagation on Twitter, including likes, retweets, comments, and posts, using the Roe v. Wade case as an example. By conducting multiple simulation experiments, we observe agents' actions, attitudes, and attitude scores over various time intervals. To

X. Tang et al. (Eds.): KSS 2024, CCIS 2269, pp. 63–77, 2025.
https://doi.org/10.1007/978-981-96-0178-3_5

understand the impact of different network structures, we employ small-world networks and scale-free networks as control variables [6–8]. Our approach allows for a controlled examination of how information disseminates through these networks, providing insights into interaction patterns and potential echo chamber effects [9].

Our study includes experiments to explore the dynamics of information spread and echo chamber formation in detail. By employing 15 agents over 10 timesteps, with each timestep representing half a day, we compare results to identify key differences in interaction patterns and how network structures affect information dissemination.

Understanding how public opinion spreads through different networks is significant for policymakers and social media platforms, as it can inform strategies to manage information and reduce misinformation [2,10]. Integrating advanced NLP models into social network simulations offers new interdisciplinary research opportunities that combine computer science, psychology, and sociology [11]. Hence, this paper not only advances theoretical understanding of social networks but also provides practical applications to improve information management in social networks.

This paper is structured as follows: Sect. 2 reviews the relevant literature on social network analysis. Section 3 details the methodology, including the simulation frameworks and the incorporation of cognitive modules in LLM agents. Section 4 describes the experimental setup, network initialization, agent interactions, and simulation parameters. Section 5 presents the results of our simulations, including visualizations of opinion dynamics, echo chamber effects, and attitude changes. Section 6 discusses the findings and their implications for information dissemination in social networks.

2   Related Work

Recent advancements in LLMs have significantly expanded the capabilities of agent-based modeling (ABM). The integration of LLMs into ABM frameworks has enabled the simulation of complex agent behaviors and interactions within various domains, enhancing the realism and applicability of these models.

Brown et al. (2020) further showcased the potential of LLMs with their work on GPT-3, highlighting the models' capabilities in few-shot learning scenarios [3]. This advancement has created new opportunities for developing agents capable of understanding and producing human-like text, making them more effective in simulating realistic interactions. Park et al. (2023) in their innovative study present a groundbreaking application of LLMs in creating a simulated environment known as "Stanford Town" [12]. This environment features generative agents that exhibit human-like behaviors and interactions, grounded in contextually rich scenarios.

Building on these foundational advancements, modern research has increasingly used LLMs to simulate complex agent behaviors and interactions in diverse fields. Notably, Yukhymenko et al. (2022), in their work present a synthetic dataset, SynthPAI, designed to address privacy risks in personal attribute inference by employing LLMs to generate synthetic profiles and interactions on Reddit [13]. Similarly, Li et al. (2023) evaluate the proficiency of LLMs in generating counterfactual scenarios across various natural language understanding tasks [14].

The potential of LLMs for large-scale social simulations is shown in Mou et al. (2023) with their study [15]. This work uses LLM-powered agents to simulate social movement dynamics, giving insights into how social change and mobilization happen on digital platforms. Further advancing this field, Vezhnevets et al. (2023) uses LLMs to create actions that fit various environments [16]. Concordia demonstrates how using LLMs in agent-based modeling (ABM) can make simulations richer, providing a flexible tool for studying complex systems and agent behavior.

Furthermore, Gao et al. (2023) provide an extensive review of the current state and future directions of integrating LLMs into ABM [17]. The challenges and opportunities in areas such as environment perception, human alignment, action generation, and evaluation are highlighted by the survey. The latest works in various domains–including network, physical, social, and hybrid environments–are also reviewed, emphasizing the potential of LLMs to enhance the realism and applicability of ABM in both real-world and virtual settings.

Compared to previous related work, we mainly focus on the application value of LLM in social network simulation. By creating relevant agents and various social network structures, we apply LLM simulations to topics in social sciences such as information dissemination and the echo chamber effect. We provide several well-designed control experiments and visualizations of simulations, offering new insights for social network simulation.

3   Methodology

3.1   Overview

In this study, we utilize advanced NLP models and simulation frameworks to investigate the propagation of public opinion on social networks. Our approach builds on the HiSim[1] and AgentVerse[2] frameworks [18], which provide robust platforms for simulating multi-agent interactions and social network. This integration allows us to simulate agents with sophisticated cognitive abilities, including memory, reflection, and chain of thought, enhancing the realism and depth of our simulations.

3.2   Large Language Model Agents

Based on the work by Cheng et al. (2024), we incorporate LLM agents into our simulations [19]. These agents are designed to exhibit advanced cognitive

---

[1] https://github.com/xymou/HiSim.
[2] https://github.com/OpenBMB/AgentVerse.

functions, which enhance their ability to simulate human-like interactions in social networks. The key cognitive modules integrated into these agents include:

Memory: This module enables agents to retain and recall past interactions and information, which influences their future decisions and behaviors. By simulating memory, agents can develop more consistent and context-aware responses, closely mimicking human behavior.

Reflection: Reflection allows agents to evaluate their actions and adjust their strategies based on outcomes and feedback from the environment. This capability is crucial for modeling adaptive behaviors and understanding how individuals learn and evolve over time.

Chain of Thought: This module facilitates complex reasoning processes, enabling agents to make more informed decisions by considering multiple factors and potential consequences. It allows for more sophisticated decision-making, reflecting the nuanced thought processes humans use in real-world scenarios.

These cognitive capabilities, powered by LLMs like GPT-4o, significantly enhance the realism and depth of our simulations, providing deeper insights into the dynamics of public opinion propagation on social networks.

3.3   Simulation Frameworks

HiSim. HiSim is a social media simulation framework that categorizes users into two types: core users driven by LLMs and numerous ordinary users modeled by deductive agent-based models. HiSim allows for the customization of agent behaviors and network structures, providing a versatile platform for studying the dynamics of information dissemination. In our study, HiSim is used to create various network configurations, including independent agents, small-world networks, and scale-free networks, to examine how these structures influence public opinion propagation.

AgentVerse. AgentVerse is designed to facilitate the deployment of multiple LLM-based agents in various applications, primarily providing two frameworks: task-solving and simulation. It extends the capabilities of HiSim by integrating advanced NLP models, enabling the simulation of agents with more complex cognitive functions. This framework supports the implementation of LLMs like GPT-4, allowing agents to engage in more realistic and context-aware interactions. By leveraging AgentVerse, we can model agents' decision-making processes more accurately, reflecting real-world scenarios where individuals' opinions and actions are influenced by their cognitive abilities and social contexts.

4   Experimental Design and Procedure

The flowchart (Fig. 1) provides a visual representation of each step involved in the experiment, ensuring a clear understanding of the process.

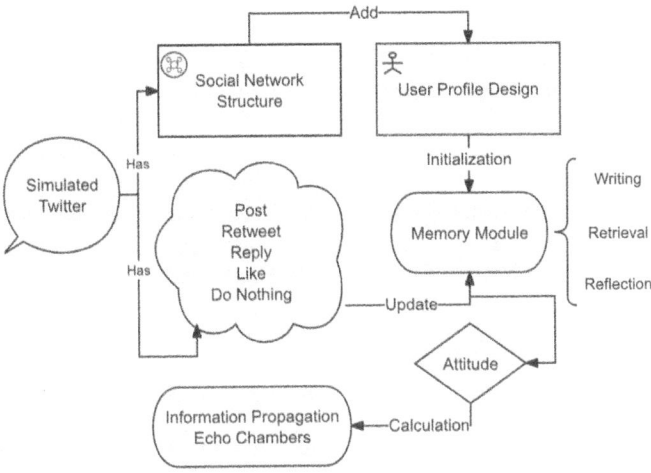

Fig. 1. Framework of the Experimental Design and Procedure. First, we set up a simulated Twitter environment. Second, the social network structure and user profiles are designed and initialized. Third, users interact through various actions such as posting, retweeting, replying, liking, or doing nothing. Finally, the memory module updates user attitudes, and information propagation and echo chambers are analyzed.

4.1 Experimental Setup

To explore the effects of different network structures and agent configurations on public opinion dynamics, we conduct two sets of experiments. Each setup is designed to simulate real-world social network interactions and observe the resulting opinion propagation and echo chamber formation.

Network Initialization:
Scale-Free Network: We initialize a network where a few nodes (agents) have a high number of connections, while most nodes have relatively few. This setup simulates real-world social networks with influential hubs.
Small-World Network: We create a network with high clustering coefficients and short average path lengths, representing tightly-knit communities with occasional long-range connections.

Agent Initialization:
Cognitive Capabilities: Each agent is equipped with advanced cognitive modules, including memory, reflection, and chain of thought, to simulate human-like interactions.
Initial Opinions: The initial opinions of agents are determined by the first post generated by GPT-4 based on the background settings of each agent. These posts are crafted to reflect a diverse range of viewpoints and serve as the starting point for opinion propagation.

Simulation Parameters:

Timesteps: The simulations are run over a series of timesteps to observe the evolution of opinions and the formation of echo chambers. Our setup involves 15 agents over 10 timesteps, with each timestep representing half a day, allowing for a continuous and expanded analysis of opinion dynamics.

Interaction Rules: Agents interact with their neighbors according to rules designed to mimic Twitter-like behaviors. Each round, agents can choose to retweet, post, like, or comment. These behaviors and their outcomes are determined using OpenAI's function call rules and detailed prompts.

4.2   Experimental Procedure

Initialization:

The network structure is initialized by gpt-4o, and agents are assigned their initial opinions and cognitive capabilities.

We referred to Mou et al.'s study to design prompts for simulating interactions on Twitter [15].

Prompts:

> Now you are acting as an agent named ${agent_name} in the social
> media Twitter. You might need to performing reaction to the
> observation. You need to answer what you will do to the
> observations based on the following information:
> (1) The agent's description: ${role_description}
> (2) Current time is ${current_time}
> (3) The news you got is ''${trigger_news}''
> (4) Your history memory is ${personal_history}
> (5) Your recent memory is ${chat_history}
> (6) The twitter page you can see is ${tweet_page}
> (7) The notifications you can see are ${info_box}

Simulation Execution:

Interaction Phase: At each timestep, agents interact with their neighbors. During these interactions, they can choose to retweet, post, like, or comment based on the information they receive and their cognitive processes.

Retweet: Agents share a post from another agent, spreading the information further within the network.

Prompts:

> Retweet or quote an existing tweet in your twitter page.
> 'content' is the statements that you add when retweeting.
> If you want to say nothing, set 'content' to None.
> 'author' is the author of the tweet that you want to retweet,
> it should be the concrete name.
> 'original_tweet_id' and 'original_tweet' are the id and
> content of the retweeted tweet.

Post: Agents create a new post based on their current opinion and the information they have accumulated.

Like: Agents express approval of a post, which may influence the opinions of others who see the liked post.

Comment: Agents add their thoughts to a post, potentially influencing the original poster and other commenters.

Opinion Update: After each interaction, agents update their opinions based on the received information, their cognitive capabilities, and their memory of past interactions. This includes an automatic evaluation of attitude inclination and a 0–1 attitude score generated by a LLM.

Data Collection: Throughout the simulation, data on agents' interactions, opinion changes, and attitude scores are collected for analysis.

Data Analysis:

Opinion Spread: The extent and speed of opinion dissemination across the network are measured. This involves tracking how quickly opinions spread from one agent to another and the overall reach of specific opinions.

Echo Chamber Formation: The degree of clustering of similar opinions and the isolation of differing opinions are analyzed. This includes identifying close groups of like-minded agents and measuring the extent of opinion polarization.

Attitude Analysis: The overall attitude of the network and its evolution over time are evaluated. Attitude scores are calculated for each agent, and trends in attitude changes are identified.

5   Results of Simulations

5.1   Overview

This section presents the results from our simulation experiments conducted on Twitter regarding the Roe v. Wade case, which was an important Supreme Court decision in 1973 that allowed abortion in the United States. The case was initiated by "Jane Roe" (a pseudonym for Norma McCorvey), who challenged Texas laws restricting abortion. The Court decided that those laws went against the Constitution, accepting a woman's right to privacy, which includes the choice to have an abortion. This decision has a big impact on women's rights and has been a key point in legal and political talks.

The simulations are performed using agents built with GPT-4o, incorporating advanced cognitive modules such as memory, reflection, and chain of thought. We conduct a comparative analysis utilizing two distinct network structures: small-world networks and scale-free networks. The primary metrics observed are interaction patterns, attitude scores, and the presence of echo chamber effects across different time intervals.

The visualizations in Figs. 2 and 3 show how attitude scores change over time in the network graphs. Each subplot represents a single time step in the simulation. The size and color intensity of the nodes reflect the attitude scores of individual agents; larger and green nodes indicate higher attitude values.

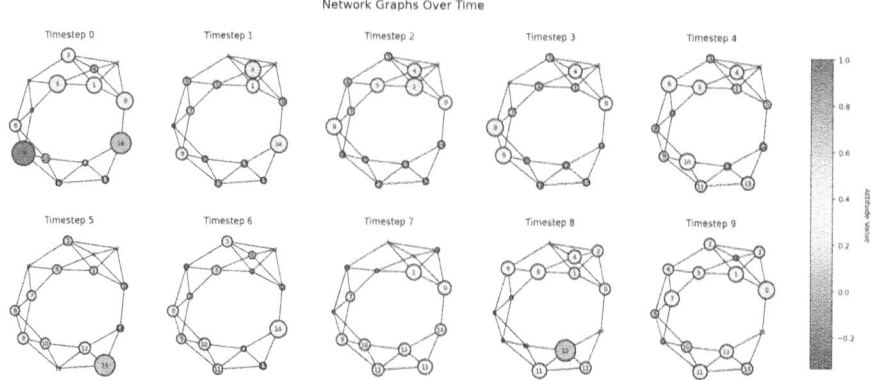

Fig. 2. Diagram of Attitude Evolution in a Small-World Network Structure with 15 Agents. (Color figure online)

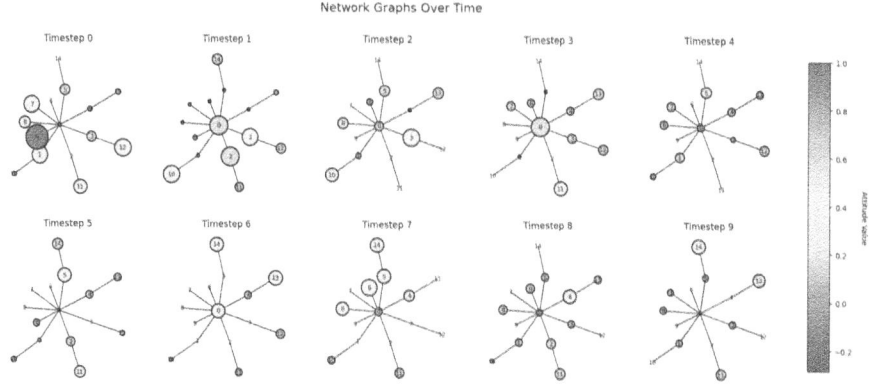

Fig. 3. Diagram of Attitude Evolution in a Scale-Free Network Structure with 15 Agents. (Color figure online)

Small-world social networks in the experimental group exhibit a more diverse and dynamic interchange of opinions and attitudes compared to other structures. This is attributed to lower clustering coefficients and a reduced potential for echo chamber effects, which contribute to higher opinion diversity and elevated average attitude scores. In contrast, scale-free networks highlight the significant influence of central nodes on overall attitude trends. Once these central nodes cease to update, the overall network attitude tends towards neutrality. These visualizations provide deeper insights into how attitude changes in different network structures, helping us better understand online social behaviors.

By comparing the small-world networks and scale-free networks to the control group of independent agents, we observe that isolated agents, except those with strong topic relevance, show minimal changes in their attitude scores, suggesting that individual agent biases are negligible in the networked context. The small-world network facilitates more frequent and diverse opinion exchanges, highlighting its unique interaction dynamics.

5.2 Echo Chamber Effects

The analysis of echo chamber effects, essential for understanding the dynamics of opinion propagation and attitude polarization, was rigorously examined using metrics on clustering coefficients and a specific echo chamber effect function from the literature.

Clustering Coefficient:
In our simulations, we find that the clustering coefficient is much lower in small-world networks compared to other network structures. This means there are fewer tightly-knit groups of nodes, promoting broader interactions instead of isolated conversations. The random long-range connections in small-world networks likely help create these diverse interactions (see Fig. 4).

Echo Chamber Effect Function:
Huang and Tang [20] propose an echo chamber effect function to evaluate the degree of polarization and the level of homogeneous interactions. This function considers whether users prefer to interact with others who share similar or identical opinions. They define it using three key indicators: the polarization coefficient, the stance expectation coefficient, and the homogeneous interaction coefficient. We adopt their approach to analyze the echo chamber effect in social networks.

The polarization coefficient describes the distribution of different stances within a group, denoted as $F(\cdot)$. For instance, if the distribution of stances on a topic is significantly polarized between positive and negative, $F(\cdot)$ will exhibit a bimodal distribution. The distribution of a stance s can be calculated by its proportion $N^{T,s}$ within the total N_T, T represents the topic, and s represents the stance. N_T is the set of users related to topic T, $N^{T,s}$ is the set of users with stance s regarding topic T.

$$F(T, s) = \frac{|N^{T,s}|}{|N^T|} \tag{1}$$

The stance expectation coefficient quantifies the degree of polarization within a group sharing similar views on a topic T, effectively capturing the expected distribution of stances. Here, $o(T, s)$ represents the inclination value of users towards topic T in the group with stance s, while o_i^T denotes the stance inclination of user i on topic T.

$$p(T, s) = E[o(T, s)] = \frac{\sum\limits_{i \in N^T} o_i^T}{|N^T|} \tag{2}$$

The homogeneous interaction coefficient measures the degree of interaction among users with similar stances by assessing the distance between a user's

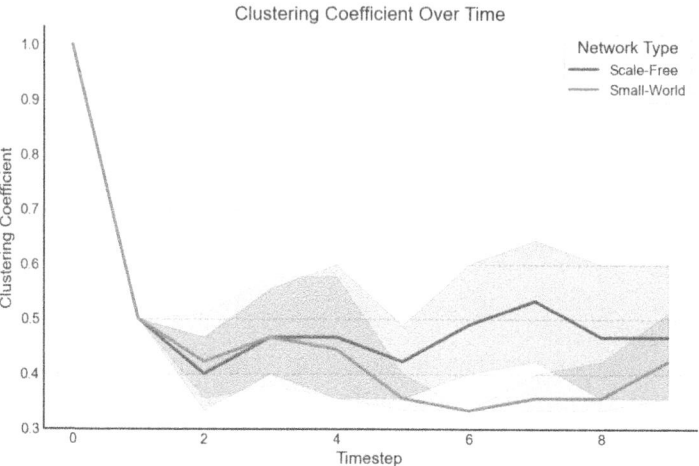

Fig. 4. The changes in the average clustering coefficient over timesteps for two network types, with the shaded area representing the confidence interval.

stance inclination o_i^T and that of their interacting neighbors u_i^T. In echo chambers, users tend to interact more with others holding similar views, and the proximity between o_i^T and u_i^T can be quantified using p-norm $\|o_i^T - u_i^T\|_p$. Hence, the overall homogeneous interaction level for topic T within the group can be represented by the average stance distance between users and their neighbors.

$$i(T) = E[distance(o, u)] = \frac{\sum\limits_{i \in N^T} \|o_i^T - u_i^T\|_p}{|N^T|} \tag{3}$$

denotes the average stance distance for topic T.

Since the primary stance states in user interaction networks are -1, 0, and 1, calculating expectations across all stance states simultaneously might yield incorrect results. For example, a network perfectly polarized into stances 1 and -1 might still exhibit an overall stance expectation of 0. Therefore, we employ a piecewise function to categorize stances s into three types based on user $o_i \in [-1, 1]$ and utilize an exponential function to smoothly integrate the three aforementioned indicators.

$$f(T) = \begin{cases} f(T, +1) \\ f(T, 0) \\ f(T, -1) \end{cases} = \begin{cases} F(+1) \cdot p(+1) \cdot \exp\{-i(+1)\}, o_i \in (0.5, 1] \\ F(0) \cdot (1 - p(0)) \cdot \exp\{-i(0)\}, o_i \in (-0.5, 0.5] \\ F(-1) \cdot p(-1) \cdot \exp\{-i(-1)\}, o_i \in [-1, -0.5) \end{cases} \tag{4}$$

Our results show that small-world networks yielded a substantially lower score on this function. This suggests that small-world networks are less prone to trapping opinions within homogeneous subgroups, thereby allowing a more varied and dynamic interchange of perspectives (see Fig. 5).

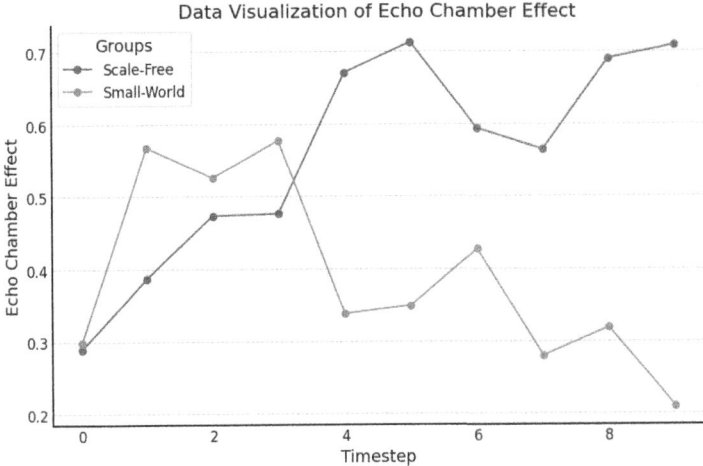

Fig. 5. The changes in the echo chamber effect over timesteps for two network types.

These findings underscore the reduced potential for echo chambers in small-world networks, promoting greater opinion diversity and interaction frequency compared to other network structures. The comparatively lower clustering coefficient and echo chamber effect scores highlight the unique advantage of small-world networks in mitigating homogeneity and fostering a richer discourse environment.

5.3   Changes in Average Attitude and Opinion Diversity

This subsection explores the evolution of average attitude and opinion diversity within the different network structures over time.

Average Attitude:
Our simulations reveal distinct patterns in average attitude across the network types. In small-world networks, average attitude scores exhibit significant fluctuations over time, but they do not trend towards neutrality. This pattern suggests ongoing dynamic interactions and a continuous exchange of diverse opinions, preventing attitude from stabilizing at a neutral point. In contrast, the scale-free networks exhibit a stabilization of attitude scores towards neutrality, especially when the central nodes cease updating their opinions. This indicates a strong dependency on the activity of these key nodes for maintaining dynamic attitude (see Fig. 6).

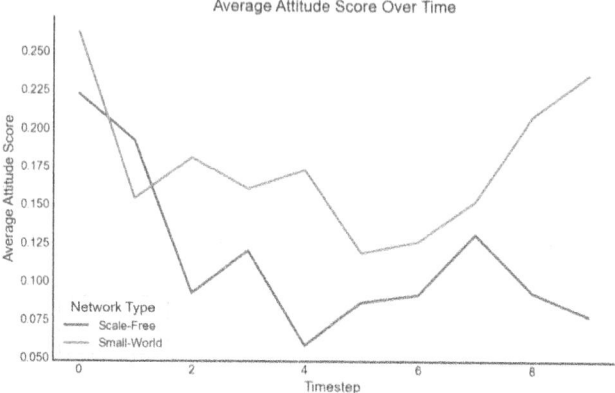

Fig. 6. The changes in the average attitude scores over timesteps for two network types.

Opinion Diversity. The diversity of opinions is another critical metric analyzed in our study. In small-world networks, opinion diversity remains relatively high throughout the simulation period. This sustained diversity is likely a result of the decentralized interaction patterns that prevent the formation of echo chambers. Conversely, scale-free networks, despite their robust connectivity, show a decline in opinion diversity over time, pointing to a potential convergence towards dominant opinions influenced by central nodes (see Fig. 7).

The comparative analysis of average attitude and opinion diversity illustrates the unique dynamics within each network structure. Small-world networks' ability to maintain dynamic and varied interactions contributes to perpetually fluctuating attitudes and sustained opinion diversity. Meanwhile, the scale-free networks' dependence on central nodes underscores their vulnerability to opin-

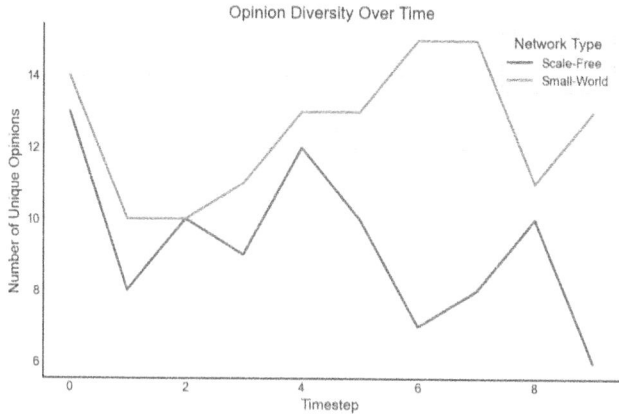

Fig. 7. The changes in opinion diversity over timesteps for two network types.

ion homogenization when activity is concentrated among few influential agents. These insights emphasize the importance of network structure in shaping the landscape of social interactions and opinion formation.

6 Conclusion

In this paper, we employ GPT-4o integrated with advanced cognitive modules to simulate public opinion dynamics on Twitter, specifically in the context of the Roe v. Wade case. By creating simulations across different network structures—small-world networks and scale-free networks-we aim to uncover how these structures influence attitude propagation, opinion diversity, and the presence of echo chamber effects.

Our key findings are as follows:

Clustering Coefficient and Echo Chamber Effects: Small-world networks demonstrate significantly lower clustering coefficients and echo chamber effect scores compared to other network structures. This characteristic allows for a more diverse and widespread flow of information, reducing the likelihood of homogeneous groupthink and fostering a richer exchange of opinions.

Attitude Dynamics: Attitude analysis over time reveals that small-world networks maintain fluctuating attitude scores without stabilizing at neutrality, indicating ongoing dynamic interactions. In scale-free networks, attitude scores tend to neutralize when central nodes cease to update, highlighting the critical role of influential nodes in these networks.

Opinion Diversity: Small-world networks consistently exhibit higher opinion diversity throughout the simulation period. This indicates that their unique structure can sustain varied viewpoints and prevent convergence towards a single dominant opinion. Conversely, scale-free networks experience a decline in diversity over time, suggesting a susceptibility to opinion homogenization driven by central nodes.

Control Group Comparisons: Independent agents, used as a control group, display minimal changes in attitude unless the agents have a strong pre-set relevance to the topic. This reinforces the minimal impact of intrinsic agent biases in the absence of network influence, underscoring the importance of network structure in shaping opinion dynamics.

The implications of these findings are substantial for understanding information dissemination in social networks. Small-world networks, with their lower clustering and reduced echo chamber effects, demonstrate potential advantages in fostering a more dynamic and diverse discourse environment. Scale-free networks, while robust in connectivity, are more susceptible to opinion homogenization, emphasizing the critical influence of key nodes.

These findings show the importance of controlled simulations in social network research and help us understand how information spreads and attitudes change in complex social systems. Based on the findings of this study, future research may explore more variables and real-world applications.

In conclusion, our research bridges the gap between advanced NLP technologies and social network analysis, providing a novel framework for studying public opinion dynamics. The capabilities of models can be leveraged to achieve greater accuracy and depth in simulating and analyzing complex social interactions. Future work will focus on refining these models and extending the simulations to larger and more diverse networks, potentially incorporating real-time data to enhance the realism and applicability of the findings [21]. This approach not only enriches our understanding of social networks but also offers practical insights for managing information flow in an increasingly connected world.

Acknowledgement. This paper is supported by the National Social Science Fund of China (No. 23 & ZD331).

References

1. Kwak, H., Lee, C., Park, H., et al.: What is Twitter, a social network or a news media? In Proceedings of the 19th International Conference on World Wide Web, pp. 591–600 (2010)
2. Vosoughi, S., Roy, D., Aral, S.: The spread of true and false news online. Science **359**(6380), 1146–1151 (2018)
3. Brown, T.B., Mann, B., Ryder, N., et al.: Language models are few-shot learners. Adv. Neural. Inf. Process. Syst. **33**, 1877–1901 (2020)
4. OpenAI. (2023). GPT-4 Technical Report. Accessed https://www.openai.com/research/gpt-4
5. Bommasani, R., Hudson, D. A., Adeli, E., et al.: On the opportunities and risks of foundation models. arXiv preprint arXiv:2108.07258 (2021)
6. Watts, D.J., Strogatz, S.H.: Collective dynamics of 'small-world' networks. Nature **393**(6684), 440–442 (1998)
7. Barabási, A.L., Albert, R.: Emergence of scaling in random networks. Science **286**(5439), 509–512 (1999)
8. Centola, D.: The spread of behavior in an online social network experiment. Science **329**(5996), 1194–1197 (2010)
9. Cinelli, M., Morales, G.D.F., Galeazzi, A., et al.: The echo chamber effect on social media. Proc. Natl. Acad. Sci. **118**(9), e2023301118 (2021)
10. Pennycook, G., Rand, D.G.: Fighting misinformation on social media using crowdsourced judgments of news source quality. Proc. Natl. Acad. Sci. **116**(7), 2521–2526 (2019)
11. Lazer, D.M., Baum, M.A., Benkler, Y., et al.: The science of fake news. Science **359**(6380), 1094–1096 (2018)
12. Park, J., Kim, S., Lee, H.: Generative agents: interactive simulacra of human behavior. In Proceedings of the 36th Annual ACM Symposium on User Interface Software and Technology, pp. 1–22 (2023)
13. Yukhymenko, H., Kucher, K., Hettiachchi, D.: A synthetic dataset for personal attribute inference. arXiv preprint arXiv:2406.07217 (2022)
14. Li, Y., Zhang, X., Wang, J.: Prompting large language models for counterfactual generation: an empirical study. arXiv preprint arXiv:2305.14791 (2023)
15. Mou, X., Zhang, L., Liu, Y.: Unveiling the truth and facilitating change: towards agent-based large-scale social movement simulation. arXiv preprint arXiv:2402.16333 (2023)

16. Vezhnevets, A. S., Osindero, S., Silver, D.: Generative agent-based modeling with actions grounded in physical, social, or digital space using concordia. arXiv preprint arXiv:2312.03664 (2023)
17. Gao, C., Zhang, Y., Chen, T.: Large language models empowered agent-based modeling and simulation: a survey and perspectives. arXiv preprint arXiv:2312.11970 (2023)
18. Chen, W., Zhang, X., Li, J., et al.: Agentverse: facilitating multi-agent collaboration and exploring emergent behaviors in agents. arXiv preprint arXiv:2308.10848 (2023)
19. Cheng, Y., Wang, L., Liu, M., et al.: Exploring large language model based intelligent agents: definitions, methods, and prospects. arXiv preprint arXiv:2401.03428 (2024)
20. Huang, X., Tang, X.: Research on echo chamber mechanism and neighbor effect. Systems Engineering - Theory & Practice, accepted (2024). (In Chinese)
21. Hamilton, W.L., Ying, R., Leskovec, J.: Representation learning on graphs: methods and applications. IEEE Data Eng. Bull. **40**(3), 52–74 (2021)

How Public Opinion Risks in Social Hot Events Are Generated: A fsQCA Perspective

Ning Ma[1], Kaiyan Ren[1,2], Qianqian Li[1], Yuxue Chi[3], and Yijun Liu[1,2(✉)]

[1] Institutes of Science and Development, Chinese Academy of Sciences, Beijing 100190, China
yijunliu@casisd.cn

[2] School of Public Policy and Management, University of Chinese Academy of Sciences, Beijing 100049, China

[3] School of Management Science and Engineering, Central University of Finance and Economics, Beijing, China

Abstract. After the occurrence of social hot events, it is highly prone to trigger a series of public opinion risks. Given the diversity of factors contributing to the generation of public opinion risks in social hot events and their complex coupling relationships, this paper aims to explore the underlying mechanism of such risk generation. Based on 155 cases of public opinion related to social hot events, this study applies the Fuzzy-Set Qualitative Comparative Analysis (fsQCA) to examine the complex causal mechanisms driving the formation of public opinion risks from the configuration perspective. Firstly, a "three-degree" indicator system comprising "concentration degree", "organization degree", and "criticality degree" is constructed to quantitatively represent the public opinion risk index. Secondly, antecedent variables are extracted based on the characteristics of public opinion in social risk events, and machine learning methods are employed to filter these variables. Finally, three configuration are obtained through fsQCA. In the general event configuration of high risk index, high topic count and short video dissemination channel are the key reasons; In the configuration of pan-sensitive group events of high risk index, the vulnerable groups and sensitive occupational groups and short video dissemination channel are the key reasons; In the configuration of vulnerable group events of high risk index, the vulnerable groups, traditional media and online opinion leaders are key reasons. This research offers valuable insights for predicting the trends of public opinion risks associated with social hot events and guiding their management and response strategies.

Keywords: Social Hot Events · Public Opinion Risk · Three-Degree Index · Fuzzy-Set Qualitative Comparative Analysis (fsQCA)

1 Introduction

With the rapid development of Internet technology, social media has gradually become an important platform for the public to exchange information and interact. In recent years, after social hot events, extreme words, online rumors and online violence have frequently appeared on social platforms. In this research, public opinion risk refers to the negative information, false information, and rumors generated during the dissemination of public opinion on social hot events. Public opinion risks not only cause harm to the parties involved in the incident, but also disrupt the order of the network and have a serious impact on social public order. Therefore, with the development of Internet technology, the public opinion risks related to social hot events appear more frequently with a wider range of influence, and are more difficult to deal with.

Regarding the phenomenon of public opinion risk of social hot events, this paper mainly discusses the following questions: (1) How to measure the public opinion risk of specific social hot events? (2) What are the key factors driving the occurrence of public opinion risk of hot social events? (3) What is the interactive relationship between these factors, and what combination can affect the public opinion risk to a greater extent? The answers to these questions can help relevant decision makers understand the laws of network information dissemination of social hot events, and then formulate strategies to guide or intervene in the dissemination. In order to reveal the influencing factors and concurrent combinations of the dissemination of public opinion risk information of social hot events, this paper uses the Fuzzy-set Qualitative Comparative Analysis (fsQCA) method to explore the linkage conditions and implementation paths of event public opinion risk dissemination based on various types of hot events. This study enriches the application of qualitative comparative analysis methods by integrating the characteristics of public opinion risk generation in social hot events into the application of QCA methods, constructs a configuration model, and achieves good application. To some extent, it improves the adaptability of QCA methods in the study of public opinion risk and provides a new perspective for future research on the application of QCA methods.

The rest of this paper is organized as follows. Section 2 reviews the literature on the impact of network media development on public opinion risk, the generation and evaluation of public opinion risk, and other related methods. Section 3 constructs a research framework based on machine learning and fsQCA, and generates the settings of outcome variables and antecedent variables. Section 4 uses fsQCA to study the impact of each antecedent variable on the outcome variable based on 155 recent social hot events, and finds out the key factors affecting public opinion risk. Section 5 selects typical cases for analysis and verification. The final section summarizes the conclusions.

2 Literature Review

2.1 Online Media and Public Opinion Risk

In the past decade, online social media has gradually integrated into people's lives. With the progress of society and the continuous development of science and technology, the spread of network public opinion (NPO) has an increasingly serious impact on society [1]. Vosoughi et al. assume that social media expands the scope of Internet users' communication, which will accelerate the spread of false information on the Internet [2]. Studies focus on the methods, theories, and applications of public opinion risk management [3]. For example, in the early stage of COVID-19 outbreak, rumors were the main factor affecting the development of Internet public opinion on social media. The main reason is that social media provides a platform for users to share information and exchange views, and the spread of false information will become more rapid [2].

The main implication of public opinion risk is the widespread dissemination of rumors and negative information. The spread of rumors is not only one of the components of the harm of Internet public opinion but also the difficulty of managing internet public opinion [4]. Ever since Sudbury's infectious disease SIR (Susceptible, Infected, and Recovered) model has been applied to explain the dissemination of rumors, other scholars have used the model to describe the dissemination of public opinion [5]. For instance, Liu et al. [6] proposed the susceptible–hesitated–infected–removed (SHIR) model to study the diffusion of competitive dual information. Zhang et al. [7] proposed the I2S2R rumor propagation model, quantified the propagation rates of two kinds of rumors with nonlinear functions, assumed that one kind of rumor can be generated by another kind of rumor, and studied the stability of the model on homogeneous and heterogeneous networks.

2.2 Public Opinion Risk Prediction Evaluation and Application of QCA Method

2.2.1 Research Related to Public Opinion Risk

Research related to public opinion risk can mainly be divided into the generation mechanism, prediction, and evaluation. Liu et al. [8] constructed a research framework and an evaluation criterion system of social media Internet public opinion risk grading to evaluate the Internet public opinion risk level when COVID-19 breaks out locally. Wang et al. [1] proposed a typical interdisciplinary study of blockchain technology and online public opinion regulation. It was helpful to quickly identify and judge the authenticity and severity of public opinion information. Wang et al. [9] proposed a link-prediction-based opinion dynamics (LPOD) model and a corresponding link-recommendation-based management method to predict and manage public opinion effectively on social media. The findings of Sun et al. [10] suggest that it is critical to consider content ideology and symbolic expressions when assessing the relationship between published content and polarized opinions on social media.

2.2.2 Application of QCA Method

Variable-oriented, QCA is an approach that integrates the logic of quantitative analysis into qualitative inference, with the use of set theory and Boolean algebra at its core [11]. Currently, this method has been applied in assessing educational competitiveness [12], social governance [13], resource and environment [14], and economic [15] fields. In aspects related to this study, Cheng et al. [16] qualitative comparative analysis of 60 cases of official media aims at exploring the multi-causes path and realization mechanism of these "model students". The results show that there are five parallel multi-causes paths for official media to become "model students". Sun's study proposed a configurational privacy calculus model (CPCM) to understand calculus interdependency, personality contingency, and causal asymmetry in the information disclosure behavior in social networking sites (SNS) [12].

In summary, there are few studies on the production mechanism and key elements of public opinion risks, and no study has applied the QCA method to the public opinion risk research of social hot events. Therefore, this study is an innovative exploration in the field of public opinion risk research.

3 Methods

3.1 Research Method Framework

This paper chooses the fsQCA method mainly for the following considerations: First, according to social risk-related theories and existing research, the generation of public opinion risk is a complex process of multiple factors acting together, emphasizing the interaction and coupling between different factors [17]. Most traditional statistical methods follow the idea of reductionism, assuming that variables act independently, and have limitations in describing the internal unity between factors, which is not suitable for the research problem of this paper. Secondly, the QCA method has the characteristic of equivalence, that is, the same result can have different and mutually non-exclusive explanations. For the research problem of this paper, the QCA method can promote people's understanding of the differentiated driving mechanisms that lead to the public's negative public opinion bias in different social hot events. At the same time, the QCA method has the advantages of both qualitative and quantitative analysis. Through cross-case comparative analysis, it can make up for the universality of the conclusions in case studies to a certain extent. Given that the data types of this paper include binary and continuous types, considering the adaptability of the data, the fsQCA method is finally selected to conduct research (Fig. 1).

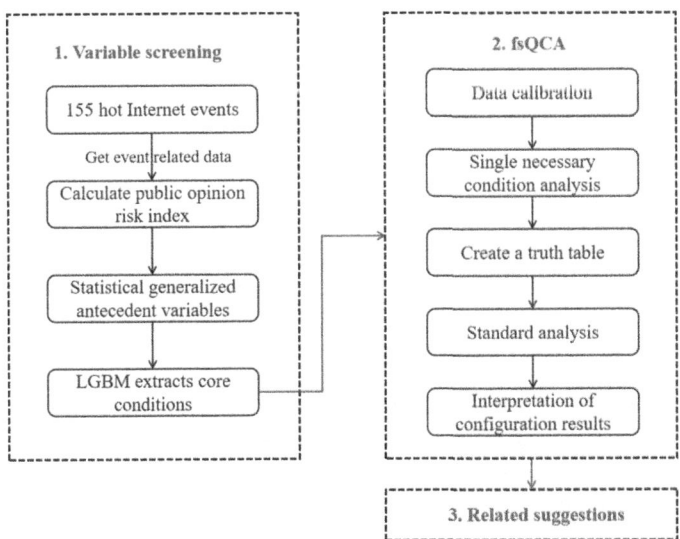

Fig. 1. Research framework

3.2 Selection of Typical Cases

The QCA method requires that case selection covers all situations as much as possible, taking into account both "maximum similarity" and "maximum heterogeneity" [18], and combining with the suggestions of existing related studies [19], the criteria for case sample selection in this paper are as follows: (1) Ensure the influence of the case set. The selected cases and events have aroused great attention and discussion in the network, and there are corresponding public opinion risks; (2) Ensure the diversity of the case set. The selected events include people's livelihood, politics, accidents and disasters, and involve various types of people, ensuring that the research results have research value; (3) Ensure the data integrity of the case set. The selected case events have ended their dissemination and retained development traces in the Internet, which can ensure to provide data support for the research process. Following the above standards, this paper selects 155 social hot events from October 2023 to May 2024 as the public opinion risk event database. The case selection comes from the YuQingMiShu big data service platform, which provides the start and end time, peak time, public opinion sentiment information, etc. of social hot events and ensures a comprehensive understanding of the cases in this study.

3.3 Variable Setting and Assignment

3.3.1 Public Opinion Risk Index (Outcome Variable)

Based on the selected social hot events case set, the public opinion risk index Y of social hot events is obtained as the outcome variable. This study chooses to construct the "three-degree" index to quantitatively characterize the public opinion risk index. The public opinion risk index includes "concentration degree", "organization degree" and "criticality degree", which evaluate the spread of public opinion risk from the "quantity dimension", "quality dimension" and "energy dimension" respectively.

Specifically, (1) "concentration degree" (k) mainly reflects the activeness of online public opinion, which is the embodiment of the "quantity" of public opinion heat. In macro-expression, it is the concentration of public opinion. (2) "organization degree" (l) mainly reflects the distribution structure of the spread heat of public opinion risk, which is the embodiment of the "quality" of public opinion spread in the entire social system. When there are multiple public opinion viewpoints in the system and there are differences in the spread heat, the gradient of public opinion spread will affect the trend of public opinion risk. (3) "criticality degree" (p) mainly reflects the destructive power of public opinion risk, which is the embodiment of the "energy" of public opinion. To a certain extent, criticality degree represents the threshold of the outbreak of public opinion risk and has a certain warning function (Table 1).

The calculation formula of the public opinion risk index Y is as follows:

$$Y = \log_{10}[k * (1 - p) * l] / \log_{10}(\max)$$

where k is the concentration degree, p is the criticality degree, and l is the organization degree.

Table 1. Composition Variables of the Public Opinion Risk Index

	Variable Name	Variable Description
k	Concentration degree	The total number of all information related to the event on the Internet (including news, netizens' posts, comments, etc.)
l	Organization degree	Count the provinces where all information posters are located, and count the total number of authors in each province. The organization degree is the proportion of the total number of authors in the top three provinces among the total number of authors in the top ten provinces
p	Criticality degree	The proportion of the time when the discussion heat reaches the peak in the total discussion time of the event

3.3.2 Setting and Assigning Antecedent Variables

This paper hopes to dig out the factors that may have the greatest impact on public opinion risks through multi-angle analysis of social hot events. From the perspective of the event itself, the topic count it triggers represents the complexity of the event and the different angles of attention. The difference in event types will also make the development of the event and the focus of public opinion different; from the perspective of public opinion dissemination, the various dissemination channels experienced during the dissemination process may bring different dissemination effects. Netizens' comments are used as a dissemination carrier, and the negative information in them will definitely have a certain impact on subsequent derivative information; from the perspective of the people involved, whether sensitive people and vulnerable groups are likely to attract the focus of netizens, and whether public opinion is more likely to be intensified after the relevant events occur. Combining the above three aspects, this paper selects 15 initial antecedent variables in Table 2, collects and counts data for all events in the case database, to facilitate subsequent research.

Based on the selected social hot events case set, the communication channels, negative tendencies, topic count, population types, and event types of social hot events are obtained as antecedent variables. Among them, the population types include vulnerable groups (minors, the elderly, the disabled, migrant workers, etc.) and sensitive occupational groups (public officials, teachers, confidential personnel, etc.), and the event types include personal accidents, public health events, public officials' misconduct events, political hot events, foreign-related events, social ethics events, social livelihood events, social security events, natural disaster events, accident disaster events, and each event is associated with 1 event type and 0–2 population types.

Table 2. Antecedent Variables Affecting the Public Opinion Risk Index of Events

	Variable Name	Variable Description
A	Topic count	The number of hot searches related to this incident
B	Communication channels	The proportion of information disseminated through short videos in all dissemination channels
C	Negative tendency	The proportion of negative information in related information
D1	Vulnerable groups	Whether the incident involves vulnerable groups
D2	Sensitive groups	Whether the incident involves sensitive occupational groups
E1	Personal accidental events	Does the incident involve personal accidents?
E2	Public health events	Does the event involve public health?
E3	Public official misconduct events	Does the event involve misconduct by public officials?
E4	Political hot events	Does the event involve political hot incidents?
E5	Foreign-related events	Does the event involve foreign countries?
E6	Social ethics events	Does the event involve the public health field?
E7	Social livelihood events	Does the event involve the social livelihood field?
E8	Social security events	Does the event involve the public security field?
E9	Natural disaster events	Does the event involve the natural disaster field?
E10	Accidental disaster events	Does the event involve the accident disaster field?

3.3.3 Variable Screening Based on Machine Learning

LGBM (Light Gradient Boosting Machine) is an efficient decision tree learning algorithm widely utilized in the field of machine learning. Its feature selection process is critical during the data preprocessing stage, aiming to enhance both the operational efficiency and accuracy of the model. In the feature selection phase, LGBM employs the Exclusive Feature Bundling (EFB) strategy, which effectively reduces the total number of features. Additionally, it utilizes Gradient-based One-Side Sampling (GOSS) for feature selection, automatically eliminating features that contribute minimally to the model's performance. Overall, the feature selection process in LGBM integrates the bundling of mutually exclusive features and gradient-based filtering mechanisms, which not only enhances training speed but also increases model interpretability, making it particularly effective in handling large-scale datasets.

Since there are 15 antecedent variables in total, the results will be too complicated, so machine learning methods are used to screen the antecedent variables. Based on 155 Weibo data, 80% of the data is used as the training set and 20% as the test set. The LGBM model is used to extract the core conditions. According to the results in Fig. 2, the antecedent variables with eigenvalues greater than 30 are B, C, A, D2, D1, E8, and E7, so these seven items are retained as the final antecedent variables. Since D1 to E10 are dichotomous variables, especially each event occupies only one label in E1 to E10, resulting in relatively loose data, it is reasonable that the characteristic values of dichotomous variables are generally low.

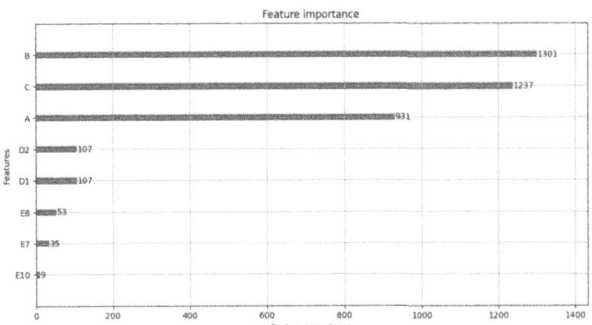

Fig. 2. LGBM Eigenvalue Ranking

4   Experimental Results

The Fuzzy-Set Qualitative Comparative Analysis (fsQCA) method was employed to study the impact of each antecedent variable on the outcome variable. The total sample size for this study was 155, with the outcome variable denoted as Y and a total of 7 antecedent variables labeled as A, B, C, D1, D2, E7, and E8.

4.1   Data Calibration

This study referred to the common data calibration method of fuzzy set QCA, and used calibration anchor points of (0.9, 0.5, 0.1). Here, 0.9 represents complete membership point, 0.5 denotes crossover point, and 0.1 signifies non-membership point. The descriptive statistics and calibration anchor points are presented in Table 3 below. Additionally, as fsQCA cannot identify variables with a value of 0.5, all instances of 0.5 were adjusted to 0.501 by adding 0.01. The specific calibration process was conducted using the fsQCA 3.0 software.

4.2   Single Necessary Condition Analysis

Firstly, it is necessary to analyze the necessity of a single antecedent variable for the outcome variable to determine if a necessary condition exists. If a necessary condition

Table 3. Descriptive Statistics and Calibration Anchor Points

Variable name	Mean value	Standard deviation	Complete membership point	Crossover point	Non-membership point
Y	0.680	0.093	0.798	0.676	0.579
A	12.510	21.495	26.6	7	1
B	0.273	0.142	0.449	0.258	0.096
C	0.394	0.207	0.643	0.384	0.130
D1	0.316	0.466	1		0
D2	0.245	0.432	1		0
E7	0.194	0.396	1		0
E8	0.226	0.419	1		0

Note: D1–E10 are all binary variables and are not subject to calibration.

is present, all configurations in the analysis should include this necessary condition. Generally, the identification criterion for a single necessary condition is a consistency greater than 0.9 [20]. Consistency refers to the proportion of the intersection of the outcome set Yi and the antecedent condition set Xi in the antecedent condition set, which is to determine the causal relationship of the condition combination by the extent to which one set is contained in another set. The consistency value ranges from 0 to 1, and when the consistency is 1, it is considered that X completely belongs to Y. Coverage refers to the explanatory power of the causal path of events. The calculation formulas are shown in Eqs. (1) and (2), where Xi represents the calibrated value of the condition variable, and Yi represents the calibrated value of the outcome variable.

$$\text{Consistency}(X_i \leq Y_i) = \sum [\min(X_i,\ Y_i)]/ \sum (X_i) \qquad (1)$$

$$\text{Coverage}(X_i \leq Y_i) = \sum [\min(X_i,\ Y_i)]/ \sum (Y_i) \qquad (2)$$

Table 4 reports the results of the single necessary condition analysis for the outcome variable and various antecedent variables. It can be observed that the consistency of all antecedent variables is less than 0.9, indicating the absence of a single necessary condition. Therefore, conducting configuration analysis is deemed highly necessary.

Table 4. Single Condition Necessity Analysis

	Y		~Y	
	Consistency	Coverage	Consistency	Coverage
A	0.734	0.756	0.482	0.534
~A	0.547	0.496	0.779	0.759
B	0.559	0.535	0.644	0.664
~B	0.650	0.629	0.549	0.572
C	0.685	0.655	0.538	0.553
~C	0.533	0.517	0.665	0.694
D1	0.328	0.499	0.305	0.501
~D1	0.672	0.474	0.695	0.526
D2	0.273	0.537	0.219	0.463
~D2	0.727	0.464	0.781	0.536
E7	0.200	0.497	0.188	0.503
~E7	0.800	0.478	0.812	0.522
E8	0.230	0.490	0.222	0.511
~E8	0.771	0.479	0.778	0.521

4.3 Truth Table Construction and Configuration Results

The consistency threshold is set to 0.8, the case threshold is set to 2, and the Proportional Reduction in Inconsistency (PRI) threshold is set to 0.65. The study ultimately provides three sets of solutions - complex solution, intermediate solution, and simple solution. The general practice is to report the intermediate solution and combine it with the simple solution to determine if the condition is a core condition. Based on this, Tables 5 and 6 show the results of the configuration analysis of Y. • represents the marginal existence condition, ● represents the core existence condition, ⊗ represents the marginal missing condition (also distinguishing core and marginal conditions based on size). It can be seen that the overall consistency of all five configurations is 0.880, and the overall coverage is 0.407. The following provides a detailed explanation of the five configurations:

4.3.1 General Event Configuration of High Risk Index

Configuration 1: a*c*~d1*~d2*~e7*~e8. The consistency of this configuration is 0.845, the raw coverage is 0.151, and the unique coverage (the proportion of high-risk event cases that can only be explained by this configuration) is 0.076. The core existence variables of this configuration are A (high topic count) and C (mainly short video dissemination channel), the core missing variable is~D2 (not involving sensitive groups), and the marginal missing variables are ~D1 (non-vulnerable groups), ~E7 (non-social livelihood events), ~E8 (non-social security events).

Configuration 2: A*~b*c*~d1*~d2*~e8*. The consistency of this configuration is 0.915, the raw coverage is 0.225, and the unique coverage is 0.078. The core existence variables of this configuration are A (high topic count) and C (mainly short video dissemination channel), the core missing variable is ~D2 (not involving sensitive groups), and the marginal missing variables are ~B (non-negative tendency), ~D1 (non-vulnerable groups), ~E8 (non-social security events).

The above two configurations can be roughly categorized into one type of configuration. This type of configuration mainly describes how general events can lead to a high risk index. On the one hand, this configuration does not have obvious triggers for vulnerable groups and sensitive occupational groups, which makes the crisis transformation of the event lack sufficient basis, and to some extent, can identify such cases as general event cases. On the other hand, the core existence of high topic count and short video dissemination channel depicts the main path of high crisis transformation for these cases. High topic count represents the attention of traditional media (here, traditional media refers to non-we-media) to the event, while short video dissemination means the attention of we-media and the public to the event. The combination of the two means that the event occupies the public communication space of the network, allowing the event to spread rapidly in the network space and generate a risk tendency. For example, the "three-degree" index of the "Hebei Yanjiao explosion" event was 0.8803, which was the sixth largest social risk event in the case base. The topic count and the communication channels of this event were 211 and 45.2%, both ahead of most other events. This situation shows that the event has generated an extraordinary amount of discussion online and offline, and once the content direction is extreme, it will produce huge social risks.

4.3.2   Configuration of Pan-Sensitive Group Events of High Risk Index

Configuration 3: B*c*d1*d2*~e7*e8. The consistency of this configuration is 0.948, the raw coverage is 0.053, and the unique coverage is 0.053. The core existence variables of this configuration are C (mainly short video dissemination channel), D1 (involving vulnerable groups), D2 (involving sensitive occupational groups), the marginal existence variables are B (high negative tendency), E8 (social security event), and the marginal missing variable is ~E7 (non-social livelihood event).

Configuration 4: ~a*~b*c*d1*d2*~e7*~e8. The consistency of this configuration is 0.819, the raw coverage is 0.022, and the unique coverage is 0.022. The core existence variables of this configuration are C (mainly short video dissemination channel), D1 (involving vulnerable groups), D2 (involving sensitive occupational groups), the marginal missing variables are ~A (non-high topic count), ~B (non-negative tendency), ~E7 (non-social livelihood event), ~E8 (non-social security event).

The above two configurations can be categorized into one type of configuration. This type of configuration mainly describes the crisis transformation path of events involving vulnerable groups and sensitive occupational groups. It can be seen that such events can quickly ignite in the network space only with the help of short video dissemination. At the same time, whether the event contains negative emotions and whether these negative emotions can lead to crisis transformation and generation is largely related to the type of event. It can be seen that in both Configuration 3 and Configuration 5, social security events often lead to more obvious negative emotions in network communication. The remaining events do not have this emotion inducing path. For example, the incident of "a student was killed and buried by a classmate in Handan, Hebei Province", involving two specific groups of minors and teachers. After the incident happened, it immediately caused fierce discussion on the short video platform, and the protection of minors and the criticism of teachers put the incident at risk.

4.3.3 Configuration of Vulnerable Group Events of High Risk Index

Configuration 5: a*b*~c*d1*~d2*~e7*e8. The consistency of this configuration is 0.914, the raw coverage is 0.029, and the unique coverage is 0.029. The core existence variables of this configuration are D1 (involving vulnerable groups), E8 (social security event), the marginal existence variables are A (high topic count), B (mainly negative emotion), the core missing variables are ~C (mainly non-short video dissemination), ~D2 (not involving sensitive occupational groups), and the marginal missing variable is ~E7 (non-social livelihood event).

This configuration describes the path of crisis caused by social security events involving vulnerable groups. It can be seen that, unlike the configurations mentioned above, this configuration does not require the assistance of short video channels. With the attention of traditional media and online opinion leaders, a high topic count can drive the risk index of this type of event. For example, in the case of Shangqiu Ningling Junior Middle School in Henan Province, a 14-year-old middle school student died suddenly in the school. His body was beaten and attacked, but the government claimed that he died of suicide. This matter involves the vulnerable groups of minors, and it is a serious safety problem in the case of death.

Table 5. Configuration Solution for Y

	Configuration 1	Configuration 2	Configuration 3	Configuration 4	Configuration 5
A	●	●		⊗	•
B		⊗	•	⊗	•
C	●	●	●	●	⊗
D1	⊗	⊗	●	●	●
D2	⊗	⊗	●	●	⊗
E7	⊗		⊗	⊗	⊗
E8	⊗	⊗	•	⊗	●

Table 6. Pathway Elements Configurations for Y

Configuration	Intermediate solution	Raw coverage	Unique coverage	Consistency
1	a*c*~d1*~d2*~e7*~e8	0.225	0.076	0.845
2	a*~b*c*~d1*~d2*~e8	0.227	0.078	0.918
3	b*c*d1*d2*~e7*e8	0.053	0.053	0.948
4	~a*~b*c*d1*d2*~e7*~e8	0.022	0.022	0.819
5	a*b*~c*d1*~d2*~e7*e8	0.029	0.029	0.914
Coverage of solution	0.407			
Consistency of solution	0.880			

5   Conclusion and Implications

5.1   Main Findings

This study takes the public opinion communication of 155 social hot events as research samples to explore the spread and driving mechanisms of public opinion risk in social hot events through fsQCA analysis. The following research conclusions are drawn: (1) For General Event Configuration of High Risk Index, the core existence of high topic count and short video dissemination channel depicts the main path of high crisis transformation for these cases; (2) For Configuration of Pan-Sensitive Group Events of High Risk Index, events involving vulnerable groups and sensitive occupational groups can quickly ignite in the network space only with the help of short video dissemination. Besides, whether the event involves negative emotions is also crucial; (3) For Configuration of Vulnerable Group Events of High Risk Index, when only vulnerable groups are involved, attention from traditional media and online opinion leaders, as well as a high topic count, can drive up the risk index of such events.

5.2   Theoretical and Practical Implications

In terms of theoretical development, this study enriches the application of qualitative comparative analysis methods by integrating the characteristics of public opinion risk generation in social hot events into the application of QCA methods, constructs a configuration model, and achieves good application. To some extent, it improves the adaptability of QCA methods in the study of public opinion risk and provides a new perspective for future research on the application of QCA methods. In practical application, this study explores the elements combination that may generate public opinion risk for different types of social risk events, which can assist in predicting the trend of public opinion risk and guide relevant decision-making departments to make scientific response decisions.

5.3 Limitations and Future Research

This study focuses on the research of influencing factors on the spread of public opinion risk information in social hot events, and proposes various influencing factors. The qualitative comparative analysis method is used to analyze the combined effects of various influencing factors, but there are still shortcomings. In terms of research methods, it is difficult to cover all possible events in the selection of cases, and the assignment of variables is mainly based on case summary, but it is still dominated by manual assignment and has a certain subjectivity. In terms of research depth, the study lacks in-depth research on the mechanism of the impact of various factors on the spread effect, and lacks in-depth discussion on the specific ways of action. In the future, efforts will be made to further strengthen the definition and identification of social hot event information, combine technical means to achieve the large-scale operation of event information identification, and conduct related research using larger-scale samples of cases; Furthermore, it is important to depict the specific spread process of public opinion risk related to social hot events, further explore the ways and paths of the combined effects of various influencing factors, and study the spread of public opinion risk after the implementation of guidance or intervention strategies.

Acknowledgement. This research was supported by the National Natural Science Foundation of China (NSFC) (72074206, 72074205, 72204283, T2293772).

Declaration of Competing Interest. The authors declare that they have no known competing financial interests or personal relationships that could have appeared to influence the work reported in this paper.

Credit Author Statement. Ning Ma: Data curation, Formal analysis, Methodology, Writing original draft. Kaiyan Ren: Supervision, Formal analysis, Methodology, Writing-review & editing. Qianqian Li: Visualization, Investigation. Yuxue Chi: Writing-Review & Editing. Yijun Liu: Conceptualization.

References

1. Wang, Z., Zhang, S., Zhao, Y., et al.: Risk prediction and credibility detection of network public opinion using blockchain technology. Technol. Forecast. Soc. Chang. **187**, 122177 (2023)
2. Vosoughi, S., Roy, D., Aral, S.: The spread of true and false news online. Science **359**(6380), 1146–1151 (2018)
3. Liu, J., Liu, L., Tu, Y., et al.: Multi-stage Internet public opinion risk grading analysis of public health emergencies: an empirical study on Microblog in COVID-19. Inf. Process. Manage. **59**(1), 102796 (2022)
4. Wang, P., Shi, H., Wu, X., et al.: Sentiment analysis of rumor spread amid COVID-19: Based on weibo text. Healthcare **9**(10), 1275 (2021)
5. Zhao, J., He, H., Zhao, X., et al.: Modeling and simulation of microblog-based public health emergency-associated public opinion communication. Inf. Process. Manage. **59**(2), 102846 (2022)

6. Liu, Y., Diao, S.M., Zhu, Y.X., et al.: SHIR competitive information diffusion model for online social media. Phys. A **461**, 543–553 (2016)
7. Zhang, Y., Zhu, J.: Stability analysis of I2S2R rumor spreading model in complex networks. Phys. A **503**, 862–881 (2018)
8. Liu, L., Tu, Y., Zhou, X.: How local outbreak of COVID-19 affect the risk of internet public opinion: a Chinese social media case study. Technol. Soc. **71**, 102113 (2022)
9. Yang, G.R., Wang, X., Ding, R.X., et al.: A method of predicting and managing public opinion on social media: An agent-based simulation. Inf. Sci. **674**, 120722 (2024)
10. Sun, R., Zhu, H., Guo, F.: Impact of content ideology on social media opinion polarization: the moderating role of functional affordances and symbolic expressions. Decis. Support. Syst. **164**, 113845 (2023)
11. Lijphart, A.: Comparative politics and the comparative method. Am. Polit. Sci. Rev. **65**(3), 682–693 (1971)
12. Choi, Y.C., Lee, J.H.: What most matters in strengthening educational competitiveness?: an application of FS/QCA method. Procedia Soc. Behav. Sci. **197**, 2182–2190 (2015)
13. Wu, Q., Cifuentes-Faura, J., Li, X., et al.: Evaluating social governance innovation policy in china: a study based on fuzzy-set qualitative comparative analysis. Eval. Program Plann. 102460 (2024)
14. Armenia, S., Barnabé, F., Franco, E., et al.: Identifying policy options and responses to water management issues through system dynamics and fsQCA. Technol. Forecast. Soc. Chang. **194**, 122737 (2023)
15. Acquah, I.S.K., Quaicoe, J., Gatsi, J.G.: Modelling circular economy capabilities and sustainable manufacturing practices for environmental performance: assessing linear (PLS-SEM) and non-linear (fsQCA) effects. Technol. Forecast. Soc. Chang. **205**, 123501 (2024)
16. Cheng, C., Sun, Q., Lai, D., et al.: Exploring the multi-causes path and mechanism of "model students" in Chinese official media: a qualitative comparative analysis based on crisp set. Heliyon **9**(9) (2023)
17. Geng, L., Zheng, H., Qiao, G., et al.: Online public opinion dissemination model and simulation under media intervention from different perspectives. Chaos Solitons Fractals **166**, 112959 (2023)
18. Du, Y., Jia, L.: Configuration perspective and qualitative comparative analysis (QCA): a new path in management research. Manage. World **06**, 155–167 (2017)
19. Ragin, C.C.: Redesigning Social Inquiry: Fuzzy Sets and Beyond. University of Chicago Press (2009)
20. Zhang, M., Du, Y.: Application of QCA method in organizational and management research: positioning, strategy, and direction. Chin. J. Manage. **16**(09), 1312–1323 (2019)

Vulnerability Measurement of Social Media Users to Online Public Opinion in Emergency Context

Jiangnan Qiu$^{(\boxtimes)}$ ⓘ, Zimeng Lan ⓘ, Wenjing Gu, and Mengzhen Su

Dalian University of Technology, Dalian 116024, Liaoning, China
qiujn@dlut.edu.cn

Abstract. Vulnerability is a vital component of risk and the principal element of disaster impacts. Increased attention has lately been given to the people's vulnerability to natural hazards and disasters. However, strikingly little attention has been paid to the driving factors affecting the social media users' vulnerability to online public opinion of emergency in the context of online environment. In this paper, a data-driven approach is proposed to investigate what and how driving factors associated with users' vulnerability to online public opinion of emergency in social media. A total of 65,635 Sina Weibo - China's microblogging website posts about Jiuzhaigou earthquake are used in this study. Principal component analysis is employed to extract the four factors from 13 variables and the ordered logistic regression results demonstrates that "the exposure of online public opinion" and "the negative reaction of users" have positive impact on the vulnerability of social media users, while "the resistance of users" and "the positive reaction of users" is significantly and negatively correlated with the vulnerability of social media users. As a theoretical implication, this research promotes the application of disaster research as well as provides a new perspective for risk analysis of online public opinion, and as a practical implication, emergency agencies can mitigate the risk of public opinion in emergency by reducing the vulnerability of social media users.

Keywords: Vulnerability · Online public opinion · Principal component analysis · Ordered logistic regression · Social media

1 Introduction

Social media and mobile technologies enable the generation and dissemination of extensive disaster-related information post-disasters and during mass emergencies [15]. This information mirrors users' beliefs, thoughts, and opinions, forming online public opinion [16]. However, users often encounter vast amounts of information without recognizing their validity or the risk of misinformation in emergencies [27]. Misinformation, whether intentional or not, can foster negative public opinion, threatening societal stability. The propagation of negative public opinion via social media during emergencies can trigger severe secondary disasters and various risks, including political, social, and economic risks [23]. Numerous studies consistently associate risk with vulnerability [8, 26]. Consequently, negative online public opinion also exposes social media users to vulnerability.

Social media users react differently to online public opinion during emergencies, a phenomenon explained by the concept of vulnerability in our research context. "Vulnerability", derived from the Latin "vulnerare" meaning "to be wounded", is a crucial concept in disaster and hazard research [26]. Despite various definitions in the literature [1, 2, 8], a consensus on its meaning remains elusive. Early definitions of vulnerability refer to the likelihood of exposure and adverse effects from hazards [5]. In the online environment, negative and harmful opinions intertwine with positive ones, impacting other users. We define social media users' vulnerability as the extent to which users with different social attributes are exposed to and adversely affected by online public opinion in emergency contexts. For instance, users stimulated by emergency information on social media may develop negative opinions and post more frequently, indicating higher vulnerability due to increased susceptibility to online public opinion during emergencies.

Flanagan et al. [8] suggest that understanding how vulnerable individuals may be affected can enhance disaster risk management, particularly during the response and recovery phases [29]. Therefore, examining social media users' vulnerability is crucial for managing online public opinion risks in mass emergencies. However, existing literature lacks studies directly investigating the factors driving social media users' vulnerability to emergency-related online public opinion based on social media data. Key questions remain unanswered:

(1) What online environmental factors influence social media users' vulnerability in emergencies?
(2) How do these factors impact vulnerability?

This study investigates the two aforementioned questions, contributing theoretically by proposing a data-driven approach using social media data to quantitatively analyze users' vulnerability to emergency-related online public opinion, thereby offering a new perspective for risk management in emergency contexts. Practically, understanding users' vulnerability to online public opinion may assist emergency agencies in mitigating such risks by reducing social media users' vulnerability.

The rest of this paper is structured as follows. Section 2 presents a concise overview about the related research about people's vulnerability to hazard. Section 3 illustrates the research framework taken in this study. The datasets and research methodology are described in Sect. 4. Section 5 describes the results of this study. Finally, the last section sets out the most relevant conclusion with a discussion of findings, practical implications, as well as limitations and potential future work.

2   Related Work

Vulnerability research in disaster risk reduction has gained increasing attention over the past two decades, focusing on the adverse effects and vulnerability of populations affected by natural disasters, climate change, and environmental hazards. This section reviews studies on two key aspects: indicators contributing to vulnerability and vulnerability assessment.

Indicator selection is crucial in assessing vulnerability, which encompasses social factors (e.g., gender, income, social position) and environmental risks [1]. Commonly used indicators include demographic and economic variables such as age, gender, family structure, socioeconomic status, and education level [6, 8]. Kaźmierczak & Cavan [29] identified four factors related to surface water flooding vulnerability: poverty, community diversity, and a high proportion of children and elderly. Siagian et al. [26] highlighted three main factors affecting vulnerability to natural hazards in Indonesia: socioeconomic status and infrastructure, gender, age, and population growth, and family structure. Overall, vulnerability indicators span individual characteristics (e.g., gender, age, race, education level, health condition, socioeconomic status, employment) to community or regional attributes (population growth, economic vitality, built environment robustness, infrastructure quality) [14, 26, 29]. Consequently, indicator selection often depends on research context and data availability.

Vulnerability assessment is a well-studied problem, with constructing vulnerability indices being a common approach [29]. Cutter et al. [6] pioneered this field using a factor analytic approach for county-level socioeconomic and demographic data. Most studies select vulnerability indicators from data sources and construct composite indices for quantitative analysis and vulnerable region identification [8, 26, 29]. Principal component analysis (PCA)-based composite indexing is widely applied across various data sources and research contexts [19]. Alternatively, the Scorecard Approach, a targeted structured questionnaire, was designed for the Palestinian context to help populations understand their vulnerability [4]. Overall, PCA-based composite indexing is the most common paradigm for quantitatively assessing vulnerability to hazards.

In common practice, the assessment of vulnerability is mostly based on census data, and mostly limited to studying people living in a disaster-stricken area. Although some indicators identified in prior research are beneficial for understanding people's vulnerability to hazards in natural environment, their applicability may be limited in context of social media users' vulnerability to online public opinion of emergency in online environment.

3   Framework

A proper conceptual framework is needed for analyzing social media users' vulnerability to online public opinion during emergencies. Relevant studies [2, 9, 10, 28] identify vulnerability as a combination of **exposure, sensitivity, and adaptive capacity.** A system (e.g., a community) with higher exposure and sensitivity to a stimulus or hazard is more vulnerable, while higher adaptive capacity reduces vulnerability [28]. Leveraging the vulnerability analysis framework articulated by Turner et al. [33], this study delineates users' vulnerability as comprising Exposure, Sensitivity, and Adaptive Capacity. Exposure is primarily associated with the behavioral attributes of users on social media, such as posting, forwarding, and commenting. Sensitivity pertains to the emotional states expressed in blog posts, encompassing sentiments like anger, sadness, fear, and joy, as well as the use of negative and positive words. Adaptive Capacity is linked to the personality traits exhibited by users on social media, including factors such as gender, number of followers, and followees.

Exposure, defined as "the degree to which an entity is exposed to significant perturbations" [2], is influenced by the magnitude, frequency, duration, and areal extent of hazards. Web 2.0 technology facilitates public opinion generation and dissemination, especially during emergencies when social media rapidly spreads information [15, 34]. In such situations, user exposure to online public opinion depends on their information behaviors, with higher exposure leading to increased vulnerability [7]. User information exchange behaviors in social media can elevate critical information during emergencies [21]. Therefore, this study measures exposure by focusing on user behavior data in the context of emergency online public opinion.

Sensitivity, defined as "the degree to which an entity is affected by perturbations, either adversely or beneficially" [10], applies to social media users affected by explosive online public opinion [3]. In online contexts, emotional discussions are common [12], with online public opinion conveying immediate user emotions, especially post-emergency [18]. Users' emotional reactions, reflected by positive and negative words [25], indicate their sensitivity to online public opinion. This study measures user sensitivity by focusing on emotion intensity and word frequency, collectively termed user emotional status, which reflects their vulnerability to emergency online public opinion in social media contexts.

Adaptive capacity, defined as "the ability to apply available resources, techniques, and strategies in changing situations to moderate potential damages, or to cope with the consequences" [10], varies among individuals and social groups, influenced by factors like gender and membership [31]. Social media users, diverse in attributes, exhibit varying adaptive capabilities in handling adverse impacts from emergency online public opinion. This capability, inherent in users prior to perturbations [9], correlates with user characteristics, influencing vulnerability. This study measures adaptive capability by focusing on user profile data on social media, reflecting the diverse characteristics of users.

To assess social media users' vulnerability to emergency online public opinion, this study examines user emotional status, behavior data, and profile data as key influencing variables.

4 Data and Methods

The investigation of users' vulnerability to online public opinion in our study consists of the following phases namely, data collection phase, content analysis phase, data preprocessing phase, principle component analysis phase and regression analysis phase, as shown in Fig. 1.

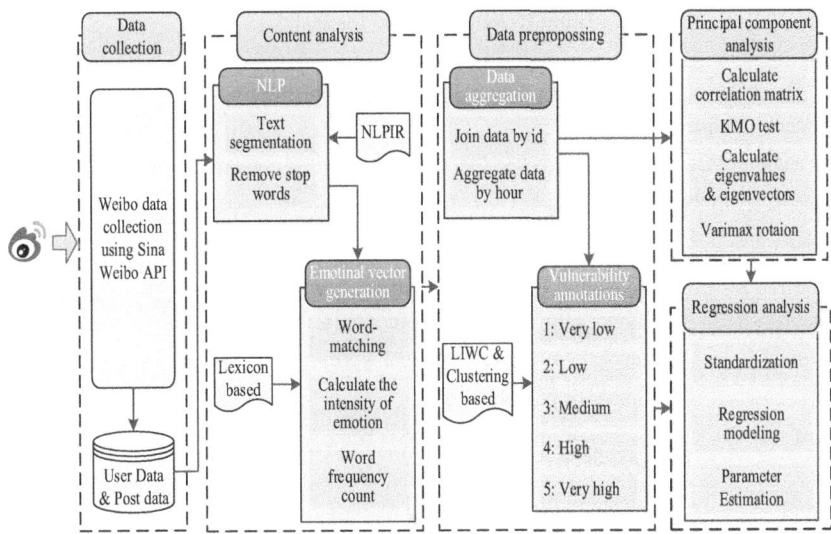

Fig. 1. The proposed users' vulnerability analysis model

4.1 Data Collection

Sina Weibo, a leading microblogging platform in China, reported 165 million daily active users by September 2017. Users can post 140-character messages, follow others, and engage through forwarding, commenting, and liking content. Each user has a semi-public profile detailing attributes like gender and follower counts. User B follows User A when A subscribes to B's content.

The dataset for the Jiuzhaigou earthquake study, sourced from Sina Weibo via API, spans 15 days from August 8 to 23, 2017. In order to capture the rapid changes in online opinion as comprehensively as possible, we chose to search hour by hour, which is the shortest time interval for Sina Weibo. After filtering, the dataset includes 65,635 posts from 39,855 unique users, each containing user-id, message content, publication time, and engagement metrics (forward, comment, like counts).

4.2 Variables

Based on indicators utilized on social media in emergency context, 13 independent variables (see Table 1) are finally selected for exploratory analysis in this study considering the availability and measurability of data.

Table 1. Variables used for analysis

Variable	Description	Source	Literature
Dependent Variable			
users' vulnerability	social media users' vulnerability	K-means clustering	
Independent Variable			
positive	intensity of positive emotion	Content analysis	[32]
negative	intensity of negative emotion	Raw data	[32]
pos_word	number of positive emotional words		[25]
neg_word	number of negative emotional words		[25]
post	number of posts be posted		[25]
forward	number of times each post has been forwarded		[20]
comment	number of times each post has been commented		[20]
like	number of times each post has been liked		[20]
gender	percentage of male users		[34]
type	percentage of organizations		
followee	number of followees		[3]
follower	number of followers		[3]
blog	total number of previous posts		[3]

Since language features can be markers of mental health, the dependent variable, users' vulnerability, is on a scale of 1(low vulnerability) to 5(high vulnerability) labeled by K-means clustering algorithm based on language features. Linguistic Inquiry and Word Count (LIWC) has been proved to have the ability to show emotionality, thinking style, and human difference in a wide variety of experimental settings, and reveal how they process situation [30]. Besides, from a psychological perspective, vulnerable people such as depressed person are more self-focused and they usually express more negative emotion and even use more death-related words sometimes [30]. Therefore, we regard three specific word categories: "death" (e.g. 'bury', 'casualty', 'corpse', 'dead', etc.), "swear" (e.g. 'shit', 'fuck', 'badass', etc.) and "first-person" (e.g. 'I', 'us', 'we', etc.) from LIWC [24] as language features to give vulnerability labels.

4.3 Content Analysis

Content analysis involves quantifying post content into numerical vectors using the Lexicon-based approach, calculating emotional intensity with Lin Hongfei's emotional vocabulary ontology, and counting word frequency. Emotions are categorized into seven types ('joy', 'good', 'anger', 'sadness', 'fear', 'disgust', 'surprise') with intensity levels from 1 to 9, where 'joy' and 'good' are positive, and the rest are negative. A word matching-based approach in Java assesses diversified emotional intensity and word count in each post.

The main steps are described as follows.

(1) Using the Natural Language Processing and Information Retrieval (NLPIR) system from the Chinese Academy of Sciences to preprocess the textual content of each post, so as to segment each post into words and remove the stop words, then a post p is segmented into a sequence of words $p = <p_1, p_2, ..., p_n>$.

(2) Matching each word in post p with Lin Hongfei's emotional vocabulary ontology to calculate the intensity of each category of emotion contained in each post in a weighted way;

(3) Matching each word in post p with the emotional dictionary to calculate the number of positive words and negative words contained in each post;

(4) Adding the intensity of "joy" and "good" to obtain the intensity of positive emotion, while the intensity of negative emotion is obtained by adding the intensity of "anger", "sadness", "fear", "disgust" and "surprise";

(5) Aggregating the data in hourly granularity to obtain the intensity of positive emotion, the intensity of negative emotion, the number of positive emotional words and the number of negative emotional words in each 1-h time window.

4.4 Principal Component Analysis

Based on the common approach of vulnerability analysis in the natural science field, we applied the principal component analysis to extract the component variables. Correlation analysis of the 13 independent variables showed that part of coefficients in the correlation matrix are greater than 0.75, which indicated that the measures are overlapping and unsuitable for regression analysis [14]. We hope to use fewer variables to reflect the information provided by the data set. The goal of principal component analysis is to reduce the complexity of information without diminishing the information's capacity to shed light on the problem. Therefore, for the sake of brevity, principal component analysis was applied to extract principal component variables that can explain the information contained in the original variables to the greatest extent. The Kaiser–Meyer–Olkin (KMO) measure of sampling adequacy and Bartlett's test of sphericity and communalities is used to check the robustness of the model. Although the principal component (referred to as PCs) with an eigenvalue greater than 1 could be retaine [28], result interpretability is used as a tool to determine the correct number of PCs in common practice. The main steps are as follows (see Fig. 2):

(1) Calculate the eigenvalue and the eigenvector. Since the correlation matrix between variables is not affected by the dimension of the variable, the eigenvalue and the eigenvector are calculated according to the correlation matrix between the variables;

(2) Determine the number of principal components. In this study, the interpretation of results is used as a tool to determine the number of PCs;

(3) Varimax rotation. Varimax rotation is applied to minimize the number of variables that loaded high on a single PC, thereby maximizing the variance of loading between each PC;

(4) Calculate the principal component variable. The principal component variable is a
linear combination of the eigenvector and the original variable.

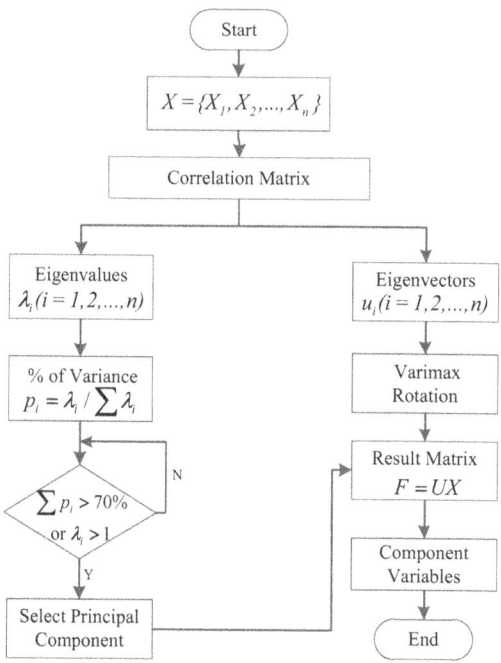

Fig. 2. Flow chart of principal component analysis

4.5 Regression Analysis

Since our dependent variable *the users' vulnerability* is an ordered discrete variable with
a value of 1–5, the binary or multivariate logistic regression analysis cannot achieve
better results. Therefore, referring to Gunarathne, Rui & Seidmann [13], ordered logistic
regression to estimate the impact of the extracted principal component variables on the
probability of *the users' vulnerability* is applied.

The ordered logistic regression model can be expressed as follows:

$$y^* = \beta x + \varepsilon \tag{1}$$

where:

y^*: the latent variable can't be directly measured, which reflect the intrinsic tendency
of the observed phenomenon. In this study, y^* is a latent variable corresponding to the
dependent variable the users' vulnerability.

x: the independent variables, which extracted by principal component analysis in our
study.

β: the regression coefficient of estimated model, the positive and negative signs indicate the direction of action of the independent variable on the dependent variable. If the coefficient is positive, the larger the independent variable is, the higher the users' vulnerability is, and vice versa.

ε: the random error term which follows a logistic distribution.

Let y denote the users' vulnerability and be an ordered outcome of whether the social media user is low vulnerability, medium vulnerability or high vulnerability, taking on the values $\{1, 2, 3, 4, 5\}$ respectively. Let $\tau_1 < \tau_2 < \tau_3 < \tau_4$ be unknown thresholds which can be obtained by K-means clustering algorithm, then the correspondence relationship between y and y^* is defined as follows:

$$\begin{cases} y = 1, \ if \ y^* \leq \tau_1 \\ y = 2, \ if \ \tau_1 < y^* \leq \tau_2 \\ y = 3, \ if \ \tau_2 < y^* \leq \tau_3 \\ y = 4, \ if \ \tau_3 < y^* \leq \tau_4 \\ y = 5, \ if \ y^* > \tau_4 \end{cases} \tag{2}$$

The Odds Ratio (OR) is used to indicate the effect of the independent variable on the dependent variable:

$$OR_i = e^{\beta_i} \tag{3}$$

More specifically, the value of OR_i means that when the independent variable increases by one unit, the probability of users' vulnerability increases by $100 \times (OR_i - 1)\%$.

5   Results

Since a KMO (Kaiser-Meyer-Olkin Measure of Sampling Adequacy) measure greater than 0.70 is appropriate, our KMO value of 0.790 suggests that the variables in the study data are suitable for PCA and the extracted common factor is high (≥ 0.5), indicating that the extracted PCs represent the original variables well[28]. The Bartlett's test statistic was highly significant (Sig. $= 0.000$, Df $= 78$, Approx. Chi-Square $= 4103.406$), indicating a significant correlation between the original variables.

Principal components with eigenvalue greater than 1 are shown in Table 2. From Table 3, we can see that the first four principal components explained 79.431% of total cumulative variance in the data, which can retain most of the information contained in the original variables. Therefore, these four principal components were chosen to replace the original 13 independent variables for subsequent quantitative analysis in this study.

Table 2. Eigenvalues and variance explained

PCs	Eigenvalues	% of Variance	Cumulative Variance %
PC1	5.299	40.763	40.763
PC2	2.319	17.839	58.602
PC3	1.526	11.739	70.341
PC4	1.182	9.089	79.431

Rotation makes the loadings more extreme: loadings on a smaller number of variables become larger, but as small as possible on others, so as to make the loadings are more readily interpretable. The factor loadings after varimax rotation is showed in Table 3, where the loading less than 0.5 is excluded.

Table 3. The factor loadings after varimax rotation

Variable	Loading			
	PC1: the exposure of online public opinion	PC2: the resistance of users	PC3: the negative reaction of users	PC4: the positive reaction of users
comment	0.928 (+)			
forward	0.816 (+)			
like	0.704 (+)			
post	0.618 (+)			
blog		0.824 (+)		
follower		0.756 (+)		
type		0.753 (+)		
followee		0.620 (+)		
gender		0.503 (+)		
neg_word			0.936 (+)	
negative			0.918 (+)	
pos_word				0.888 (+)
positive				0.854 (+)

Note: + (−) in parentheses means the variable is positively (negatively) loaded.

The PCA results in Table 3 show that PC1 reflects users' information exchange behavior and represents online opinion exposure. PC2 represents users' resistance. PC3 contains neg_word and negative variables, reflecting users' negative reactions, and PC4 contains pos_word and positive variables, reflecting users' positive reactions. As mentioned earlier, sensitivity is the degree to which an entity is negatively or favorably affected. Thus PC3 and PC4 represent the sensitivity of social media users. These components collectively define user sensitivity to public opinion, aligning with disaster research theories on vulnerability [2, 10], encompassing exposure, sensitivity, and adaptive capacity.

In order to further investigate how these factors extracted by PCA affect the users' vulnerability, we took the users' vulnerability as the dependent variable and the exposure of online public opinion, the resistance of users, the negative reaction of users and the positive reaction of users as the independent variables to estimate an ordered-logit model. The regression results are reported in Table 4.

Table 4. The parameter estimation of ordered logistic regression

Component Variable	Coefficient	Odds Ratio
the exposure of online public opinion	0.702^{***} (0.116)	2.018^{***} (0.235)
the resistance of users	-1.382^{***} (0.135)	0.251^{***} (0.034)
the negative reaction of users	0.729^{***} (0.122)	2.073^{***} (0.252)
the positive reaction of users	-1.048^{***} (0.127)	0.351^{***} (0.045)
Cut 1 Constant	-2.183 (0.173)	
Cut 2 Constant	0.018 (0.126)	
Cut 3 Constant	1.128 (0.140)	
Cut 4 Constant	3.022 (0.230)	
Prob > chi2	0.000	
Pseudo R2	0.192	

Notes: Standard errors are in parentheses. * Significant at the 10 percent level; ** significant at the 5 percent level; *** significant at the 1 percent level.

Analysis from Table 4 indicates that exposure to online public opinion (coefficient $0.702, p < 0.01, OR = 2.018$) and negative user reactions (coefficient $0.729, p < 0.01, OR = 2.073$) significantly increase vulnerability, with negative reactions having a greater impact. Conversely, user resistance (coefficient $-1.382, p < 0.01, OR = 0.251$) and positive reactions (coefficient $-1.048, p < 0.01, OR = 0.351$) significantly decrease vulnerability, with resistance having a more substantial effect. Notably, negative emotional reactions are more strongly linked to vulnerability than both exposure and positive reactions. Further investigation, detailed in Fig. 3, examines the frequency of negative words to understand these reactions.

Obviously, from Fig. 3, we can see that users' negative reaction to online public opinion is mainly caused by these two negative emotion categories: sadness-related (e.g. 'grief', 'cry', 'suffer', etc.), and disgust-related (e.g. 'rumor', 'liar', 'lies', etc.) words.

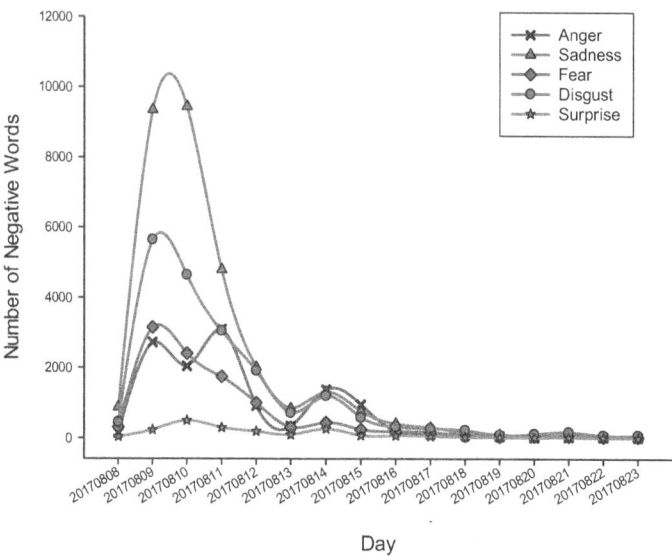

Fig. 3. Number of negative words in daily granularity

6 Conclusion

6.1 Discussion

Leveraging a vulnerability analysis framework from disaster research, this study employed a data-driven approach using Sina Weibo data from the Jiuzhaigou earthquake to quantitatively assess factors influencing user vulnerability to emergency-related online public opinion. Results indicate that vulnerability is impacted by user behavior, social attributes, and emotional reactions (both positive and negative).

(a) Users' vulnerability positively correlates with online public opinion exposure, encompassing variables such as comment, forward, like, and post. These metrics reflect the frequency, severity, and scope of public opinion [7]. Notably, forwarding accelerates information dissemination rapidly [22]. Consequently, greater exposure to online public opinion increases the likelihood of user engagement, thereby enhancing vulnerability.

(b) Negative user reactions, comprising variables negative and neg_word, positively correlate with vulnerability, whereas positive reactions (positive and pos_word) are negatively associated with vulnerability in emergency contexts. Negative reactions, often seeking emotional support, indicate heightened vulnerability [11]. Notably, negative reactions are more influential on vulnerability than public opinion exposure, frequently manifesting as sadness and disgust, particularly in natural disasters. This aligns with findings that sadness-related words outnumber anger-related words in disaster contexts [12], suggesting that natural disasters, unlinked to human intent, elicit predominantly sorrowful responses.

(c) User resistance, encompassing variables followers, followees, type, gender, and blogs, negatively impacts social media vulnerability. Specifically, users with more followers, followees, and previous posts exhibit lower vulnerability, aligning with the Theory of Planned Behavior, where resources and sociability mitigate vulnerability [34]. Followers symbolize social resources and popularity [17], while followees and posts reflect social activeness, reducing vulnerability to negative public opinion impacts. Organizational users also demonstrate lower vulnerability due to greater trust and social support [30]. Additionally, male users are less vulnerable than female users, consistent with research indicating higher female vulnerability during disasters [26].

6.2 Theoretical Contributions

This study extends vulnerability literature by integrating disaster research concepts into online public opinion contexts, focusing on social media users' vulnerability during emergencies. Unlike traditional vulnerability studies based on census data and physical disaster areas, this research explores vulnerability in online environments. Leveraging existing vulnerability analysis methods, it identifies and examines factors influencing social media users' vulnerability, thereby broadening the application of disaster research to digital social spaces.

This research extends online public opinion literature by adapting vulnerability analysis frameworks from disaster research to assess social media users' vulnerability in emergency contexts, marking a novel approach in this domain. Despite the established link between risk and vulnerability in literature, previous studies on online public opinion risk during emergencies have overlooked users' vulnerability. This study introduces a fresh perspective to risk research in online public opinion emergencies, contributing to the development of a more robust risk analysis system for online public opinion.

6.3 Practical Implications

This research offers practical insights for emergency agencies by elucidating social media users' vulnerability, enhancing the efficiency of situational awareness and disaster management. Specifically, mitigating users' vulnerability to online public opinion can be strategically integrated into managing the risk associated with mass emergencies.

(a) The study indicates that heightened exposure to public opinion exacerbates users' vulnerability. Consequently, government and emergency agencies must adeptly discern and swiftly counteract negative online narratives, such as rumors, crafted by extremists within a deluge of disaster-related information. This proactive approach aims to curtail the dissemination of harmful content and safeguard the public from detrimental effects associated with such malicious opinions.
(b) The empirical analysis underscores the significance of emotional reactions in social media, particularly during emergencies. Government and emergency agencies are advised to intensify monitoring of public emotional responses, especially extreme negative reactions, through content analysis. Additionally, disseminating credible information imbued with positive emotions during crises can mitigate public sensitivity to negativity.

(c) The research suggests that government and emergency agencies should prioritize enhancing social media engagement and follower counts to bolster resistance to online public opinion. This strategy is integral for establishing an effective feedback and response mechanism in disaster scenarios.

6.4 Limitations and Future Work

Our study, limited by its focus on Sina Weibo data related to the Jiuzhaigou earthquake, may face challenges in generalizability. Future research should explore online public opinion across various emergencies, including man-made disasters. Additionally, constraints imposed by the Sina Weibo API and language measures restrict comprehensive user profile analysis; subsequent studies should incorporate broader variables such as user tags, account age, and education. Despite these limitations, our research offers insights into social media user vulnerability and presents a novel perspective for managing online public opinion risks during emergencies.

Acknowledgments. This study was funded by Liaoning Province Economic and Social Development Research Project (grant number 2024lslybkt-049).

Disclosure of Interests. The authors have no competing interests to declare that are relevant to the content of this article.

References

1. Adger, W.N.: Social vulnerability to climate change and extremes in coastal Vietnam. World Dev. **27**(2), 249–269 (1999)
2. Adger, W.N.: Vulnerability. Glob. Environ. Chang. **16**(3), 268–281 (2006)
3. Burnap, P., et al.: Tweeting the terror: modelling the social media reaction to the Woolwich terrorist attack. Soc. Netw. Anal. Min. **4**(1), 1–14 (2014)
4. Cerchiello, V., Ceresa, P., Monteiro, R., Komendantova, N.: Assessment of social vulnerability to seismic hazard in Nablus, Palestine. Int. J. Disast. Risk Reduct. **28**, 491–506 (2018)
5. Cutter, S.L.: Living with Risk: The Geography of Technological Hazards. E. Arnold (1993)
6. Cutter, S.L., Boruff, B.J., Shirley, W.L.: Social vulnerability to environmental hazards. Soc. Sci. Q. **84**(2), 242–261 (2003)
7. Du, Z., Xie, X.: The establishment of public opinion forecasting and early-warning model with the methods of grey forecasting and pattern recognition. Libr. Inf. Serv. **57**(15), 27–33 (2013)
8. Flanagan, B.E., Gregory, E.W., Hallisey, E.J., Heitgerd, J.L., Lewis, B.: A social vulnerability index for disaster management. J. Homeland Secur. Emerg. Manage. **8**(1) (2011)
9. Gallopín, G.C.: Linkages between vulnerability, resilience, and adaptive capacity. Glob. Environ. Chang. **16**(3), 293–303 (2006)
10. Gerlitz, J.Y., Banerjee, S., Brooks, N., Hunzai, K., Macchi, M.: An approach to measure vulnerability and adaptation to climate change in the Hindu Kush Himalayas. In: Handbook of Climate Change Adaptation, pp. 151–176 (2015)
11. Gilbert, E., Karahalios, K.: Predicting tie strength with social media. In: Proceedings of the SIGCHI Conference on Human Factors in Computing Systems, pp. 211–220. ACM (2009)

12. Greving, H., Kimmerle, J., Oeberst, A., Cress, U.: Emotions in Wikipedia: the role of intended negative events in the expression of sadness and anger in online peer production. Behav. Inf. Technol. 1–11 (2018)
13. Gunarathne, P., Rui, H., Seidmann, A.: Whose and what social media complaints have happier resolutions? Evidence from Twitter. J. Manag. Inf. Syst. **34**(2), 314–340 (2017)
14. Holland, I.S., Lujala, P., Rød, J.K.: Social vulnerability assessment for Norway: a quantitative approach. Norsk Geografisk Tidsskrift-Norwegian J. Geogr. **65**(1), 1–17 (2011)
15. Imran, M., Castillo, C., Diaz, F., Vieweg, S.: Processing social media messages in mass emergency: Survey summary. In: Companion of the the Web Conference 2018 on the Web Conference 2018, pp. 507–511. International World Wide Web Conferences Steering Committee (2018)
16. Jamali, M., Nejat, A., et al.: Social media data and post-disaster recovery. Int. J. Inf. Manage. **44**, 25–37 (2019)
17. Jin, X., Jin, K., Tang, Z., Zhou, Z.: Factors influencing content forwarding behavior in microblog during emergency events: a perspective of information source. J. China Soc. Sci. Tech. Inf. **34**(8), 809–818 (2015)
18. Jin, X., Fang, Y., Zhou, Z.: Understanding user-generated information sharing in Microblog-based on impulsive behavior perspective. J. China Soc. Sci. Tech. Inf. **35**(7), 739–748 (2016)
19. Kaźmierczak, A., Cavan, G.: Surface water flooding risk to urban communities: analysis of vulnerability, hazard and exposure. Landsc. Urban Plan. **103**(2), 185–197 (2011)
20. Kim, J., Bae, J., et al.: Emergency information diffusion on online social media during storm Cindy in U.S. Int. J. Inf. Manage. **40**, 153–165 (2018)
21. Li, L., Zhang, Q., Tian, J., Wang, H.: Characterizing information propagation patterns in emergencies: a case study with Yiliang earthquake. Int. J. Inf. Manage. **38**(1), 34–41 (2018)
22. Li, Q., Liu, Y.: Exploring the diversity of retweeting behavior patterns in Chinese microblogging platform. Inf. Process. Manage. **53**(4), 945–962 (2017)
23. Lu, Y.: Analysis on the forms of internet public opinion risk and coping strategies. New Media Res. **24**, 1–3 (2018)
24. Pennebaker, J.W., Booth, R.J., Boyd, R.L., Francis, M.E.: Linguistic inquiry and word count: LIWC2015. Pennebaker Conglomerates, Austin (2015)
25. Ragini, J.R., Anand, P.R., Bhaskar, V.: Big data analytics for disaster response and recovery through sentiment analysis. Int. J. Inf. Manage. **42**, 13–24 (2018)
26. Siagian, T.H., Purhadi, P., Suhartono, S., Ritonga, H.: Social vulnerability to natural hazards in Indonesia: driving factors and policy implications. Nat. Hazards **70**(2), 1603–1617 (2014)
27. Simon, T., Goldberg, A., Adini, B.: Socializing in emergencies—a review of the use of social media in emergency situations. Int. J. Inf. Manage. **35**(5), 609–619 (2015)
28. Smit, B., Wandel, J.: Adaptation, adaptive capacity and vulnerability. Glob. Environ. Chang. **16**(3), 282–292 (2006)
29. Solangaarachchi, D., Griffin, A.L., Doherty, M.D.: Social vulnerability in the context of bushfire risk at the urban-bush interface in Sydney: a case study of the Blue Mountains and Ku-ring-gai local council areas. Nat. Hazards **64**(2), 1873–1898 (2012)
30. Tausczik, Y.R., Pennebaker, J.W.: The psychological meaning of words: LIWC and computerized text analysis methods. J. Lang. Soc. Psychol. **29**(1), 24–54 (2010)
31. Thathsarani, U.S., Gunaratne, L.H.P.: Constructing an index to measure the adaptive capacity to climate change in Sri Lanka. Procedia Eng. **212**, 278–285 (2018)
32. Torkildson, M.K., Starbird, K., Aragon, C.R.: Analysis and visualization of sentiment and emotion on crisis tweets. In: Luo, Y. (ed.) CDVE 2014. LNCS, vol. 8683, pp. 64–67. Springer, Cham (2014). https://doi.org/10.1007/978-3-319-10831-5_9

33. Turner, B.L., Kasperson, R.E., Matson, P.A., McCarthy, J.J., Corell, R.W., Christensen, L., et al.: A framework for vulnerability analysis in sustainability science. Proc. Natl. Acad. Sci. **100**(14), 8074–8079 (2003)
34. Xie, Y., Qiao, R., et al.: Research on Chinese social media users' communication behaviors during public emergency events. Telemat. Inform. **34**(3), 740–754 (2017)

Analyzing Replies and Interactions Among Users with Different Stances: A Case Study of the Russia-Ukraine Conflict

Xiaohui Huang[1,3] and Xijin Tang[1,2(✉)]

[1] Academy of Mathematics and Systems Science, Chinese Academy of Sciences, Beijing 100190, China
huangxiaohui@amss.ac.cn, xjtang@iss.ac.cn
[2] University of Chinese Academy of Sciences, Beijing 100049, China
[3] Department of Statistics and Data Science, College of Science, Southern University of Science and Technology, Shenzhen 518055, China

Abstract. Since its outbreak, the Russia-Ukraine conflict has been extensively discussed on social media. Analyzing the replies and interactions across different stances is significant for comprehending the political, economic, military, and diplomatic factors, as well as the increasing polarization that has been widely observed among online discussions. This paper analyzes the replies from Reddit discussions, focusing on user replies to understand sentiment, topics, and interaction patterns among users with Pro-Russia, Neutral, or Pro-Ukraine stances. Descriptive statistical analysis is employed to study word frequency, entity frequency, and emotion distribution. Topic modeling is utilized to extract topics from replies with different stances, revealing differences in the content and providing insights into stances formation. User interaction analysis indicates that replies between opposing stances tend to exhibit stronger negative sentiment, and different stances also reduce the likelihood of interaction. Pro-Ukraine users prefer interacting with like-minded users, while neutral users interact more equally with different stances, aiming to ease tensions and foster discussion.

Keywords: Russia-Ukraine conflict · User stance · Topic analysis · Reply network · User interaction

1 Introduction

Since Russia initiated its "special action" in Ukraine on February 24, 2022, the Russia-Ukraine conflict has drawn significant international attention. This conflict has impacted geopolitics [1,2] and led to extensive discussions on social media like Twitter and Reddit [3–5]. These discussions reflect a wide range of perspectives, emotions, and narratives, shaping public understanding and sentiment about the conflict [6,7]. Analyzing the discussions on the Russia-Ukraine

X. Tang et al. (Eds.): KSS 2024, CCIS 2269, pp. 109–123, 2025.
https://doi.org/10.1007/978-981-96-0178-3_8

conflict provides a detailed understanding of public sentiment, discourse dynamics, and interactive patterns, contributing to academic research, media strategies, policy-making, and societal understanding. Therefore, this paper focuses on analyzing Reddit discussions to explore how users with different stances engage in discussions about this conflict, offering insights into their interaction and the underlying public discourse.

Many researchers have employed various text mining methods to explore the sentiments and content from online discussions about the conflict. Chen and Ferrara [8] collected nearly 500 million tweets using specific keywords since the conflict broke out, and developed an effective dataset for studying social media's impact on cyber information warfare. Aslan [9] discussed how sentiment analysis effectively captures people's thoughts on the Ukraine-Russia conflict. He proposed a BiLSTM-based method to identify sentiment and used Latent Dirichlet Allocation (LDA) for topic extraction to uncover discussion themes, revealing people's sentiment orientations and views on the conflict. Sazzed [10] further analyzed emotional and thematic differences among various demographic groups regarding the conflict, providing insights into people's perceptions across the globe. While these researches focus on sentiment and topic analysis, their results are often limited and overlook some important details, such as group differences in the discussions.

Additionally, some researchers have considered interaction relationships within reply networks to analyze differences among users with various characteristics (such as stances). Xi and Tang [11] analyzed posts on the debate about traditional Chinese medicine on the Tianya forum as an example, combining user stances to study interaction behaviors in online debate networks. They found that online social platforms promote dialogue between individuals with different stances, with replies between opposing stances happening faster than those between same stances. Lai et al. [12] examined Twitter discussions on the Italian constitutional referendum, analyzing interactions between users with similar and differing opinions across multiple social network structures, such as friendships, retweets, quotes, and replies. They found that different types of social relationships significantly affect network structures, with users tending to express differing opinions by replies. Evkoski et al. [13] analyzed the replies on social media in the ex-Yugoslavia regarding Russia's invasion of Ukraine, comparing topological interactions and content-based similarities between communities. Their results suggested that different types of communities, such as political and non-political ones, react differently toward the same invasion-related content but show high consistency on specific topics. These studies suggest that focusing on the interactions between users with different stances in the context of the Russia-Ukraine conflict may provide valuable insights into people's views and behavior patterns regarding this topic.

This paper conducts analysis of Reddit discussions about the Russia-Ukraine conflict by integrating text analysis, reply networks, and user stances. The differences in sentiment, topics and interactions are examined among users with different stances, revealing insights into public opinions and interaction behaviors specific to this geopolitical issue. We uniquely focus on how different stances, Pro-Russia, Neutral, and Pro-Ukraine, shape the discourse and interactions, uncovering echo chambers and polarized communication. The distinct themes among users with varying stances and emphasizes the role of neutral users in fostering balanced discussions, are explored to offer novel perspectives on the online conflict-related discussions and practical applications for social media management and policy-making.

The remainder of this paper is organized as follows. Section 2 gives the data processing and statistical analysis of the replies. In Sect. 3, LDA topic modeling and topic analysis are presented. Section 4 shows reply network and user interaction analysis. Finally, conclusions are given in Sect. 5.

2 Data Processing and Statistical Analysis

This section presents the dataset generated and details the word frequency analysis, entity analysis, and emotion analysis performed on the data.

2.1 Data Preprocessing

Since the outbreak on February 24, 2022, the Russia-Ukraine conflict has been widely discussed. This paper examines the content and behavior of users with different stances by analyzing a dataset from Reddit discussions on the conflict [14]. Data are collected from relevant subreddits, including 'UkraineConflict,' 'RussiaUkraineWar2022,' 'UkrainianConflict,' 'UkraineWarReports,' 'ukraine,' and 'UkraineWarVideoReport.' As shown in Table 1, the dataset includes 29,844 posts and 226,187 replies. A directed reply network, which comprises 131,627 nodes (users) and 220,730 edges (reply connections), is built based on the users' replies. If user i replies to user j, a directed edge is drawn from user i to user j. Since posts are undirected, we only discuss their text content in statistical analysis (Sect. 2) and topic analysis (Sect. 3), but do not consider their direction in the reply network analysis in Sect. 4.

Table 1. Dataset. This table shows the number of posts, replies, users, and edges.

Num of Posts	Num of Replies	Num of Users	Num of Edges
29,844	226,187	131,627	220,730

Figure 1 depicts the trend in total discussion on the conflict from February 24, 2022, to March 10, 2022. Discussions soon became very hot within the first week of the conflict and decline gradually. This suggests that as the conflict dragged on, public interest dropped. Thus, analyzing the data from the first two weeks after the outbreak provides insights into public reactions and reply behaviors while reducing the influence of irrelevant and redundant data.

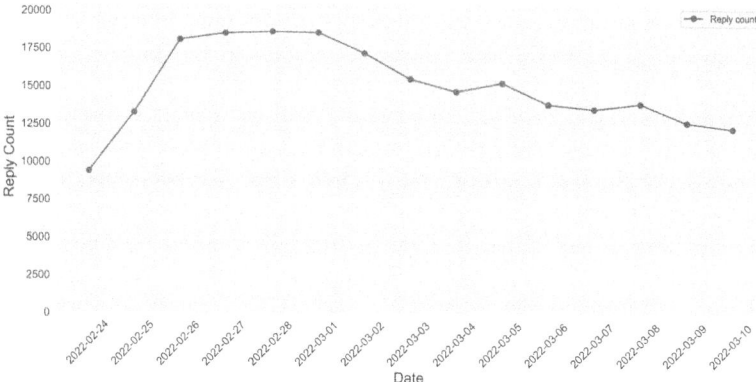

Fig. 1. The trend chart of daily reply counts. Figure shows the trend in total discussion on this conflict from February 24, 2022, to March 10, 2022.

The main objective of this paper is to identify differences in replies from users with different stances. To determine the stances of users, we build a sample dataset with 2,000 randomly selected replies for labeling. Based on the content, we classify the replies as Pro-Russia, Neutral, or Pro-Ukraine. The labeling principle focuses on whether the replies express support or opposition to the Russia-Ukraine conflict or related events. Replies that did not clearly fall into these categories are labeled as neutral. The labeled sample dataset is split into 80% for training and 20% for validation to fine-tune a pre-trained BERT model for predicting the stance of replies. The model, achieving 80% prediction accuracy, is then used to determine the stance of the remaining replies.

The results show that out of the total replies, 16,762 (7.4%) are Pro-Russia, 113,985 (50.4%) are Pro-Ukraine, and 95,503 (42.2%) are neutral. Related researches [15,16] on online social media like Twitter and Facebook indicate that most users are Pro-Ukraine, with fewer Pro-Russia supporters. As Reddit is banned in Russia, this also results in a small number of Pro-Russia supporters.

2.2   Word Frequency Analysis

Figure 2 presents the word cloud generated by the Python tool WordCloud[1] after removing common stop words. The most commonly used words are shown in larger fonts to highlight their importance across different stances. Across all stances, both 'Ukraine' and 'Russia' are the most frequently mentioned words, indicating their importance in the discourse. Similarly, words like 'war' and 'invasion' also appear frequently in all replies. Among Pro-Ukraine replies, phrases such as 'Glory Ukraine' and 'Together victory' are frequently used, reflecting the desire for a Ukrainian victory and opposition to Russian actions. The frequent appearance of words like 'Kyiv,' 'Moscow' and 'eastern Ukraine' in Pro-Russia

---

[1] https://pypi.org/project/wordcloud/.

replies suggests a focus on the conflict's events and locations, possibly justifying actions and emphasizing regional elements. The neutral replies include words like 'people,' 'Nazi,' and 'government,' reflecting concerns about the conflict's impacts on people and its political justification.

(a) Pro-Russia   (b) Neutral   (c) Pro-Ukraine

Fig. 2. Word clouds of replies from different stances. The most commonly used words are highlighted in larger fonts across different stances.

2.3   Entity Analysis

The Spacy-transformers library[2] in Python is used to extract the entities from replies with different stances. By combining transformer models like Bert and RoBERTa with the Spacy framework, this tool improves the efficiency of tasks such as text classification, named entity recognition, and sentiment analysis. Figure 3 shows the counts of some key personal names and countries or location entities across replies with various stances. In Fig. 3, the vertical axis represents entities, the horizontal axis represents the corresponding count, and the legend represents different stances. The markers with different shapes in the figure represent the count of each entity across different stances.

As shown in Fig. 3, the entities 'Ukraine' and 'Russia', are mentioned most in all replies, reflecting their central role in discussions related to the conflict. There is a clear polarization of entity mentions based on the stance. Pro-Ukraine replies prominently feature entities such as 'USA,' 'NATO,' 'UK,' 'EU,' and so on, emphasizing allies and opposition to Russia. In contrast, Pro-Russia replies frequently mention entities like 'Kyiv,' 'Moscow,' 'Kherson,'and 'Donetsk,' highlighting Russian interests and supportive narratives. Neutral replies show a more balanced distribution but still lean towards the entities more frequently mentioned in Pro-Ukraine or Pro-Russia. The high frequency of these entities highlights their influence in shaping the narrative and bias within each stance, emphasizing the importance of critically evaluating media representations in geopolitical conflicts. Neutral replies often serve as mediators, striving to clarify and ease tensions between opposing viewpoints.

---

[2] https://spacy.io/.

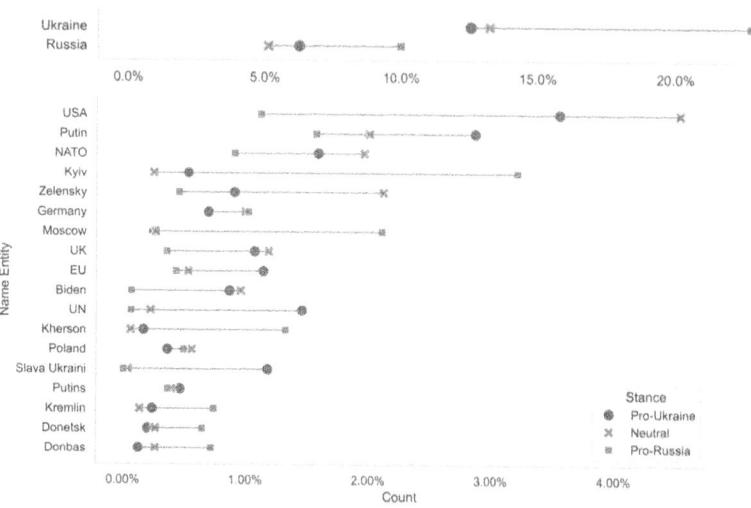

Fig. 3. The top entity words with the highest proportions in different stances. The vertical axis represents entities, the horizontal axis represents the corresponding count, and the legend represents different stances. The markers with different shapes in the figure show the count of each entity across different stances. The figure is split into two parts due to the significant difference in proportions between the top two entities and the others, which would make it difficult for comparison in a single scale.

2.4   Emotion Analysis

In psychology, Plutchik et al. [17, 18] categorize human emotions into eight types: anger, anticipation, disgust, fear, joy, sadness, surprise, and trust. The emotional characteristics of responses from different stances are reflected through the distribution of emotions. The NRC Emotion Lexicon [19], which contains 14,200 words, each with a tendency score for each emotion, is used to obtain the distribution of the above eight emotions.

Figure 4 shows the emotion distribution for three stances: Pro-Russia, Neutral, and Pro-Ukraine. Fear is the dominant emotion among all stances. In Pro-Ukraine responses, fear is particularly high at 28.55%, reflecting significant anxiety and concern about the conflict's results and impacts on each side. Pro-Ukraine replies exhibit higher anger levels, reflecting strong emotional reactions against perceived aggression. In contrast, Pro-Russia replies show a notable amount of anticipation, joy and trust, potentially indicating confidence in their stance. Sadness and disgust are consistently expressed across all replies, with a higher presence in Pro-Russia replies, potentially indicating disapproval and sadness regarding the progression of the conflict. Replies with Neutral stance show a more balanced distribution of emotions, particularly focusing on trust and fear, which suggests an effort to maintain an impartial perspective, acknowledging the various risks and expectations linked to the conflict.

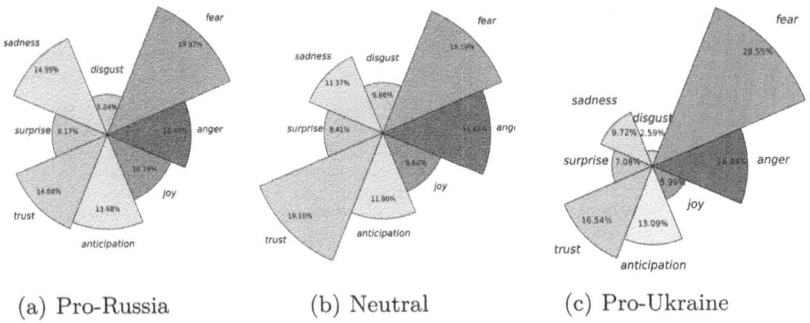

(a) Pro-Russia   (b) Neutral   (c) Pro-Ukraine

Fig. 4. Emotion distribution of replies from different stance. The larger the sector area, the higher the proportion of that emotion in replies with that stance.

3   Topic Modeling and Topic Analysis

In this section, we apply topic modeling on replies with different stances and compared the differences among them.

3.1   Topic Model

To compare the differences in reply content across various stances, we apply the BERTopic model [24] to extract the topics from the replies. Traditionally, the LDA model has been widely used for topic extraction [25, 26]. However, LDA works well with long text but struggles with short social media texts like tweets. BERTopic uses a pre-trained model to convert documents into embeddings, effectively handling short texts, and then applies dimensionality reduction and clustering techniques to achieves semantically consistent topic clusters. BERTopic also allows users to choose different embeddings, dimensionality reduction techniques, or clustering models, making it adaptable to different types of data.

This paper uses the sentence-transformers model[3] for document embeddings, UMAP [27] for dimensionality reduction, and HDSCAN [28] for clustering. Different with LDA and other clustering method, HDSCAN automatically determines the optimal number of clusters. As summarized in Table 2, the topics under different stances are obtained through the BERTopic model. The number of topics identified for Pro-Russia, Neutral, and Pro-Ukraine stances are 15, 214, and 245, respectively. Based on topic size, probability and coherence, we present several representative topics and their representative replies in Table 2.

3.2   Topic Analysis with Different Stances

The topic modeling results reflect the distinct thematic foci across the Pro-Russia, Neutral, and Pro-Ukraine stances. Pro-Russia replies mainly discuss the

---

[3] The all-MiniLM-L6-v2 pre-trained model in sentence-transformers library is chosen. https://huggingface.co/sentence-transformers/all-MiniLM-L6-v2.

Table 2. Sample topics of each stance. The table lists major topics among replies of different stances.

Topic label	Size	Representative replies
Pro-Russia		
Justification	24.1%	["A memorandum was held in Crimea and the decision of the people was to join Russia but how else to protect people from the fate of Donbass without introducing a military when no one wants to recognize the will of the people of Crimea." "Ukraine is currently winning the information war exposing itself only as a victim of bloodthirsty Russia What would have happened if by that time Ukraine was part of NATO 3 world war."]
War	12.4%	["Russian jets can fly most of the time out of reach of MANPADS and dive down only for attack thus minimizing exposure to MANPADS." "Russian still have the missiles that targeted and destroyed so much of the Ukrainian Airbases."]
NATO	11.4%	["Today we need to admit that there is a crisis in European and global politics. One of the reasons is a lack of desire on the part of our Western partners to take Russias point of view and legal interests in security into consideration." "I am referring primarily to NATO expansion missile defense plans."]
Neutral		
Oil	13.6%	["I think youre right If the west stops buying Russian oil someone out there will still buy it The difference will be oil prices will have gone up and Russia will be making more money from their oil exports." "You are mixing gas and oil together Europe cant really cut off gas but can cut off oil."]
Bioweapons	10.4%	["Washington has been moving its bioweapons research out of Ukraine. The head of Russias Nuclear Biological and Chemical Defense Troops has claimed."]
UN	8.9%	["Vote would be taken by the Security Council not the general membership. And being that Russia has a Veto vote in that council. They would not allow it. They veto it the second when it came up for discussion."]
Pro-Ukraine		
Peace	12.7%	["If Russia's invasion succeeds, other countries might follow suit, leading to more conflicts and wars. Supporting Ukraine is crucial for maintaining global peace."]
Sovereignty	11.3%	["Ukraine is a sovereign nation with the right to determine its own future. Russia's aggression is a severe violation of this sovereignty." "Just Putin and Sadam have broken these international sovereignty laws so blatantly in modern history."]
Continuing		
Sanctions	7.1%	["Its either war or more sanctions More sanctions it is." "if anything that would make more sanctions against them."]
NW	5.1%	["So here is my hope when this all ends and lets assume Putin is no longer in power that the folly of nuclear weapons is realized and countries around the world rid themselves of these weapons." "If one good thing can come from this disaster lets hope banning nuclear weapons is it."]

NATO: North Atlantic Treaty Organization.
UN: the United Nations.
NW: nuclear weapons.

justification of Russia's actions (24.15%) and the threats of NATO's expansion (12.4%). As shown by the representative replies in the Table 2, these topics cover users' opinions on the Crimea region and NATO's plans, thereby expressing their support for Russia's actions. Furthermore, 12.4% of the replies analyze and discuss some actions related to Russia in this conflict.

Neutral replies center on the economic and political impact, such as oil exports, bioweapons, and the role of the United Nations in mediating this conflict. 13.6% of the neutral replies focus on the conflict's impacts on oil prices and supply. Discussions about the bioweapons labs that Russia claims were built by the U.S. in Ukraine make up 10.4%. Meanwhile, 8.9% of the replies discuss the influence of the United Nations on this conflict. Neutral replies objectively discuss events and impacts of the Russia-Ukraine conflict, without clearly supporting either Russia or Ukraine.

Pro-Ukraine replies focus on the topics of peace, sovereignty, sanctions, nuclear weapons, and so on, expressing users' opposition to Russian actions and support for Ukraine. 12.7% of the replies call for ending the conflict and maintaining peace, while 11.3% express dissatisfaction with Russia's actions from the perspective of national sovereignty. Discussions about sanctions on Russia make up 7.1% of the replies. Additionally, 5.1% of the replies mention the threat of Russia's nuclear weapons and the concerns about their use in the conflict.

The topic extraction results across different user stances reflect the diversity of public participation in the Russia-Ukraine conflict discussions on Reddit. The differences in topics among the three stances show the varying events users focus on in the conflict, illustrating how Pro-Russia and Pro-Ukraine users justify their stances, and how these opposing stances are formed.

4   Reply Network and User Interaction

In this section, we analyze the user reply network and discuss the interactions between users with different stances.

4.1   Reply Network

As shown in Fig. 5a, the directed user reply network is created based on the interactions between users. In this network, each node represents a user, and each edge represents at least one reply from one user to another. If user i replies to user j, a directed edge is drawn from user i to user j. The color of each node represents users' stance. The reply network contains 131,627 nodes (users) and 220,730 edges (reply relationships). Previous studies indicate that individuals' stances on a topic are stable over short periods [20,21]. Thus, a user's stance on the Russia-Ukraine conflict is represented by the average stance of all his replies over the period.

Figure 5(a) indicates that most users are Pro-Ukraine or neutral and tend to cluster together, while Pro-Russia users are relatively fewer. Our earlier work [22] shows that this reply network exhibits an echo chamber structure, with users of same stance more likely to interact. This homogeneity is observed through the network's community structure. We use the Louvain algorithm [23] to identify all community structures in the network and calculate their average stance. Figure 5(b) illustrates the community sizes and average stance within the reply network. The x-axis represents the community id sorted by average stance, and

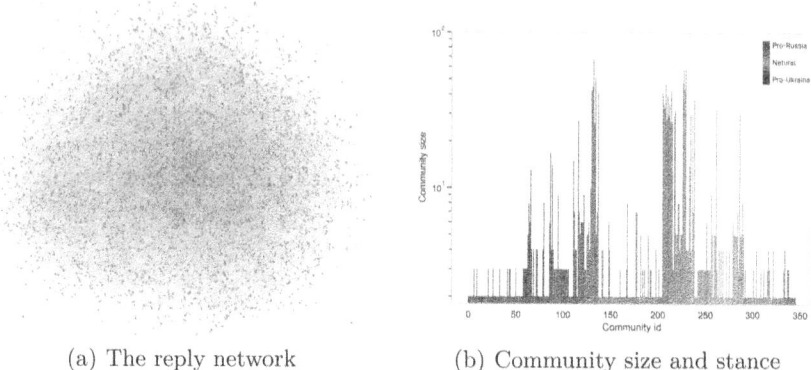

(a) The reply network   (b) Community size and stance

Fig. 5. The reply network and communities.(a) Each node represents a user, and each edge represents a reply between users. If user i replies to user j, a directed edge is drawn from user i to user j. The node colors represent users' stances, with green indicating Pro-Russia, blue for neutral, and orange for Pro-Ukraine. (b) The x-axis represents the community id sorted by average stance, and the y-axis represents the community size, i.e., the number of nodes.

the y-axis represents the community size, i.e., the number of nodes. The statistics show more Pro-Ukraine and neutral communities with larger sizes, while Pro-Russia voices are scarce, appearing only in small communities.

This distribution suggests a clear echo chamber effect, where users with same stances cluster and reinforce each other's opinions. Larger Pro-Ukraine and neutral communities may amplify shared narratives, while smaller Pro-Russia groups remain more insulated. However, the diversity in community sizes and stances suggests potential cross-stance interactions. Although larger communities dominate discussions, the presence of smaller groups allows for cross-stance engagement, offering opportunities to challenge or influence entrenched positions and fostering a more dynamic discourse.

4.2   User Interactions with Different Stances

The aforementioned analysis shows differences in reply content across stances and suggests potential interactions between these communities. Here, we conduct a deeper analysis of interactions between users with different stances. We utlize the Apriori algorithm [29,30] to mine the association relationships within user interactions. The Apriori algorithm determines if users with stance X have specific reply tendencies toward users with stance Y. Association rules mining is usually screened using three statistical measures: support, confidence, and lift ratio. Support is an indication of how often the association rule appears in the dataset. Confidence reflects the reliability of the rule calculation. Lift indicates the relationship between the antecedents and consequences of the association rule. The three statistical measures are calculated as shown in Eqs. (1) to (3),

$$\text{Support}(X \rightarrow Y) = P(X \cap Y) \tag{1}$$

$$\text{Confidence}(X \rightarrow Y) = \frac{P(X \cap Y)}{P(Y)} \tag{2}$$

$$\text{Lift}(X \rightarrow Y) = \frac{P(X \cap Y)}{P(X) \times P(Y)} \tag{3}$$

where $P(X \cap Y)$ is the probability that X and Y occur together in the dataset, $P(X)$ is the probability that X appears, and $P(Y)$ is the probability that Y appears in the dataset.

Besides, identifying sentiment from user interactions is also essential for uncovering biases and offering deeper insights into engagement and conflict within the discourse. Comparison with emotion analysis in Sect. 2.4, sentiment is preferred in explaining user interactions as it reflects broader attitudes, while emotion focuses on specific micro-level expressions within the text [31]. To explore the sentiment in their replies, we employ a RoBERTa pre-trained model[4] that has been trained and fine-tuned on over 100 million tweets, achieving high accuracy in sentiment analysis of short social media texts. We obtain the sentiment for all replies and summarize the average during interactions between users with different stances.

Applying the Apriori algorithm with a minimum support threshold of 0.8, we discover that all item pairs across the three stances occur frequently. Their high frequency of co-occurrence indicates a high level of interactions between different stances. Based on the co-occurrence of replies between different stances, the conditional probability $P(X|Y)$ of a user with stance X replying to a user with stance Y, is calculated.

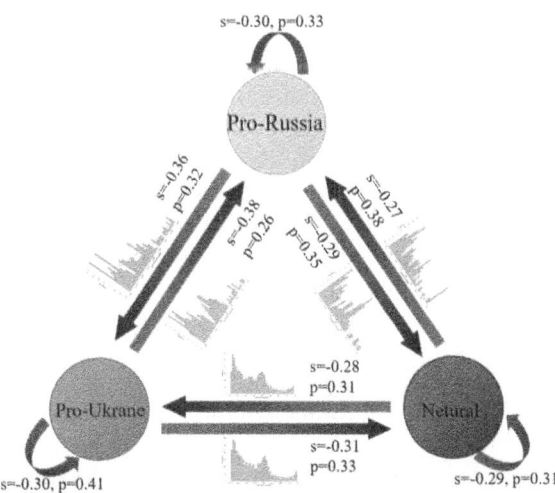

Fig. 6. Interaction relationships between users with different stances. In user interactions, s represents the sentiment of text and t represents the reply probability.

---

[4] https://huggingface.co/cardiffnlp/twitter-roberta-base-sentiment-latest.

Figure 6 displays the average sentiment, sentiment distribution, and reply probabilities in interactions between users with different stances. Negative sentiment dominates across all replies, with the average sentiment in all interactions being negative, and the sentiment distribution skewing toward the negative end. We find that the proportion of negative sentiment is higher in replies between Pro-Russia and Pro-Ukraine users. The average sentiment for (Pro-Russia $\rightarrow$ Pro-Ukraine) and (Pro-Ukraine $\rightarrow$ Pro-Russia) are -0.36 and -0.38, respectively, significantly more negative than other interactions, indicating that negative, even aggressive, content is more frequent in interactions between users with opposing stances. Other replies also show negative sentiment, ranging from -0.31 to -0.27, but with less intensity than those between Pro-Russia and Pro-Ukraine users. Discussions related to the conflict are inherently more adversarial and emotional, leading supporters on both sides to react more strongly and sharply. Neutral users are more likely to focus on facts and analysis, resulting in relatively calmer emotional reactions.

As for the reply interaction pattern, we discover that Pro-Ukraine users have the highest probability (0.41) of interacting within the group that have the same stance, indicating strong internal communication and solidarity. Such a pattern, known as homogeneous interaction, is identified and confirmed in our previous work [22]. When Pro-Ukraine users are trapped in an echo chamber, their exposure to users with opposite views decreases, explaining why they are less likely to interact with Pro-Russia users compared to neutral users. The interaction probabilities of (Pro-Russia $\rightarrow$ Pro-Ukraine), and (Pro-Ukraine $\rightarrow$ Pro-Russia) are lower, at 0.32 and 0.26 respectively, suggesting that conflicting stances may also decrease the willingness of them to communicate. Even if they reply to each other, the sentiment is likely to be negative. Interestingly, different from these findings, previous research on debates about traditional Chinese medicine [11] found that users with opposing stances interacted more frequently, revealing how echo chambers influence user interactions and restrict open discussion. Besides, Pro-Russia users, as confirmed in previous work [22], are not in the echo chamber. As shown in Fig. 6, they are more likely to interact with users of different stances, resulting in similar interaction probabilities with the other two stances. Neutral users reply equally to those with different stances, using more rational and positive words, indicating their dedication to reducing conflict and fostering discussion from a neutral stance.

5   Conclusions

Since its outbreak, the Russia-Ukraine conflict has become a major topic of discussion in the international community and on social media. This paper aims to analyze the Reddit discussions related to the conflict, focusing on user replies to understand sentiment, topics, and interactions among users with different stances. Understanding public sentiment, discourse content, and interactions is essential for policymaking and maintaining a healthy cyber space.

We collect replies related to the Russia-Ukraine conflict on Reddit from February 24 to March 10, 2022. After data processing, we identify the stance of each reply as Pro-Russia, Neutral, or Pro-Ukraine. Word and entity frequency analysis identify the most frequently occurring keywords and entities in replies with different stances. Emotion analysis is used to examine the emotional differences in the replies. We employ topic modeling to uncover the topics of concern in replies with different stances, providing insights into the formation of these stances. The analysis of the reply network and user interactions, including sentiment and reply probabilities, reveal sentiment differences and reply tendencies among users with different stances.

Topic analysis explains the topics that users are concerned about. We observe that neutral users tend to focus on economic topics like oil, while Pro-Russia and Pro-Ukraine users prefer topics that align with their stances. These observations provide insights into how Pro-Russia and Pro-Ukraine users justify their stances and how these opposing stances are formed. By the analysis of user interactions, we discover that replies between users with opposing stances exhibit higher levels of negative sentiment. Pro-Ukraine users are more likely to interact with users of same stance and less with those of opposing stance, while neutral users have more balanced interactions. Comparison with previous research shows how echo chambers shape user interactions and limit open discussion.

This paper provides a comprehensive analysis of Reddit discussions on the Russia-Ukraine conflict, revealing significant insights into public sentiment, topic preferences, and interactions among users with different stances. Different from previous research, we explore the semantic differences in reply content and interactions among users with varying stances. We highlight the polarized nature of online discourse and the role of neutral users in fostering balanced discussions. For social media platforms and community managers, our findings into the echo chambers and neutral users help guide the strategies to foster more balanced and inclusive discussions. Promoting cross-stance interactions and amplifying neutral voices contributes to mitigating polarization and fostering healthier online environments. For policymakers, understanding the dynamics of online discourse surrounding contentious geopolitical issues can inform more effective communication strategies, helping to bridge divides and reduce social tensions. The methodology we used, including BERTopic and network analysis, provides a versatile tool to analyzing online discourse. More research expanding to additional platforms and timeframes is desired in the future to validate our findings.

Acknowledgement. This paper is supported by the National Social Science Fund of China (No. 23 & ZD331).

References

1. Kotoulas, I.E., Pusztai, W.: Geopolitics of the war in Ukraine. Foreign Affairs Inst. 41–54 (2022)
2. Moisio, S.: Geopolitics of explaining Russia's invasion of Ukraine and the challenge of small states. Polit. Geogr. **97**, 102683 (2022)
3. Sufi, F.: Social media analytics on Russia-Ukraine cyber war with natural language processing: perspectives and challenges. Information **14**(9), 485 (2023)

4. Zasiekin, S., Kuperman, V., Hlova, I., et al.: War stories in social media: personal experience of Russia-Ukraine war. East Eur. J. Psycholinguist. **9**(2), 160–170 (2022)

5. Babacan, K., Tam, M.S.: The information warfare role of social media: fake news in the Russia-Ukraine war. Erciyes Iletişim Dergisi **3**, 75–92 (2022)

6. Kusa, I.: Russia-Ukraine war. Policy Perspect. **19**(1), 7–12 (2022)

7. Orhan, E.: The effects of the Russia-Ukraine war on global trade. J. Int. Trade Logist. Law **8**(1), 141–146 (2022)

8. Chen, E., Ferrara, E.: Tweets in time of conflict: a public dataset tracking the twitter discourse on the war between Ukraine and Russia. In: Proceedings of the International AAAI Conference on Web and Social Media, vol. 17, pp. 1006–1013 (2023)

9. Aslan, S.: A deep learning-based sentiment analysis approach (MF-CNN-BILSTM) and topic modeling of tweets related to the Ukraine-Russia conflict. Appl. Soft Comput. **143**, 110404 (2023)

10. Sazzed, S.: The dynamics of Ukraine-Russian conflict through the lens of demographically diverse twitter data. In: IEEE International Conference on Big Data, pp. 6018–6024 (2022)

11. Xi, Q., Tang, X.J.: Analysis of interactive behavior of online debate network. J. Syst. Sci. Math. Sci. **39**(9), 1361 (2019). (In Chinese.)

12. Lai, M., Tambuscio, M., Patti, V., et al.: Stance polarity in political debates: a diachronic perspective of network homophily and conversations on Twitter. Data Knowl. Eng. **124**, 101738 (2019)

13. Evkoski, B., Kralj Novak, P., Ljubešić, N.: Content-based comparison of communities in social networks: ex-Yugoslavian reactions to the Russian invasion of Ukraine. Appl. Netw. Sci. **8**(1), 40 (2023)

14. Zhu, Y., Haq, E., Lee, L.H., et al.: A reddit dataset for the Russo-Ukrainian conflict in 2022. arXiv preprint arXiv: 2206.05107, 2022

15. Zia, H.B., Haq, E.U., Castro, I., et al.: An Analysis of Twitter Discourse on the War Between Russia and Ukraine. arXiv preprint arXiv: 2306.11390, 2023

16. Pathinayake, N., Kulatileke, L., Hettiarachchi, C., et al.: Sentiment analysis of tweets on the Russia-Ukraine war. In:4th International Conference on Advanced Research in Computing (ICARC), pp. 143–148 (2024)

17. Plutchik, R.: Emotions: a general psychoevolutionary theory. Approaches Emot. **197–219**, 2–4 (1984)

18. Plutchik, R.: The Psychology and Biology of Emotion. HarperCollins College Publishers, Gurugram (1994)

19. Mohammad, S.M., Turney, P.D.: NRC emotion lexicon. Natl. Res. Counc. Can. **2**, 234 (2013)

20. Hyland, K., Jiang, F.: Change of attitude? A diachronic study of stance. Writ. Commun. **33**(3), 251–274 (2016)

21. Alkhalifa, R., Zubiaga, A.: Capturing stance dynamics in social media: open challenges and research directions. Int. J. Digit. Humanit. **3**(1), 115–135 (2022)

22. Huang, X.H., Tang, X.J.: Research on echo chamber mechanism and neighbor effect. Syst. Eng.- Theory Pract. (2024). (In Chinese.)

23. De Meo, P., Ferrara, E., Fiumara, G., et al.: Generalized Louvain method for community detection in large networks. In: 2011 11th International Conference on Intelligent Systems Design and Applications, pp. 88–93 (2011)

24. Grootendorst, M.: BERTopic: neural topic modeling with a class-based TF-IDF procedure. arXiv preprint arXiv: 2203.05794, 2022

25. Blei, D.M., Ng, A.Y., Jordan, M.I.: Latent dirichlet allocation. J. Mach. Learn. Res. **3**(Jan), 993–1022 (2003)
26. Huang, X., Tang, X.: Understanding of the party's construction and governing philosophy by an analysis of the reports of successive CPC's congresses. In: Chen, J., Huynh, VN., Tang, X., Wu, J. (eds.) Knowledge and Systems Sciences. KSS 2023. CCIS, vol. 1927, pp. 215–229. Springer, Singapore (2023). https://doi.org/10.1007/978-981-99-8318-6_15
27. Vermeulen, M., Smith, K., Eremin, K., et al.: Application of uniform manifold approximation and projection (UMAP) in spectral imaging of artworks. Spectrochim. Acta Part A Mol. Biomol. Spectrosc. **252**, 119547 (2021)
28. McInnes, L., Healy, J., Astels, S.: HDBSCAN: hierarchical density based clustering. J. Open Source Softw. **2**(11), 205 (2017)
29. Borgelt, C., Kruse, R.: Induction of Association Rules: Apriori Implementation. In: Hardle, W., Ronz, B. (eds.) Compstat, pp. 395–400. Physica, Heidelberg (2002). https://doi.org/10.1007/978-3-642-57489-4_59
30. Santoso, M.H.: Application of association rule method using apriori algorithm to find sales patterns case study of indomaret tanjung anom. Brill. Res. Artif. Intell. **1**(2), 54–66 (2021)
31. Sailunaz, K., Alhajj, R.: Emotion and sentiment analysis from Twitter text. J. Comput. Sci. **36**, 101003 (2019)

Explainable Machine Learning-Based Research on Key Factors in the Formation of Public Opinion on Similar Events

Yuxue Chi[1], Ning Ma[2], and Yijun Liu[2,3]([envelope])

[1] School of Management Science and Engineering, Central University of Finance and Economics, Beijing 100081, China
[2] Institutes of Science and Development, CAS, No. 15 ZhongGuanCunBeiYiTiao Alley, Haidian District, Beijing 100190, China
Yijunliu@casisd.cn
[3] University of Chinese Academy of Sciences, No. 19A Yuquanlu, Beijing 100049, China

Abstract. Nowadays, the situation of cumulative public opinion risks caused by repeated occurrences of similar events is becoming increasingly common. Therefore, it is necessary to study the key factors in the formation of public opinion on similar events. Currently, there are fewer studies on public opinion on similar events, and the model of the public opinion evolution mechanism is getting more complicated, which raises the cost of making suggestions for public opinion response. In response, integrating the machine learning explanation model, text mining algorithms, and complex network, this study constructed an algorithm for mining key factors of public opinion evolution that considers the characteristics of similar events. The effectiveness of the algorithm was demonstrated through specific cases.

Keywords: Explainable Machine Learning · Similar Events · Public Opinion · Public Opinion Formation

1 Introduction

Influenced by the "memory function" of online social platforms, the situation of cumulative public opinion risks caused by repeated occurrences of similar events is becoming increasingly common. For example, the Snow Country, which is still mired in the "rip-off" dilemma due to the "rip-off" event in 2017. The "long-tail effect" of negative impacts from historical hot events cannot be ignored. It has been found that the serial presentation of hot events might lead to the deterioration and solidification of online public opinion [1]. Therefore, research on the key factors in formation of public opinion on similar events is necessary, which is helpful to quickly anticipate potential public opinion risks at the beginning of an event and minimize negative impacts.

Public opinion researches on similar events can be summarized as follows. Wang et al. [2] pointed out that the phenomenon of hotspot event public opinion association is common in the Internet era, and the connection between events is mainly formed by

© The Author(s), under exclusive license to Springer Nature Singapore Pte Ltd. 2025
X. Tang et al. (Eds.): KSS 2024, CCIS 2269, pp. 124–136, 2025.
https://doi.org/10.1007/978-981-96-0178-3_9

common subjects, themes or emotions. In addition, with the depth of research, some scholars have tried to construct a netizen attention transfer model for two public opinion events with short interval [3]. Liang et al. [4] studied the topic resonance of Micro-blogs on two similar events. Current researches on similar events still focuses on qualitative analysis and case studies.

The research on opinion formation and diffusion can be traced back to some classical models, such as infectious disease model, Hegselmann-Krause model [5]. Recently, researchers have optimized models based on classic models through refinement of public opinion dissemination scenarios [6, 7] and the integration of multiple models [8, 9]. On the other hand, the development and widespread use of technologies such as complex network and machine learning have provided strong support for the construction and validation of public opinion evolution models, making the construction of these models increasingly complex.

Taken together, existing studies provide rich research foundation for this study. Increasingly complex public opinion evolution models help to analyze the vein of public opinion propagation and evolution in a more detailed and precise way. However, at the same time, it also raises the cost of making suggestions for public opinion response. In addition, most of the related researches propose general models, with few studies considering the characteristics of similar events in the modeling process. In response, this study tried to apply the machine learning explanatory model to the mining of key factors in the evolution of public opinion. In details, SHAP (SHapley Additive exPlanations) [10], which can be used to quantify the contribution of each feature in the machine learning model by calculating the SHAP value, was used in this study.

Integrated the machine learning explanation model, text mining algorithms, and complex network, this study constructed an algorithm for mining key factors of public opinion evolution that considers the characteristics of similar events. Compared to existing studies, the algorithm is more general in identifying the key factors affecting the evolution of public opinion, and more efficient in proposing targeted suggestions for public opinion response. By analyzing historical data of similar events, the algorithm could quickly extract key topics or emotional factors affecting the formation of public opinion on specific type of events, which would help to quickly form effective public opinion response suggestions.

2 Literature Review

2.1 Research on Public Opinion Evolution

Public opinion evolution has always been one of the focuses of public opinion researches, and researchers have tried to analyze the interactions among features such as heat, consensus, sentiment, and topic in the process of public opinion evolution, to discover the laws of public opinion evolution, as well as construct models of public opinion evolution.

Related researches can be traced back to the classical infectious disease model, Hegselmann-Krause model [5], the voter model [11], etc. These classical models provide an important research foundation for the study of public opinion evolution, and many researchers have conducted more in-depth studies on public opinion evolution based on these models. Yuan et al. [7] constructed a model about network public opinion polarization based on SIR model. Some researchers have constructed opinion evolution models that further consider conflict sources [12] and value co-creation [13] on the basis of the HK model. There are also researchers studied the effect of mass media on opinion evolution based on the nonlinear q-voter model on the fully connected network [14]. In addition, Li et al. [9] attempted to integrate the advantages of multiple models and proposed the HK-SEIR model.

Moreover, complex network models and machine learning algorithms are widely used in opinion evolution researches. For complex network, because of the advantages in portraying interactions and social networks, opinion evolution models are often modelled with network models as the skeleton [15–18]. In these types of studies, network models are mostly used to portray the propagation and interaction of information during the evolution of public opinion. Recently, researchers have increasingly started to use higher-order network models such as multilayer networks [15] and coupled networks [18]. Machine learning algorithms not only provide powerful tools for mining features such as emotions and topics in the evolution of public opinion, but in recent years, the development of deep learning technologies has greatly contributed to the improvement of the accuracy of the prediction of public opinion dynamics. The application of deep learning models such as generative adversarial network [19], graph neural network [20], bidirectional encoder representation from transformer (BERT) [21] and so on is becoming more and more common.

Taken together, current researches on public opinion evolution are mostly about the improvement of model fitting effect or the refinement of public opinion evolution scenarios with few studies considering the characteristics of similar events in the modeling process. The development of complex networks, machine learning and other technologies has provided strong support for the construction and testing of public opinion evolution models. The construction of opinion evolution models is becoming more and more complicated.

2.2 Research on Key Factors of Public Opinion Evolution and Machine Learning Explanatory Model

Research on key factors in the evolution of public opinion can not only help us comprehend laws of public opinion evolution, but also provide targeted suggestions on public opinion response.

Current related studies can be divided into those based on opinion evolution models [7, 22–24] and case-specific studies [25, 26]. The former mostly analyses specific rules according to the characteristics of effect factors and incorporates them into the model construction. The factors of these researches are relatively specific, such as dynamic network structure [12], mass media [22], information interaction between government and media [23], social bots [24], and so on. Case-specific studies typically propose specific factors based on attributes of those events. For example, Wang et al. [25] found that the severity of the epidemic, the number of Internet users, the number of media reports and the region's attributes jointly affect the spatial and temporal evolution pattern of online public opinions about the epidemic.

As for now, the specificity of the influencing factors proposed in these studies is obvious, which is not conducive to propose public opinion response recommendations efficiently. If common factors (e.g., emotions, topics, etc.) are chosen to be used in these researches, the identification of the key factors becomes a new challenge.

In response, considering the advantages of machine learning explanatory model in identifying key factors, this study tries to apply it to the identification of factors influencing public opinion evolution. In details, SHAP (SHapley Additive exPlanations) [10], which can be used to quantify the contribution of each feature in the machine learning model by calculating the SHAP value, was used in this study. For its good performance, SHAP is currently used in many fields such as transportation [27], medical [28], aerospace [29] and so on.

3   The Key Factors Mining Algorithm for the Formation of Public Opinion on Similar Events

The key factors mining algorithm for the formation of public opinion on similar events consists of three parts: construction of features, selection of the best algorithm and mining of key explanatory factors.

The structure of the algorithm is shown in the Fig. 1.

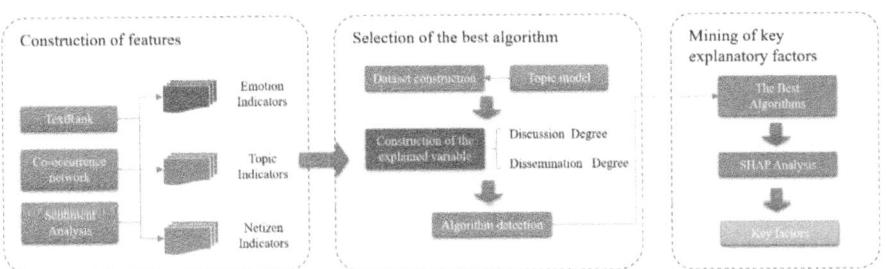

Fig. 1. Algorithm flow chart.

3.1   Construction of Features

It has been pointed out that the public opinion association of hot events mainly comes from common subjects, themes or emotions [2]. Accordingly, in this study, the features of similar events are composed of emotion indicators, topic indicators and netizen indicators.

Emotion Indicators
These indicators characterize the proportion of each type of emotion in the discussion of historical events, which are divided into eight categories: neutral, sympathy, surprise, optimism, sadness, fear, anger and disgust.

In details, the emotion classification is implemented by integrating the UIE (Universal Information Extraction) model [30] and the UTC (Universal Text Classification) model proposed by Baidu. Baidu, as the largest Chinese search engine service provider, has accumulated rich Chinese text data and strong technical strength. The UTC proposed by Baidu has won the top spot in the ZeroCLUE[1]. The effect of sentiment classification task based on UTC is guaranteed.

Topic Indicators

These indicators characterize the proportion of hot topics in the historical discussion of historical events. Topic Indicators vary by event type, and could be constructed in the following steps.

Firstly, the TextRank [31] is used to extract the representative keywords in the discussion of each event. Then, the keyword co-occurrence network is constructed based on the co-occurrence of the extracted keywords in different events. Subsequently, representative keywords for similar events are obtained by calculating the Pagerank value of each node in the keyword network. Finally, topic indicators can be obtained by categorizing top-ranked keywords.

Netizen Indicators

These indicators characterize discussants and consists of two indicators, the percentage of authenticated official accounts (c_off) and the percentage of authenticated individual accounts (c_ind).

3.2 Selection of the Best Algorithm

The performance of the commonly used algorithms is examined in similar events dataset, so as to select the optimal algorithm for the subsequent analysis of the key elements. The specific steps include the following.

Construction of the Dataset

To refine the research granularity, introduce the topic model to split the discussion of each event into different themes. In order to determine the optimal number of topics, in this paper, the perplexity [32] is chosen as the evaluation metric to determine the optimal number of topics of each event. Perplexity is a common metric used in the field of natural language processing to measure the effectiveness of a language model, lower value indicates that the model is more effective.

The corresponding explanatory variables and event indicators will be extracted from records of each theme.

Construction of the Explained Variable

Based on the data structure, construct indicators representing the degree of discussion and dissemination for public opinion formation and diffusion.

---

[1] https://www.cluebenchmarks.com/zeroclue.html.

In details, the discussion degree of theme i could be calculated by the Eq. (1).

$$DiscussionDegree_i = (\sum_{j=1}^{Num} Comment_i + Num)/Num \tag{1}$$

$Comment_i$ denotes the number of comments on each record, and Num denotes the total number of record in the theme i.

The dissemination degree of theme i could be calculated by the Eq. (2).

$$Dissemination\ Degree_i = (\sum_{j=1}^{Num} Forward_i + Num)/Num \tag{2}$$

$Forward_i$ denotes the number of forwards on each record, and Num denotes the total number of record in the theme i.

Algorithm Detection

In this part, the MSE (mean-square error) is used as the evaluation metric. By comparing the prediction performance of different algorithms on the similar events dataset, the optimal prediction algorithm could be identified for subsequent SHAP analysis.

3.3 Mining of Key Explanatory Factors

SHAP, a machine learning explanation model, is introduced in this part, which can be used to quantify the contribution of each feature in the machine learning model by calculating the SHAP value.

By performing SHAP analysis on the optimal algorithm selected in the previous step, the key factors influencing the formation and diffusion of public opinion for specific type of events can be identified.

4 Experimental Results

4.1 Data Set

In order to test the efficiency of the algorithm, this study took the school bullying events that have occurred frequently in recent years as an example for experimental analysis.

Specially, 25 representative school bullying events (Table 1) during the period 2019–2024 were selected to construct the dataset. We then collected data from Sina Weibo, one of the largest Chinese online social platforms, about these 25 events (8128 records). To ensure that the collected microblog data are public discussions related to the events, we first collected the microblog topics corresponding to these events (microblog topics are bracketed by #, e.g., #Sichuan Guangyuan school bullying#). Then, we obtained the microblog data under each topic to construct the dataset.

Table 1. Details of Events.

No	Time	Location of events
1	2019.10	Jiangsu Province
2	2021.3	Beijing
3	2021.4	Guizhou Province
4	2021.6	Sichuan Province
5	2021.7	Sichuan Province
6	2021.9	Yunnan Province
7	2021.11	Jiangxi Province
8	2022.4	Anhui Province
9	2022.7	Gansu Province
10	2022.10	Guangdong Province
11	2022.11	Henan Province
12	2023.2	Guizhou Province
13	2023.3	Hainan Province
14	2023.3	Hebei Province
15	2023.5	Jiangxi Province
16	2023.6	Beijing
17	2023.9	Shanxi Province
18	2023.9	Hebei Province
19	2023.10	Beijing
20	2023.10	Hebei Province
21	2023.10	Hunan Province
22	2023.10	Fujian Province
23	2023.10	Hebei Province
24	2024.3	Hunan Province, Jishou
25	2024.3	Hunan Province, Changsha

4.2 Construction of Features

In the part of the construction of features, five topic indicators were analyzed by TextRank and co-occurrence network analysis. The correspondence between these indicators and keywords is shown in the Table 2.

Table 2. Topic indicator.

Index	Keywords
School-related	'campus', 'school', 'education', 'dropout', 'teacher'
Victim-related	'child', 'minor', 'student', 'girl'
Home-related	'parent', 'father and mother'
Process-related	'hope', 'follow-up', 'attention', 'voice out', 'justice', 'end without resolution'
Event-related	'violence', 'perpetrator', 'video'

All indicators constructed for school bullying events are shown in Table 3.

Table 3. Indicators for school bullying events

Emotion Indicators	Topic Indicators	Netizen Indicators
Neutral	School-related	c_off
Sympathy	Victim-related	c_ind
Surprise	Home-related	
Optimism	Process-related	
Sadness	Event-related	
Fear		
Anger		
Disgust		

4.3 Selection of the Best Algorithm

In the part of selection of the best algorithm, using the topic model, the records of 25 events were specifically divided into 99 themes.

Considering the size of the dataset, predictors that perform well on small-scale data were selected when constructing the set of alternative algorithms. Alternative algorithms include predictors built on Support Vector Machine (SVM), k-Nearest Neighbor (KNN) algorithm, Random Forest (RF) [33], and Light Gradient Boosting Machine (LightGBM) [34]. The parameters are set as follows. For SVM, "kernel" is set to "rbf". For KNN, "n_neighbors" is set to 3. For Random Forest, "n_estimators" is set to 10. For LightGBM, "metric", "boost-ing_type", and "learning_rate" are set as "mse", "gbdt" and 0.01. To ensure the stability of the results, each group of experiments was repeated 500 times to take the average value of mean-square error (MSE).

In addition, the calculation of the standard deviation of the MSE values was added to analyze the stability of these algorithms. The performance of each algorithm is shown in Table 4.

Table 4. Performance alternative algorithms.

Index	Dissemination Degree		Discussion Degree	
Model	MSE	S.D	MSE	S.D
SVM	**0.954832287**	0.089370555	0.914572812	0.101694765
RF	1.305782996	0.435437137	1.022234897	0.320346563
KNN	1.240296638	0.358351374	1.05050741	0.298891454
LGBM	0.974998571	**0.078642379**	**0.894463318**	**0.096555556**

According to Table 4, it can be seen that the SVM and LightGBM performed best in predicting both Dissemination Degree and Discussion Degree, respectively. The Light-GBM is chosen by considering the performance of the algorithm and the stability of the results.

4.4 Mining of Key Explanatory Factors

In the part of key explanatory factors mining, the LightGBM was selected for SHAP analysis based on the experimental results in the Table 4. The experimental results of SHAP analysis on LightGBM are shown in the Figs. 2 and 3.

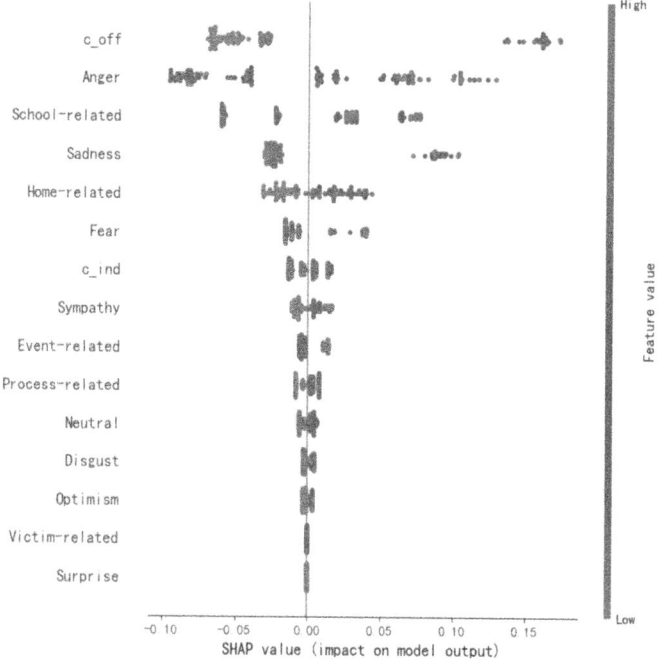

Fig. 2. Results of SHAP analysis (Dissemination Degree).

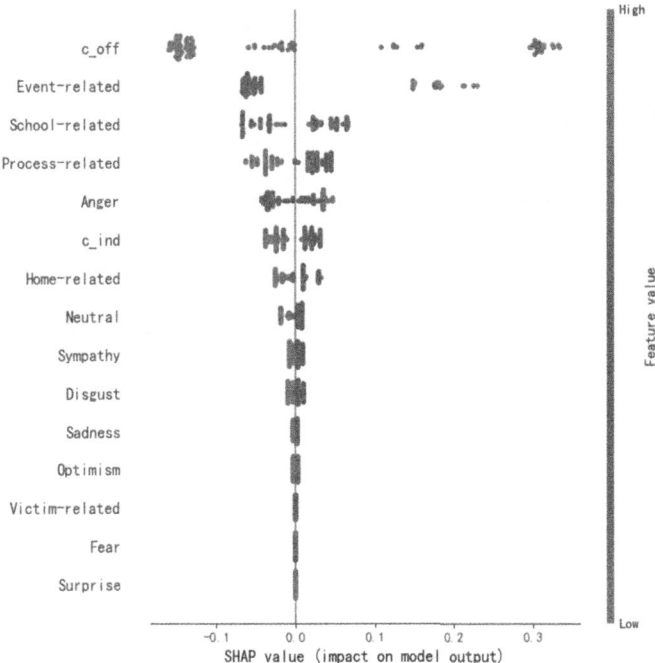

Fig. 3. Results of SHAP analysis (Discussion Degree).

Figures 2 and 3 show that for both the dissemination degree and discussion degree, the percentage of certified official accounts plays a key role. In other words, media coverage takes on a crucial position in shaping public opinion on such events.

There is a difference in the impact of topic indicators and emotion indicators on dissemination degree and discussion degree. For dissemination degree, emotion indicators have a more pronounced effect than topic indicators, whereas the opposite is true for discussion degree. Specifically, the effect of more intense emotions such as anger, sad and sympathy on the dissemination degree is more significant, which is in line with the findings of existing researches, where it has been noted that negative emotions such as anger are more spreadable than joy [35]. Comparatively, the impact of the topic indicators is weaker, with only school-related topics playing a more pronounced role.

For discussion degree, except for school-related topics, both the event-related and process-related topics play significant roles in shaping the discussion. In terms of emotion, only the indicator about anger shows a significant explanatory effect.

Accordingly, it can be known that when dealing with public opinion on school bullying incidents, it is important not only to report on the handling of the incident on official media accounts and respond to the public's concerns about the duty of the school, but also to take care to appease the public's anger when reporting on the incident.

5 Conclusion

For the study of public opinion on similar events, this study first integrated text mining and complex network algorithms to construct three-dimensional feature indicators including emotion, topics and netizens. Further, the machine learning explanatory model is introduced to implement the key factors mining algorithm for the formation of public opinion on similar events. Subsequently, the effectiveness of the algorithm is demonstrated through specific cases.

In the future, the key factors for different types of incidents can be compared. By analyzing the key elements of different types of events, diversified decision-making suggestions can be provided for public opinion management.

Acknowledgments. This research was supported by the National Natural Science Foundation of China (NSFC) (72204283, 72074205, 72074206, T2293772).

Disclosure of Interests. The authors declare that they have no known competing financial interests or personal relationships that could have appeared to influence the work reported in this paper.

References

1. Fang, F., Xiao, L., Wang, G.: Research on consecutive hot events of network public opinion. J. Intelligence **30**(02), 1–5 (2011)
2. Wang, G., Deng, H., Wang, Y.: A study on public opinion relevancy of network hot issues. J. Intelligence **31**(07), 1–5 (2012)
3. Lan, Y., Liu, B., Zhang, P.: The research on netizens attention transfer model for big data of network public opinion. J. Model Inf. **38**(10), 10–15 (2018)
4. Liang, Y., An, L., Liu, J.: Topic resonance of microblogs on similar public health emergencies. Data Anal. Knowl. Discov. **4**(Z1), 122–133 (2020)
5. Rainer, H., Krause, U.: Opinion dynamics and bounded confidence: models, analysis and simulation. J. Artif. Soc. Soc. Simul. **5**(3) (2002)
6. Dong, J., Hu, J., Zhao, Y., Peng, Y.: Opinion formation analysis for expressed and private opinions (EPOs) models: reasoning private opinions from behaviors in group decision-making systems. Expert Syst. Appl. **236**, 121292 (2024)
7. Yuan, J., Shi, J., Wang, J., Liu, W.: Modelling network public opinion polarization based on SIR model considering dynamic network structure. Alex. Eng. J. **61**(6), 4557–4571 (2022)
8. Gong, H., Guo, C., Liu, Y.: Measuring network rationality and simulating information diffusion based on network structure. Physica A **564**, 125501 (2021)
9. Li, Q., et al.: HK–SEIR model of public opinion evolution based on communication factors. Eng. Appl. Artif. Intell. **100**, 104192 (2021)
10. Lundberg, S.M., Lee, S.I.: A unified approach to interpreting model predictions. In: Proceedings of the 31st International Conference on Neural Information Processing Systems, pp. 4768–4777 (2017)
11. Clifford, P., Sudbury, A.: A model for spatial conflict. Biometrika **60**(3), 581–588 (1973)
12. Glass, C.A., Glass, D.H.: Opinion dynamics of social learning with a conflicting source. Physica A **563**, 125480 (2021)
13. Jiang, G., Luo, T., Liu, X.: Opinion evolution model for online reviews from the perspective of value co-creation. Inf. Fusion **88**, 41–58 (2022)

14. Muslim, R., NQZ, R.A., Khalif, M.A.: Mass media and its impact on opinion dynamics of the nonlinear q-voter model. Physica A Stat. Mech. Appl. **633**, 129358 (2024)

15. Geng, L., Yang, S., Wang, K., Zhou, Q., Geng, L.: Modeling public opinion dissemination in a multilayer network with SEIR model based on real social networks. Eng. Appl. Artif. Intell. **125**, 106719 (2023)

16. Ma, N., Yu, G., Jin, X.: Dynamics of competing public sentiment contagion in social networks incorporating higher-order interactions during the dissemination of public opinion. Chaos Solitons Fractals **182**, 114753 (2024)

17. Wang, G., Chi, Y., Liu, Y., Wang, Y.: Studies on a multidimensional public opinion network model and its topic detection algorithm. Inf. Process. Manage. **56**(3), 584–608 (2019)

18. Zhang, L., Su, C., Jin, Y., Goh, M., Wu, Z.: Cross-network dissemination model of public opinion in coupled networks. Inf. Sci. **451**, 240–252 (2018)

19. Haihong, E., Hu, Y., Peng, H., Zhao, W., Xiao, S., Niu, P.: Theme and sentiment analysis model of public opinion dissemination based on generative adversarial network. Chaos Solitons Fractals **121**, 160–167 (2019)

20. Zeng, Z., Sun, S., Li, Q.: Multimodal negative sentiment recognition of online public opinion on public health emergencies based on graph convolutional networks and ensemble learning. Inf. Process. Manage. **60**(4), 103378 (2023)

21. Su, M., Cheng, D., Xu, Y., Weng, F.: An improved BERT method for the evolution of network public opinion of major infectious diseases: case study of COVID-19. Expert Syst. Appl. **233**, 120938 (2023)

22. Hu, H., Chen, W., Hu, Y.: Opinion dynamics in social networks under the influence of mass media. Appl. Math. Comput. **482**, 128976 (2024)

23. Sun, L., Zhang, Y.: Research on the impact of information interaction between government and media on the dissemination of public opinion on the internet. Heliyon **9**(6), e17407 (2023)

24. Zhang, Y., Ma, J., Fang, F.: How social bots can influence public opinion more effectively: right connection strategy. Physica A **633**, 129386 (2024)

25. Wang, J., Zhang, X., Liu, W., Li, P.: Spatiotemporal pattern evolution and influencing factors of online public opinion—evidence from the early-stage of COVID-19 in China. Heliyon **9**(9), e20080 (2023)

26. Weitzman, J., Filgueira, R., Grant, J.: Identifying key factors driving public opinion of salmon aquaculture. Mar. Policy **143**, 105175 (2022)

27. Sun, H., Cheng, Q., Wang, P., Huang, Y., Liu, Z.: Lane change decision prediction: an efficient BO-XGB modelling approach with SHAP analysis. Transportmetrica A Transp. Sci. 1–38 (2024)

28. Shaon, M., Karim, T., Shakil, M., Hasan, M.: A comparative study of ma-chine learning models with LASSO and SHAP feature selection for breast cancer prediction. Healthc. Analytics **6**, 100353 (2024)

29. Khattak, A., Zhang, J., Chan, P., Chen, F.: SPE-SHAP: self-paced ensemble with shapley additive explanation for the analysis of aviation turbulence triggered by wind shear events. Expert Syst. Appl. **254**, 124399 (2024)

30. Lu, Y., et al.: Unified structure generation for universal information extraction. In: Proceedings of the 60th Annual Meeting of the Association for Computational Linguistics, pp. 5755–5772 (2022)

31. Mihalcea, R., Tarau, P.: TextRank: bringing order into text. In: Proceedings of the 2004 Conference on Empirical Methods in Natural Language Processing, Association for Computational Linguistics (2004)

32. Blei, D.M., Ng, A.Y., Jordan, M.I.: Latent dirichlet allocation. J. Mach. Learn. Res. **3**(Jan), 993–1022 (2003)

33. Breiman, L.: Random forests. Mach. Learn. **45**, 5–32 (2001)

34. Ke, G., et al.: LightGBM: a highly efficient gradient boosting decision tree. In: Advances in Neural Information Processing Systems, vol. 30 (2017)
35. Fan, R., Zhao, J., Chen, Y., Xu, K.: Anger is more influential than joy: sentiment correlation in Weibo. PLoS ONE **9**(10), e110184 (2014)

Knowledge Technologies and Systems Engineering

UMCap: User Memory Augmented Method for Personalized Image Descriptions

Dang-Man Nguyen[1], Xuan-Thang Tran[1,2], Duc-Vinh Vo[3(✉)],
and Van-Nam Huynh[1,4]

[1] Japan Advanced Institute of Science and Technology, Nomi, Japan
{mannd,txthang,huynh}@jaist.ac.jp
[2] Tay Nguyen University, Buon Ma Thuot, Vietnam
[3] Ho Chi Minh University of Banking, Ho Chi Minh City, Vietnam
vinhvd@hub.edu.vn
[4] School of Finance and Accounting, Industrial University of Ho Chi Minh City, Ho
Chi Minh City, Vietnam

Abstract. Despite significant advancements in general Image Captioning (IC), captions are most effective when they transcend obvious descriptions and engage the user's interest. Personalized image captioning (PIC) addresses this need by generating descriptions that incorporate the user's prior knowledge, such as active vocabulary, writing style, and other user-specific information. This task is complicated by the large number of users on social networks and the evolving nature of user preferences for image captions. This paper proposes a hybrid model that combines a pre-trained model with a retrieval-based memory mechanism to tackle the Personalized Image Captioning problem. Our method involves two main phases: (1) Constructing User Memory (UM), and (2) generating image descriptions using the pre-trained model and UM. We create external user memories containing word-level knowledge in textual and image contexts, referred to as user historical context. During the second phase, we utilize a retrieval method to measure the similarity between historical and current contexts, retrieving relevant knowledge from the UM to feed into the pre-trained model to generate descriptions word by word. Our framework does not require additional parameters, supports expansion, and enables explicit and interpretable memorization of user knowledge. Furthermore, our method outperforms baseline models and achieves results comparable to state-of-the-art methods.

Keywords: Personalization · Image Captioning · Retrieval Mechanism

1 Introduction

In the age of abundant digital imagery, accurately understanding and describing visual content is immensely valuable. Images have become integral to our daily lives, conveying information, emotions, and narratives in ways that text alone cannot match. Consequently, there is a growing demand for intelligent systems capable of automatically generating engaging captions for images. Recent

X. Tang et al. (Eds.): KSS 2024, CCIS 2269, pp. 139–154, 2025.
https://doi.org/10.1007/978-981-96-0178-3_10

advances in Computer Vision and Natural Language Processing have significantly improved our ability to understand visual content, particularly in the challenging task of image captioning [1,2,16]. Image captioning translates visual content into natural language sentences, bridging the gap between visual and textual information [5,7,13]. This task requires extracting semantic meaning from text and understanding object relationships within images. Despite its challenges, image captioning has numerous applications, including aiding visually impaired individuals, automating the description process for images, and enhancing machine understanding of visual data.

While existing image captioning systems [6,8,9,11] have made progress in generating descriptive and relevant captions, they often fail to capture personalized elements or consider users' unique preferences and experiences. Conventional models produce identical captions for a given image, regardless of the user. However, users have diverse backgrounds, interests, and perceptual biases that influence their interpretation of visual content. Therefore, there is a growing need for personalized image captioning approaches that account for individual user characteristics, contextual information, and preferences to generate more tailored and engaging descriptions. See an example in Fig. 1.

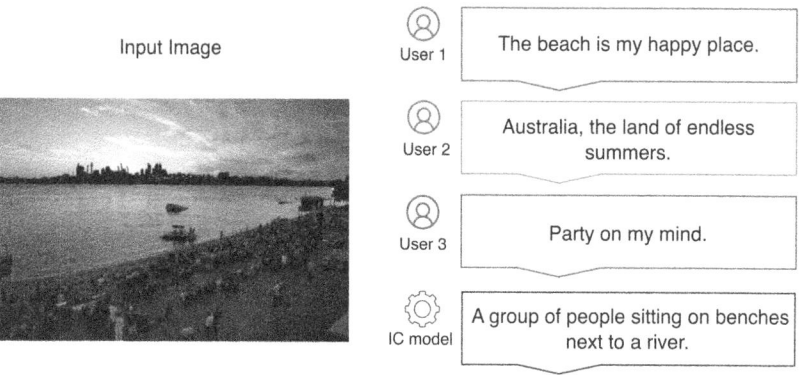

Fig. 1. Diverse user perspectives: the motivation of personalized image captioning. Users have diverse backgrounds, interests, and perceptual biases, which can greatly influence their interpretation and understanding of visual content.

The growing demand for personalized recommendations has highlighted the importance of incorporating user information in image captioning. This trend enhances captions by making them more relevant, engaging, satisfying, accessible, and interpretable, leading to a more personalized user experience with visual content. According to the Adobe Digital Index Social Intelligence Report (2014)[1], social media posts with images receive 650% more engagement than

---

[1] https://news.adobe.com/news/news-details/2014/Adobe-Report-Finds-Social-Media-Drives-Record-Revenue-Per-Visit-for-Retailers-Facebook-Competition-Heats-Up/default.aspx.

text-only posts. Personalized image captioning can amplify this engagement by tailoring captions to viewers' interests and preferences. This field has promising implications across various sectors such as e-commerce, marketing, and healthcare, where accurate and relevant descriptions of visual content are crucial. Relatively few pioneering studies [3, 4, 10, 14, 20] have explored the impact of user preferences on generating effective personalized captions. These studies use active vocabularies, self-descriptions, and unique identifiers to learn latent representations of users. By combining these user representations with visual representations from images, models generate captions tailored to individual preferences, distinguishing personalized models from traditional ones that do not consider user preferences. However, these approaches have several drawbacks:

- *Scalability of trained models:* Scalability is a challenge in prior studies, especially when incorporating new contexts and users. Expanding or updating users' prior memory and adding new users often require fine-tuning. However, finding the optimal strategy for integrating new users without overfitting or compromising generalization remains complex. The scarcity of user-specific data further complicates capturing unique preferences, impeding the model's ability to generate accurate and personalized captions for new users.
- *Limited Adaptability to Dynamic Preferences:* User preferences for image captions evolve over time, requiring models to adapt. Many current models lack the flexibility to effectively capture and model these dynamic preferences, limiting their ability to produce up-to-date and personalized captions. Updating models to accommodate new preferences while retaining previous knowledge is a complex challenge.
- *Lack of Interpretability:* Current models often use complex deep-learning architectures that are difficult to interpret. These models do not readily offer insights into the rationales behind their predictions. Understanding how and why models generate specific personalized captions is crucial for fostering user trust and ensuring transparency.

To address these challenges, we propose a hybrid model that integrates a pretrained image captioning model with a retrieval-based memory mechanism. This approach allows direct revision and expansion of 'augmented historical context,' enhancing our ability to explicitly record user knowledge. Our work leverages the extensive knowledge from a pre-trained model trained on a much larger dataset than user data alone, an advantage not present in existing PIC research. Our hybrid model does not require additional parameters while facilitating the expansion and interpretability of user knowledge. Unlike traditional approaches that incorporate user knowledge as model parameters, our model provides a more transparent representation. Our objective is to improve interpretability, simplifying the process of understanding how user knowledge integrates into personalized image captioning. We assume that users tend to use similar words in specific contexts previously encountered or related to their historical contexts. This assumption is based on consistent language patterns and preferences observed in user interactions with image captions. By leveraging this trend, we aim to enhance personalized image captioning models to generate more contextually relevant and user-tailored captions.

We summarize the contributions of this paper as follows:(1) To the best of our knowledge, this is the first attempt to construct a non-parametric, retrieval-augmented method for personalized image captioning. (2) We devised a User Memory, serving as an embedding space encapsulating historical context and relevant knowledge from the user's past posts. (3) Extensive experiments on two publicly available datasets demonstrate that our approach achieves competitive performance in personalized image captioning tasks.

2   Related Work

2.1   Image Caption

Image captioning is a challenging task that has garnered significant attention from researchers in natural language processing (NLP) and computer vision. Numerous studies have aimed to develop methods for generating accurate and meaningful image descriptions. Early methods treated image captioning as a retrieval problem [5,13] or used template-based approaches [17], retrieving pre-existing captions from a database based on similarity measures or generating captions using predefined templates. These methods often resulted in generic and rigid descriptions lacking nuanced details. Recently, deep neural networks, particularly those utilizing an encoder-decoder framework [7,9,11], have shown remarkable effectiveness and flexibility in image captioning tasks. The encoder-decoder architecture facilitates end-to-end training, enabling models to map visual features to textual descriptions. These models typically comprise a two-stage architecture: a visual encoder and a language decoder. The visual encoder, utilizing convolutional neural networks (CNNs) or transformers, extracts high-level visual features from the input image. The language decoder, employing recurrent neural networks (RNNs) or transformer-based language models, generates a coherent and descriptive caption from the encoded visual features. This method produces more contextually relevant and diverse captions. Advanced techniques such as attention mechanisms [18] have further enhanced image captioning models. Attention mechanisms allow the model to focus selectively on different regions of the image during caption generation, enabling a more detailed and aligned connection between visual content and textual descriptions.

2.2   Personalized Image Caption

Indeed, captions are perceived as more effective by humans when they extend beyond mere obviousness and are crafted in a style that engages their interest. Personalized image captioning seeks to meet this demand by generating descriptions that take into account the user's prior knowledge, including their active vocabulary, writing style, and other personal attributes. Despite substantial progress in general image captioning, personalized image captioning—which more closely aligns with users' specific needs—has only begun to receive significant attention recently. Our work shares similarities with previous personalized image captioning models that emphasized the modeling of user preferences to enhance personalization. These models utilize user-specific data, including preferences, contextual factors, and user experiences, to learn personalized models. Pioneering studies [3,14] in this field aimed to capture user personalization by extracting user-active vocabulary from their previous image descriptions. To imitate user preferences, these studies proposed context sequence models to store the historical words that users most frequently used. An alternative approach, [10] involves learning user representations by utilizing user IDs to capture latent preferences. To effectively capture the unique styles and preferences of different users, this approach adopts a user embedding scheme where each user is represented as a unique interest vector within the model, enabling the model to better incorporate the user's context and preferences. [12,15] enhance the description models by incorporating a more comprehensive set of personality traits. In these approaches, a total of 215 distinct personality traits are specified to characterize each user, with the training model designed to be dependent on these traits. these methods aim to improve the model's understanding of image content while generating engaging and accurate captions. The most recent work in this area is [20] which introduces a multi-modal Transformer network that personalizes captions based on the user's recent captions and a learned user representation. This innovative model consists of an input representation module, a hierarchical transformer encoder, and a transformer decoder. In summary, when being compared with previous PIC studies, the uniqueness of our framework is highlighted by the following points: (1) We utilize a pre-trained model combined with an argument retrieval mechanism to generate personalized captions for each user. This approach enhances the quality of generated captions by leveraging the extensive knowledge from the pre-trained model, which complements the limited information from user data while capturing user-specific preferences; (2) Our approach involves constructing a non-parametric retrieval-based memory, which does not require further training, making it particularly suitable for the dynamic data of users on social networks; (3) We propose storing and representing user historical knowledge with both textual and visual contexts at the word level. This method allows for the analysis of words in the User Memory and the augmented retrieval mechanism, making our approach more transparent than other deep learning models.

3   Proposed Approach

This section introduces our proposed method for addressing PIC tasks. The method consists of two main components: (1) constructing User Memory (UM), and (2) generating image descriptions based on a pre-trained model and the UM. Overview of the method is depicted in Fig. 2.

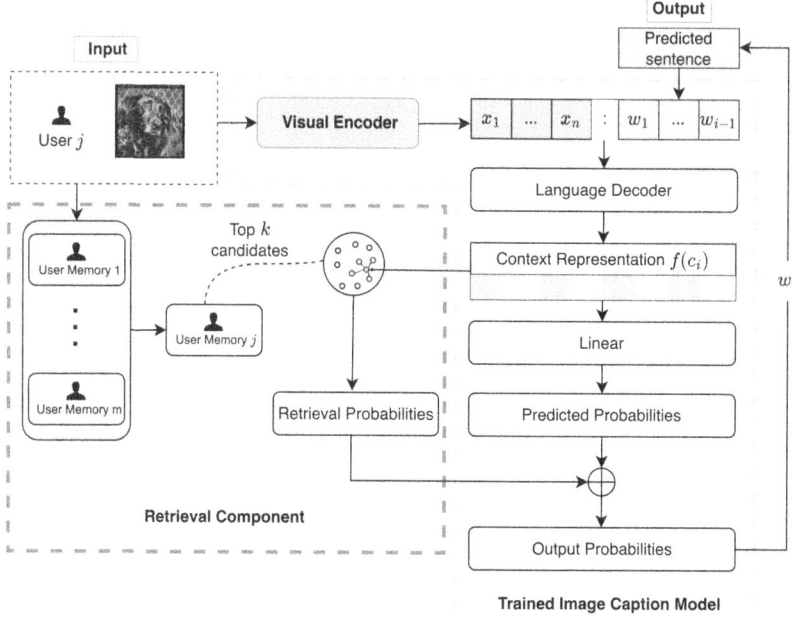

Fig. 2. The overall architecture of our Hybrid Augmented model. The hybrid model includes two main components: the Retrieval Component and the Trained Image Caption Model.

3.1 Constructing User Memory

User image-caption posts are encoded and stored as a user memory context to facilitate personalized caption generation. Particularly, an encoder-decoder architecture will be employed to create a context vector that consists of both the image context and textual context of the user's post. In this work, we used the trained image captioning model following an encoder-decoder architecture, where the encoder processes the input image, and the decoder functions as a language model. The trained model's objectives are (1) to learn general patterns and representations of visuals and language from standard image captioning data; and (2) to encode the extracted information into a context vector c_i. This model includes two primary components: a visual encoder and a language model.

Visual Encoder. To extract visual features from the input image, the Encoder utilizes the existing pre-trained visual encoder of the Contrastive Language - Image Pre-training model (CLIP) as follows:

$$X = encoder(x) = f_{\mathbf{CLIP}}(I; \theta_{enc}) \tag{1}$$

where $f_{\mathbf{CLIP}}$ is the visual encoder defined by the CLIP model with parameters θ_{enc}. X is the encoded visual feature of input image I.

The output generated by the visual encoder is regarded as the visual context of user posts. This is predicated on the assumption that images with similar visual content are likely to convey overlapping user information.

Language Model. The objective of this language model is to encode natural language into embedding vectors, which are regarded as the textual context of the associated image. Subsequently, the model predicts the probability of subsequent words in a given sentence. In this research, we employ the auto-regressive language model GPT-2 to handle both the encoding and sentence generation tasks.

$$P(w_i|X; w_1, w_2, \ldots, w_{i-1}) = g_{\mathbf{GPT\text{-}2}}(f(X; w_1, w_2, \ldots, w_{i-1}); \theta_{gpt}) \quad (2)$$

To predict captions based on an input image, the language model is tasked with generating captions sequentially, word by word. It considers both the visual information from the image and the sequence of words $(X; w_1, w_2, \ldots, w_{i-1})$. During each step of the generation process, the language model transforms the input image and tokens into a hidden representation using a function $f(.)$. This hidden representation is then utilized to predict the probability of the next token, employing the trained GPT-2 model g with parameters θ_{gpt}.

User Memory. After configuring the trained Image Captioning model, the user data are encoded by this model and subsequently stored in their respective User Memory. The User Memory is specifically designed to retain the user's prior knowledge at the word level, capitalizing on information extracted from past image-caption user posts. This accumulated knowledge is organized into a key-value map (K, V), where the value v_i represents a word, and the key k_i corresponds to the associated image and caption context of the value word:

- The value v_i in the User Memory corresponds to an actual word w_i found within an image-caption post of a user. This post, encompassing both visual and textual contexts, is considered to represent the relevant knowledge of the user associated with that specific word.
- The key k_i is the complete caption context, which includes both the image and textual context associated with the corresponding value v_i. This representation is facilitated by a function $f(.)$ that maps the current context vector c_i to a fixed-length vector, as computed by a pre-trained image captioning model. In our study, we employ a transformer-based image caption model, where the key representation $f(c_i)$ is derived from the output of the self-attention layer in the language model. Alternatively, in different architectures, this representation can be obtained from the output of the cross-attention layer in a transformer-based image caption model or from the attention strategy utilized in an LSTM decoder.

When provided with a training image-caption dataset D_j belonging to a user u_j, the User Memory can be generated through a single calculation over each sample using the pre-trained model. Finally, the complete User Memory for user j, denoted as M_j can be formulated as follows:

$$M_j = (K, V)_j = \{(f(c_i), w_i)|(c_i, w_i) \in D_j\} \tag{3}$$

The procedure for constructing User Memory for a given user, based on their dataset, is depicted in Algorithm 1.

Algorithm 1. Building User Memory for a Given User

Input: Trained image captioning model $P_M(w|c_i; \theta)$, image-caption dataset D_j belonging to a user u_j
Output: User Memory $M_j = (K^j, V^j)$

1: **for** each image-caption pair $(I_i, S_i) \in D_j$ **do**
2:    extract image features from the input image X_i via Equation 1
3:    **for** each word $w_t^i \in$ caption S_i **do**
4:       generate current context representation $f(c_t^i)$
5:       **if** $f(c_t^i) \notin K^j$ or $w_t^i \notin V^j$ **then**
6:          insert $(f(c_t^i), w_t^i)$ into $M_j = \{K^j, V^j\}$
7:       **end if**
8:    **end for**
9: **end for**
10: Return User Memory M_j of user u_j

After constructing the User Memory for each user, this module becomes instrumental in assisting the trained Image Captioning model to generate personalized image captions that are tailored to the corresponding user. Additionally, the User Memory can be incrementally updated as new image-text pairs are added, accommodating both the new data and new users.

3.2 Generating Personalize Image Description: Combination and Inference

The construction process described in the previous step culminates in the creation of User Memories and the configuration of a trained model. To leverage the valuable information contained within these User Memories for generating appropriate image descriptions, a retrieval mechanism is employed during inference time. This mechanism is specifically designed to extract relevant information from the constructed memory.

Key-Value Retrieval. Building on the previous explanation, the User Memory stores records of each user's actual words alongside their associated historical contexts within the retrieval dataset. Given the User Memory M_i for a specific user u_i and an input query representing the current context C_t, a knowledge retrieval component is utilized to perform an approximate k-nearest neighbor search. Outcomes of this retrieval are denoted as $R(c_t) = \{(v_1, d_1), \ldots, (v_k, d_k)\}$, where the top k values v_j and their respective distances d_j -closest to the query context c_t - are identified. The distance d_j is computed using the similarity function $distance(.,.)$ between the key k_j in the User Memory and the representation of the current context $f(c_t)$:

$$d_j = (distance(k_j, f(c_t))|k_j \in M_i); j \in [1, k] \qquad (4)$$

To efficiently navigate the extensive User Memory, the retrieval component utilizes the Facebook AI Similarity Search (FAISS) library. FAISS is specifically designed for indexing high-dimensional vectors, which makes it ideally suited for managing User Memory. Furthermore, FAISS facilitates rapid and efficient searching through the memory by employing a similarity measure, such as Euclidean distance or cosine similarity, denoted by $distance(.,.)$.

Algorithm 2. Hybird Augmented Model Prediction for A Given User

---

   Input: User Memory M_j of given User u_j, trained image captioning model $P_M(w|c_t; \theta)$, input image I
   Output: Image caption $Y = \{w_1, \ldots, w_t\}$

1: **for** $i = 1$ to t **do**
2:    Get current total context $c_i = \{I, w_1, \ldots, w_{i-1}\}$
3:    Calculate output distribution of trained model $p_M(w|c_i)$.
4:    Generate current context representation $f(c_i)$
5:    Calculate distance between current context with each element in User Memory M_j via Equation 4
6:    Select top k nearest candidates $R(C_i) = \{(v_1, d_1), \ldots, (v_k, d_k)\}$
7:    Normalize, aggregate the distribution of k retrieved candidates from Retrieval Component $p_{RC}(w|c_i)$ via Equation 5
8:    Generate next word based on combined distribution $p(w|c_i)$(Equation 6)
$$w_i = argmax(p(w|c_i))$$
9: **end for**

Inference. Based on the integration of the trained model and user memory, we implement the previously described knowledge retrieval strategy to generate the final output sentence. This model is designed to predict a temporal distribution across the vocabulary, taking into account the input image, the currently generated sentence, and the retrieved user memory as conditioning factors. During the inference process, given an input image I of user u_j, the trained image caption model makes predictions based on the generated context $c_i = \{X, w_1, \ldots, w_{i-1}\}$. This context encompasses both the input visual context $X = \{x_1, \ldots, x_k\}$ and the generated textual context. At inference step t, the trained model generates distribution outputs over the vocabulary $P_M(w|c_i; \theta)$ and computes the current context representation, $f(c_i)$. Subsequently, the model queries the corresponding User Memory M_j using this context representation to retrieve the top n similar

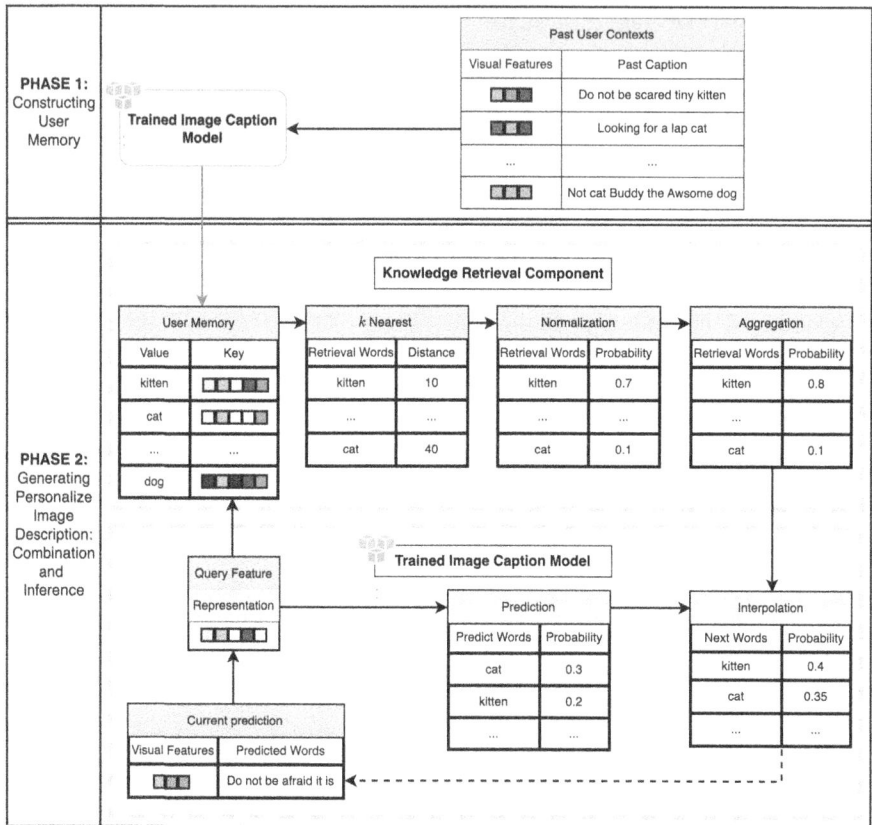

Fig. 3. The chart illustrates the process of generating captions using our hybrid augmented model for a specific user.

candidates $R^j(c_i) = \{(v_1, d_1), \ldots, (v_n, d_n)\}$ based on the values from the distance function $distance(.,.)$, as defined in Eq. 4. Afterward, our retrieval component normalizes the distribution of these retrieved k candidates by applying a softmax function to their negative distances, taking into account multiple occurrences of the same retrieval candidate. For candidates that are not among the retrieved n candidates, zero probabilities are assigned to them in the output probability of the Retrieval Component, denoted as $p_{RC}(w|c_i)$.

$$p_{RC}(w|c_i) \sim \sum_{(k_m, v_m \in R^j)} \mathbb{I}_{w=v_m} exp(-distance(k_m, f(c_i))) \qquad (5)$$

We integrate the retrieval distribution with the trained model distribution using a tuned parameter α to achieve a balanced contribution of model predictions and user retrieval knowledge in the final output of the hybrid model:

$$p(w|c_i) = \alpha p_M(w|c_i) + (1 - \alpha)p_{RC}(w|c_i) \qquad (6)$$

The process of prediction at each step is presented in greater detail in the Algorithm 2 and Fig. 3.

4   Experiments

In this section, we initially introduce the datasets and research questions. Subsequently, we conduct experiments and analyze the results obtained from both our proposed model and the baseline models. Finally, an ablation study is performed to further ascertain the contribution of User Memory to the outcomes.

4.1   Datasets and Baseline Models

In our study, we utilize two publicly available datasets: (1) the InstaPIC-1.1M dataset, which comprises image posts from real users on Instagram; and (2) the Yahoo Flickr Creative Commons 100 Million Dataset (YFCC100M), which has been reconstructed [3] to be most suitable for the PIC task. Due to the inherent characteristics of raw user data, such as brevity, noise, and non-English content, we adopted a data preprocessing approach similar to that outlined in the work by Zhang et al. [20]. This preprocessing involves the removal of duplicate posts, lengthy captions, and non-English posts to ensure the dataset's quality and suitability for our research. To implement and evaluate the baseline models as well as our proposed method, we divided each dataset into training and test sets. Specifically, the first 80% of user's posts were allocated as the training set, while the remaining 20% were used as the test set. The dataset statistics for this experiment are presented in Table 1. We evaluate our approach in comparison with the leading research and other nearest-neighbor methods within the domain of personalized image captioning, as detailed below: (1) **1NN-IM**: for a given input query image, we identify its closest match in the training set based on the l_2 distance, then return the caption of the retrieved image as final prediction; (2) **1NN-UsrIM** [14]: This model acts as a retrieval-based baseline by selecting a caption from the user's training image as the generated caption; (3) **MeaCap**: is a memory-augmented zero-shot image captioning framework developed by [19] for the IC task. Utilizing retrieved textual memory, they developed a memory-augmented visual-related fusion score to guide the generation of captions. To adapt this framework for personalized image captioning (PIC) tasks, we create memory separately based on user training data; (4) **ClipCap** [11]: is a transformer-based architecture specifically designed for image captioning tasks, utilizing the CLIP model for image encoding and the GPT-2 language model for

Table 1. Statistics of InstaPIC-1.1M and YFCC100M Dataset

Data	# Post	# User	# Posts/User	# Words/Post
Instagram	363656	2888	125.9	8.55
YFCC100M	328289	5560	60	6.3

sentence generation; (5) **BLIP-2** [8]: BLIP-2 leverages frozen pre-trained image encoders and large language models (LLMs) by incorporating a lightweight. This model achieves state-of-the-art performance on various vision-language tasks, it also exhibits emerging capabilities in zero-shot instructed image-to-text generation. We fine-tune the BLIP-2 model using the trained data without any specific input prompt as the base model; (6) **BLIP-2-Usr**: We fine-tune the BLIP-2 model using trained data with the prompt "You are user '*user ID*', write a description for this image:". Subsequently, we apply the same prompt format for the test data.

4.2 Evaluation Metrics

The experiments are designed to address the following criteria: *C1*. How is the overall performance of the proposed framework and baseline models in predicting the user image descriptions according to common evaluation metrics?; *C2*. How effective is the UM in our proposed framework? Common metrics such as Bilingual Evaluation Understudy (BLEU), METEOR, and ROUGE-L are employed to assess the linguistic similarity between the predicted sentences and the ground truths. Besides, Consensus-based Image Description Evaluation (CIDEr) is an image caption quality measure since it looks for similarities between several human-annotated captions.

4.3 Results

In this study, we utilize ClipCap [11] as our trained image caption model, which incorporates a state-of-the-art transformer-based architecture specifically designed for image captioning tasks. The model encodes images using the CLIP method, while the GPT-2 language model is utilized in a frozen state to preserve its pre-trained weights. Keys and values for each User Memory are generated through a single forward pass across the training set of all users. Additionally, during the experiments, we tuned the number of retrieved candidates, setting k to values of $\{1, 8, 256\}$, and explored various settings of the combination parameter α ranging from 0 to 1 for optimizing user knowledge retrieval.

In Table 2, we provide a summary of the experimental results for the YFCC-100M and InstaPIC-1.1M datasets, wherein we compare our approach with other models. For each model evaluated, we report outcomes based on several evaluation metrics, including the language similarity metrics of BLEU, CIDEr, METEOR, and ROUGE-L. Notably, in our approach, the User Memory is constructed using the same data split that was employed for training the compared models. In the results presented in Table 2, we observe that the retrieval-based model, 1NN-Im and 1NN-Usr, exhibit the lowest performance compared to the other generative models across both datasets. Moreover, the ClipCap model, which does not incorporate personalization, demonstrates superior modeling capabilities relative to the retrieval-based model. Among the subsequent three baselines—MeaCap and BLIP-2, BLIP-2-Usr—there is a general trend of better performance across most metrics, underscoring the importance of integrating personalization to achieve effective captioning outcomes. Upon comparing our approach with the aforementioned baselines, we observe consistent and significant improvements in performance. Our method demonstrates enhanced capabilities in personalized captioning and achieves superior results across multiple evaluation metrics in the YFCC100M and InstaPIC-1.1M datasets. Moreover, compared to the state-of-the-art BLIP-2-User, our approach provides the significant advantage of easily revising and expanding user knowledge by leveraging the augmented historical context stored in the User Memory, without the need for additional training. Additionally, our method offers a more transparent representation of users' knowledge by directly considering retrieval candidates. This feature enhances the interpretability of our model, facilitating a clearer understanding and analysis of how users' knowledge is integrated into the personalized image captioning process.

Table 2. The performance comparison of our method with various baseline methods on the InstaPIC-1.1M and YFCC100M Dataset.

YFCC100M dataset

Method	BLEU-1	BLEU-2	BLEU-3	BLEU-4	METEOR	ROUGE-L	CIDEr
1NN-Im	0.0178	0.0035	0.0008	0.0003	0.0055	0.0179	0.0065
1NN-Usr	0.0854	0.0497	0.0352	0.0270	0.0357	0.0715	0.1831
MeaCap	0.0840	0.0291	0.0115	0.0055	0.0299	0.0828	0.0884
ClipCap	0.1097	0.0471	0.0245	0.0145	0.0412	0.1124	0.2013
BLIP-2	0.1460	0.0744	0.0434	0.0282	0.0630	0.1466	0.3465
BLIP-2-Usr	0.1546	0.0804	0.0480	0.0320	0.0632	0.1515	0.4054
ClipCap + UM (Ours)	**0.1701**	**0.1186**	**0.0912**	**0.0741**	**0.0807**	**0.1544**	**0.5925**

Instagram dataset

Method	BLEU-1	BLEU-2	BLEU-3	BLEU-4	METEOR	ROUGE-L	CIDEr
1NN-Im	0.0423	0.0084	0.0010	0.0005	0.0211	0.0302	0.0104
1NN-User	0.0554	0.0197	0.0106	0.0069	0.0294	0.0478	0.0680
MeaCap	0.0745	0.0246	0.0097	0.0043	0.0387	0.0705	0.0733
Clipcap	0.0757	0.0313	0.0161	0.0098	0.0358	0.0748	0.1498
BLIP-2	**0.0924**	0.0348	0.0171	0.0095	0.0463	0.0924	0.1566
BLIP-2-Usr	0.0835	0.0357	0.0173	0.0100	0.0473	0.0938	0.1710
ClipCap + UM (Ours)	0.0883	**0.0433**	**0.0275**	**0.0196**	**0.0474**	**0.0950**	**0.2182**

4.4 Ablation Study

To assess the significance of the User Memory component in our approach, we conducted ablation experiments comparing the results of the ClipCap base model, the fine-tuned ClipCap model, and our approach (ClipCap + UM). The fine-tuned ClipCap model refers to the base model after it has been fine-tuned on the same personalized dataset, ensuring a fair comparison among the models. The results of the ablation experiment are detailed in Table 3.

Table 3. Ablation experiment of User Memory component on the InstaPIC-1.1M and YFCC100M Dataset.

YFCC100M dataset

Method	BLEU-1	BLEU-2	BLEU-3	BLEU-4	METEOR	ROUGE-L	CIDEr
Clipcap	0.0628	0.0176	0.0059	0.0021	0.0327	0.0708	0.0750
Fine-tuned Clipcap	0.1097	0.0471	0.0245	0.0145	0.0412	0.1124	0.2013
Clipcap + UM	**0.1701**	**0.1186**	**0.0912**	**0.0741**	**0.0807**	**0.1544**	**0.5925**

Instagram dataset

Method	BLEU-1	BLEU-2	BLEU-3	BLEU-4	METEOR	ROUGE-L	CIDEr
Clipcap	0.5439	0.0129	0.0037	0.0012	0.0248	0.0574	0.0625
Fine-tuned Clipcap	0.0757	0.0313	0.0161	0.0098	0.0358	0.0748	0.1498
Clipcap + UM	**0.0883**	**0.0433**	**0.0275**	**0.0196**	**0.0474**	**0.0950**	**0.2182**

Based on the results presented in Table 3, it is evident that the augmented retrieval component plays a crucial role in enhancing the model's performance. Our approach surpasses the base ClipCap model across the majority of the evaluation metrics. Additionally, it also outperforms the fine-tuned model, which utilizes the same dataset. These findings underscore the significant contribution of the augmented retrieval component to the overall effectiveness of our approach in personalized image captioning.

5   Conclusion

In this work, we introduced a hybrid model for personalized image captioning that integrates a pre-trained image captioning model with a retrieval-based memory mechanism. The method consists of two main components: (1) constructing User Memory (UM), and (2) generating image descriptions based on the pre-trained model and UM. Our framework significantly contributes to personalized image captioning by presenting the first non-parametric, retrieval-augmented method in this domain. We developed a User Memory that serves as an embedding space for historical context and relevant user knowledge. Extensive experiments on publicly available datasets demonstrate that our approach outperforms baseline models and achieves competitive results compared to state-of-the-art methods. By leveraging the augmented historical context stored in the User Memory, our method enables the revision and expansion of user knowledge, resulting in more personalized and engaging image captions.

References

1. Bai, S., An, S.: A survey on automatic image caption generation. Neurocomputing **311**, 291–304 (2018)
2. Bernardi, R., et al.: Automatic description generation from images: a survey of models, datasets, and evaluation measures. J. Artif. Intell. Res. **55**, 409–442 (2016)

3. Chunseong Park, C., Kim, B., Kim, G.: Attend to you: personalized image captioning with context sequence memory networks. In: Proceedings of the IEEE Conference on Computer Vision and Pattern Recognition, pp. 895–903 (2017)
4. Cohen, N., Gal, R., Meirom, E.A., Chechik, G., Atzmon, Y.: This is my unicorn, fluffy: personalizing frozen vision-language representations. In: Avidan, S., Brostow, G., Cissé, M., Farinella, G.M., Hassner, T. (eds.) Computer Vision – ECCV 2022. ECCV 2022. LNCS, vol. 13680, pp. 558–577. Springer, Cham (2022). https://doi.org/10.1007/978-3-031-20044-1_32
5. Farhadi, A., et al.: Every picture tells a story: generating sentences from images. In: Daniilidis, K., Maragos, P., Paragios, N. (eds.) ECCV 2010. LNCS, vol. 6314, pp. 15–29. Springer, Heidelberg (2010). https://doi.org/10.1007/978-3-642-15561-1_2
6. Hu, X., et al.: Vivo: visual vocabulary pre-training for novel object captioning. In: Proceedings of the AAAI Conference on Artificial Intelligence, vol. 35, pp. 1575–1583 (2021)
7. Karpathy, A., Fei-Fei, L.: Deep visual-semantic alignments for generating image descriptions. In: Proceedings of the IEEE Conference on Computer Vision and Pattern Recognition, pp. 3128–3137 (2015)
8. Li, J., Li, D., Savarese, S., Hoi, S.: Blip-2: bootstrapping language-image pre-training with frozen image encoders and large language models. arXiv preprint arXiv:2301.12597 (2023)
9. Li, J., Li, D., Xiong, C., Hoi, S.: Blip: bootstrapping language-image pre-training for unified vision-language understanding and generation. In: International Conference on Machine Learning, pp. 12888–12900. PMLR (2022)
10. Long, C., Yang, X., Xu, C.: Cross-domain personalized image captioning. Multimed. Tools Appl. **79**, 33333–33348 (2020)
11. Mokady, R., Hertz, A., Bermano, A.H.: Clipcap: clip prefix for image captioning. arXiv preprint arXiv:2111.09734 (2021)
12. Nguyen, T., Phung, D., Hoai, M., Nguyen, T.H.: Structural and functional decomposition for personality image captioning in a communication game. arXiv preprint arXiv:2011.08543 (2020)
13. Pan, J.Y., Yang, H.J., Duygulu, P., Faloutsos, C.: Automatic image captioning. In: 2004 IEEE International Conference on Multimedia and Expo (ICME)(IEEE Cat. No. 04TH8763), vol. 3, pp. 1987–1990. IEEE (2004)
14. Park, C.C., Kim, B., Kim, G.: Towards personalized image captioning via multimodal memory networks. IEEE Trans. Pattern Anal. Mach. Intell. **41**(4), 999–1012 (2018)
15. Shuster, K., Humeau, S., Hu, H., Bordes, A., Weston, J.: Engaging image captioning via personality. In: Proceedings of the IEEE/CVF Conference on Computer Vision and Pattern Recognition, pp. 12516–12526 (2019)
16. Stefanini, M., Cornia, M., Baraldi, L., Cascianelli, S., Fiameni, G., Cucchiara, R.: From show to tell: a survey on deep learning-based image captioning. IEEE Trans. Pattern Anal. Mach. Intell. **45**(1), 539–559 (2022)
17. Yao, B.Z., Yang, X., Lin, L., Lee, M.W., Zhu, S.C.: I2T: image parsing to text description. Proc. IEEE **98**(8), 1485–1508 (2010)
18. You, Q., Jin, H., Wang, Z., Fang, C., Luo, J.: Image captioning with semantic attention. In: Proceedings of the IEEE Conference on Computer Vision and Pattern Recognition, pp. 4651–4659 (2016)

19. Zeng, Z., Xie, Y., Zhang, H., Chen, C., Wang, Z., Chen, B.: MeaCap: memory-augmented zero-shot image captioning. arXiv preprint arXiv:2403.03715 (2024)
20. Zhang, W., Ying, Y., Lu, P., Zha, H.: Learning long-and short-term user literal-preference with multimodal hierarchical transformer network for personalized image caption. In: Proceedings of the AAAI Conference on Artificial Intelligence, vol. 34, pp. 9571–9578 (2020)

Research on the Construction Method of Island Chain Knowledge Graph

Guangfei Yang[1,2], Feifan Li[1], Xingqian Zhao[2], and Lin Tang[3(✉)]

[1] Dalian University of Technology, Dalian 116000, China
gfyang@dlut.edu.cn
[2] PLA Dalian Naval Academy, Dalian 116000, China
[3] Dalian University, Dalian 116000, China
tanglin@dlu.edu.cn

Abstract. Due to the complexity of data and the urgency of time, traditional methods can only focus on a single element when processing battlefield information, which may result in the omission of key information. Knowledge graph has the inherent advantage of integrating multi-dimensional and multi category data, providing a feasible solution for focusing on multi element processing of battlefield information. Therefore, we proposes a framework for constructing a military domain knowledge graph that integrates multi-dimensional battlefield elements, fine-tuning the CasRel model to more accurately extract entity relationships. Then, based on this framework, a semi automated island chain knowledge graph is constructed and an application example is given. The graph can comprehensively and accurately express the geographical, military, political and other aspects of the island chain region, providing important support for military decision-making of multi-dimensional battlefield elements.

Keywords: Island Chain · Battlefield Elements · Knowledge Graph · Ontology · Relation Extraction

1 Introduction

Island chain is a geographical security concept created by the United States, which defines defensive or offensive boundaries by connecting islands and other large land groups together, aiming to limit the range of activities of the Chinese Navy and play a strategic deterrent and reconnaissance role [1]. The location of islands on the island chain and military base facilities are important basis for military's strategic planning. By analyzing and processing multidimensional elements such as the layout of the island chain, key islands and military bases, China can better formulate corresponding military deployments and resource allocation to enhance its military strength. Moreover, by strengthening military deployment and patrol monitoring, China can improve its monitoring capabilities in the surrounding waters of the island chain, timely detect and respond to potential military threats.

With the continuous development of information technology, big data technology has gradually shown significant advantages in integrating battlefield information. The

X. Tang et al. (Eds.): KSS 2024, CCIS 2269, pp. 155–168, 2025.
https://doi.org/10.1007/978-981-96-0178-3_11

use of big data technology can integrate and analyze massive data from multiple sources such as sensors, satellite images and social media data [2, 3], thereby helping the military obtain comprehensive battlefield information. Artificial intelligence algorithms can automatically identify and extract key information, accelerating information collection and processing [4]. The existing work involves the following aspects: Yu Y B et al. [5] combined the credibility of different reconnaissance platforms on the battlefield and the detection results of multiple target attributes to perform information fusion and target value judgment, achieving multi-source information fusion in the field of maritime target detection; Wu M Y et al. [6] conducted theoretical research on data preprocessing, data association and information fusion in multi-sensor information fusion technology and proposed improvement ideas to improve algorithm performance, further achieving information fusion of battlefield sensor data; Chen X F et al. [7] constructed a sea battlefield situation data visualization mining platform, which can achieve data mining and visualization of weapons, equipment, battlefield environment and other elements separately, but lacks further multidimensional element fusion.

In summary, current technology mainly focuses on a single element [8, 9], making it difficult to achieve comprehensive analysis of multi-source information such as ubiquitous intelligence and geographic intelligence [10] and there is a possibility of missing key information. The knowledge graph in the field of maritime battlefields can integrate multiple sources of data, including satellite remote sensing data, intelligence, geographic information, etc., thereby more comprehensively revealing the situation of maritime battlefields [11]. By constructing a knowledge graph of multidimensional battlefield elements, it is possible for the military to conduct in-depth analysis of intelligence data based on multiple elements [12]. Although there have been relevant studies on knowledge graphs in military equipment [13] and threat intelligence [14], no knowledge graph has been constructed in the island chain field. In addition, there is a relative lack of research on the construction and application of military knowledge graphs that integrate multidimensional elements.

Therefore, based on general domain knowledge modeling and knowledge graph construction techniques, combined with the characteristics of island chain knowledge and naval warfare requirements, we design a multi element knowledge graph construction framework and specific construction methods for island chain. The framework consists of two layers: the ontology layer and the instance layer. Firstly, under the guidance of domain experts, the island chain ontology layer was constructed and then the island chain entities, relationships and attributes were extracted through the model to construct an instance layer. The multi-dimensional battlefield element fusion island chain knowledge graph constructed in this article is in line with practical military needs and provides data support for naval warfare intelligent recommendation systems [15], machine reasoning systems [16] and auxiliary decision-making systems [17].

2 Framework

2.1 The Concept and Characteristics of Island Chain Knowledge Graph

The island chain knowledge graph that integrates multidimensional battlefield elements represents and stores relevant knowledge of islands, military bases and other elements in the island chain in the form of a directed graph. Among them, the nodes and edges of the directed graph correspond to entities and relationships respectively and the node attributes correspond to entity features. Entities contain specific things in the island chain domain, entity features contain their descriptive information and relationships include spatial relationships, subordinate relationships, etc. between entities. Specifically, the island chain knowledge graph for the fusion of multidimensional battlefield elements can be defined as [18]:

$$G = \{E, R, F, A\} \tag{1}$$

Among them, G represents the island chain knowledge graph of multi-dimensional battlefield element fusion and E represents the collection of entities such as islands, military bases, ports, etc. in this knowledge graph; R represents a set of equal relationships located in or belonging to; F represents a set of entity relationship triplets, each consisting of a head entity e_+, a tail entity e_2 and a relationship r, such as the Kadena located in the Ryukyu Islands; A represents the set of attributes.

2.2 Graph Construction and Application Process

The construction and application process of the island chain knowledge graph for the fusion of multidimensional battlefield elements are shown in Fig. 1. Firstly, under the guidance of domain experts, an ontology layer was constructed. Then, entity relationship extraction and attribute extraction were performed on multi-source heterogeneous data to complete the instance layer construction. Finally, application directions such as intelligent search and intelligent recommendation were explored.

Ontology Layer Construction. Ontology layer construction refers to the creation of an ontology to formalize and clarify the description of shared concepts, including object concepts, types, relationships and attributes [19].

We refer to the seven step method and uses a combination of top-down [20] and bottom-up [21] to construct the ontology layer of the island chain knowledge graph.

Firstly, obtain data from relevant papers, existing ontology, professional books and expert knowledge in the island chain field, summarize important terms, start from the top-level concepts and gradually refine to the lowest level concepts; For entities from sources such as map images and military news, abstract the underlying concepts first and then add them to the ontology model.

Instance Layer Construction. The construction of the instance layer is guided by the ontology layer, which involves extracting entities, attributes and relationships in the island chain domain from multi-source heterogeneous data such as satellite images, professional books and military news and then importing them into the graph database to complete the construction of the graph.

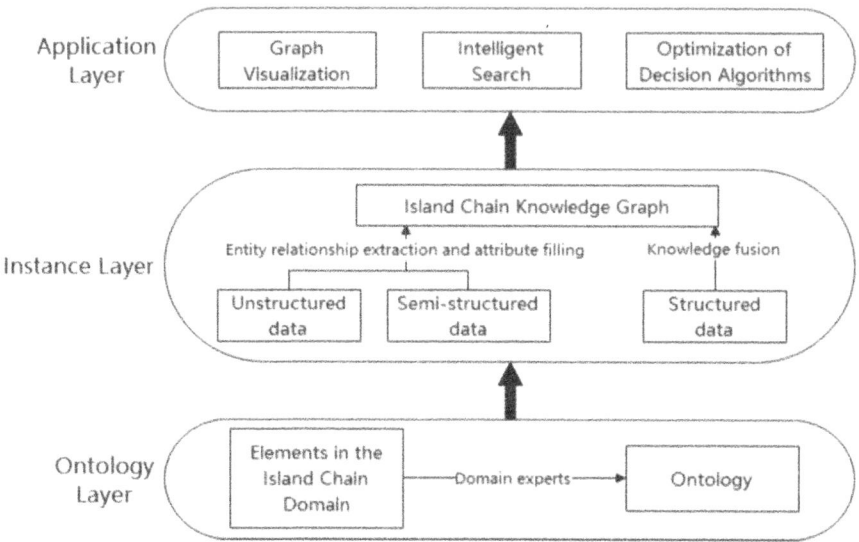

Fig. 1. Knowledge Graph Construction and Application Process

3   Construction of Island Chain Knowledge Graph

3.1   Data Entity Acquisition

We mainly studies the location of islands, military bases, military situation, etc. Different data types and sources are shown in Fig. 2. Among them, unstructured data needs to be annotated before entity relationship extraction and then unstructured data is converted into structured triplets and imported into a knowledge graph The weapon and equipment graph in structured data is fused into the knowledge graph through entity alignment and relationship mapping. The Strait Port Database mainly contains attribute information of entities such as straits and ports. After entity alignment, the entity attributes can be imported into the island chain knowledge graph.

3.2   CasRel Model Structure

CasRel [22] is a parameter sharing based entity relation joint extraction method. It first identifies all possible head entities in the text, and then extracts the tail entities that have the same relationship with the head entity in each relation category. This task decomposition method fundamentally solves the problem of extracting overlapping entity relation triplets. Therefore, it performs well in tasks involving multiple relationships between entities, with performance metrics such as accuracy, recall, and F1 score typically outperforming traditional relationship extraction models. In the field of island chains, in addition to general relationships, entities also have geographical relationships, political affiliations, and other relationships. Using the Casrel model can effectively extract various relationships between entities, providing strong support for the construction and analysis of island chain knowledge graphs.

Fig. 2. Data types and source diagrams in the island chain field

The goal of entity relationship joint extraction task is to identify all triples in a sentence. The principle of the CasRel model used in this article is as follows:

$$\prod_{(s,r,o)} p((s, r, o)|x) =$$

$$\prod_{s \in T} p((s|x)) \prod_{(r|o)} p((r, o)|x, s) = \tag{2}$$

$$\prod_{s \in T} p(s|x) \prod_{r \in T|s} p(o|x, s, r) \prod_{r \in R \backslash T|t} p(o_\phi|x, s, r)$$

In the formula, s Representing the subjects in overlapping relationships; $T|s$ is a triplet with s as the head entity in T; $(r, o) \in T|s$ is the tail relationship (r, o) in $T|s$; r is the set of all relationships; $R \backslash T|s$ Represents all relationships except for the subject in T which is s; o_ϕ indicates that except for the relationships contained in the triplet $T|s$, all other relationships have no corresponding objects in sentence $s \circ p(s|x)$ Represents the probability of the existence of s in x. Therefore, the entire solution is to first find all the head entities s and then find the tail entities o based on the relationships, in order to extract the entities and relationships (Fig. 3).

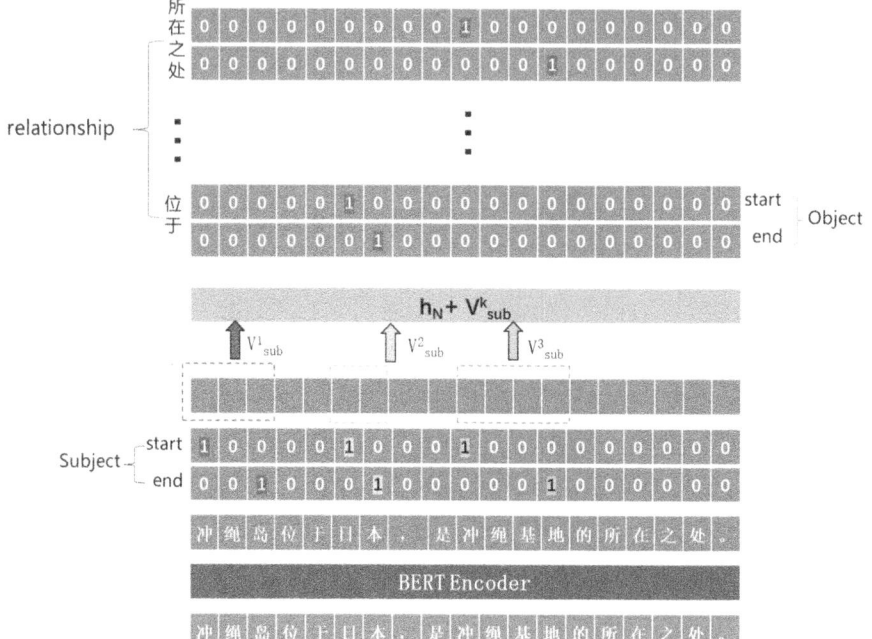

Fig. 3. Structure diagram of Casrel model.

The model consists of two parts.

1. Sentence encoding section

This section uses an encoding layer based on BERT [23] to represent characters/words, obtain contextual semantic information and further input it into the subject recognition and subject object relationship recognition modules.

2. Cascade decoding section

The cascading decoding section consists of two modules: the head entity recognition module and the tail entity recognition module for a given relationship. Firstly, through the subject recognition module, the subject in the sentence is marked and then through the subject object relationship recognition module, the object corresponding to the relationship is associated.

(1) Head entity recognition layer

The head entity recognition layer can directly decode the encoding layer results and then recognize all possible head entities. This layer is divided into two parts:

Firstly, use a linear layer and sigmoid activation function to determine whether each token is the beginning or end of the header entity;

$$p_i^{start_s} = \sigma(W_{start}x_i + b_{start}) \tag{3}$$

$$p_i^{end_s} = \sigma(W_{end}x_i + b_{end}) \tag{4}$$

Then, using the nearest matching principle, the identified start and end are paired to obtain a set of candidate head entities.

(2) Relationship and tail entity joint recognition layer

CasRel identifies relationships and tail entities through a set of relationship related tail entity recognition layers. The structure of each tail entity recognition layer is the same as that of the head entity recognition layer, with the main difference being the input: the input of the head entity recognition layer is the output of the encoding layer, while the input of the tail entity recognition layer also considers the characteristics of the head entity.

$$p_i^{start_o} = \sigma(W_{start}^r(x_i + v_{sub}^k) + b_{start}^r) \tag{5}$$

$$p_i^{end_o} = \sigma(W_{end}^r(x_i + v_{sub}^k) + b_{end}^r) \tag{6}$$

v_{sub}^k is the average of all token vectors contained in the k-th candidate header entity.

3.3 Ontology Layer Construction

We refer to the core concepts in the field of naval warfare and, under the guidance of domain experts, abstracts 7 entity types from massive multi-source heterogeneous island chain data. These entities have spatial relationships such as "located" and "adjacent", as well as subordinate relationships such as "membership". The entity types, their attributes and relationship types are shown in Tables 1 and 2, respectively.

Table 1. Entity and their attributes

Island	Total length
	Number of bases
	Strategic role
	Belonging sea area
	Total area
	Number of bases
Military base	Latitude and longitude
	Number of bases
	Latitude and longitude
	The measure of area
	Climate
	Latitude and longitude
Port	The closest distance to China
	The measure of area
	Troops
	Facilities
	Weaponry
	Latitude and longitude
	Position
Strait	The measure of area
	Water depth
	Port throughput
	Climate
	Latitude and longitude
	Connected sea areas
Country	

Table 2. Relationship types

Head entity	Relationship	Tail entity
Archipelago	be located on	Island Chain
Island	be located on	Archipelago
Military base	be located on	Island
Port	be located om	Island
Strait	near	Island
Military base	subjection	Country
Strait	subjection	Country
Island	subjection	Country
Military base	near	Country

3.4 Instance Layer Construction

The CasRel model has excellent performance in entity relation joint extraction tasks, but further fine-tuning is needed to improve performance. The specific process is to randomly select 150 entries from papers in the island chain field, import them into the annotation tool doccano, manually annotate entities and relationships, and then randomly split them into training set, development set, and testing set in a ratio of 0.70:0.15:0.15. Supervised learning is performed on the pre trained model using the training set and development set, and model parameters are adjusted. The final evaluation is fine tuned using the testing set. The process of fine-tuning the Casrel model is shown in pseudocode 1 (Table 3).

Pseudo code 1: Fine tuning Casrel model

```
Input: Island chain domain entity data and labels
TrainModel:
   TotalEpochs = Specify total number of training epochs
   LearningRate = Specify learning rate for model optimization
   For each Epoch in TotalEpochs:
    For each Batch in Data:
      1.Input, IslandChainDomainEntitiesAndLabels = Get input data and la
        bels for current batch
      2.PredictedIslandChainDomainEntityLabels = Model output(Input)
      3.Loss = Calculate loss function(IslandChainDomainEntitiesActualLa
        bels, PredictedIslandChainDomainEntityLabels)
      4.Gradient = Calculate gradient(Loss, Model parameters)
      5.Update model parameters(Model parameters, Gradient, Learning
        rate)
```

Table 3. Comparison of entity relationship extraction performance before and after fine-tuning.

Category	accuracy	recall	F1 score
Before fine-tuning	0.586	0.492	0.487
After fine-tuning	0.620	0.523	0.516

Using the fine-tuning model, entity relationship extraction was performed on the data, resulting in 3 island chains, 27 island clusters, 143 islands, 28 military bases, 48 ports, 35 straits and 13 countries. There were a total of 946 pairs of relationships between these entities. Then, further manual corrections are made to the results to improve accuracy.

We use the read_HTML method in the Pandas library to quickly retrieve attribute table data from Baidu Baike and Wikipedia. The process is shown in pseudocode 2.

Pseudocode2: Baidu Baike and Wikipedia attribute extraction

Set URL as the target webpage address
Set CSV file path as 'D:\\百科\\getContent\\baidubaike_con-
tent\\1.csv'
For each table i:
  1. Read HTML tables from the URL
  2. Get the i-th table and name it tb
  3. Append tb to the CSV file using UTF-8 encoding
  - Set the header to 1
  - Do not include row index

We choose Neo4j as the graph database system to store the data of the island chain knowledge graph.- The final nodes of the island chain knowledge graph are shown in Fig. 4:

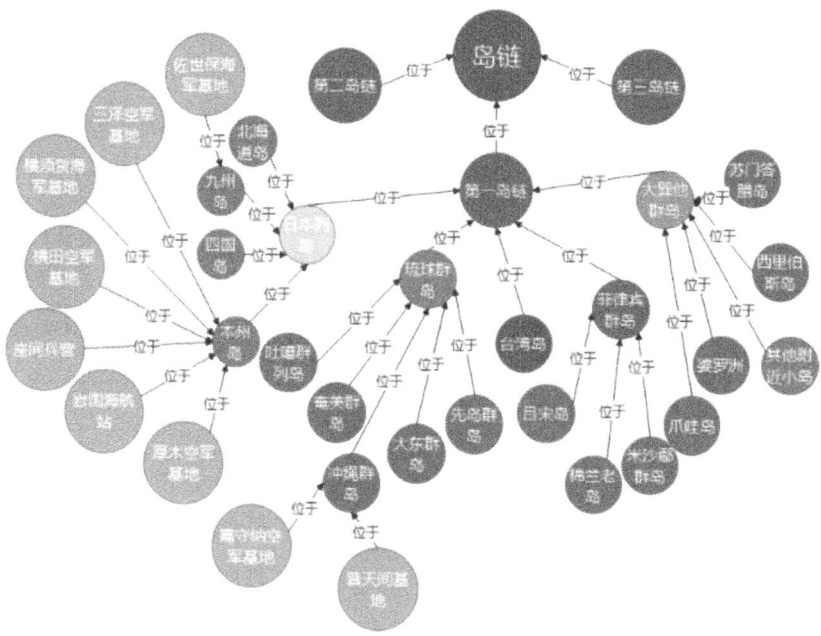

Fig. 4. Island Chain Knowledge Graph Part Node Display

4   Example of Graph Application

4.1   Construct the Island Chain Geographic Information System

After converting the csv files containing the entity relationship attributes of military bases, straits, and ports in the knowledge graph into geojson format files through applications (such as ASPOSE), you can import the map data visualization platform (taking

Google My Maps as an example) to further build an island chain geographic information system that integrates information from multiple military elements, in which users can search or click on entities to display their attributes and relationships, as shown in Fig. 1. In the figure, the blue logo represents the military base, the orange mark represents the important military port, the purple mark represents the important strait, the dialog box on the right is the attribute information of the current entity, the red line represents the closest distance between the geographic entity and China, and the gray circle represents the scope covered by the "near" relationship of the entity.

In the Island Chain GIS, users can view a panoramic view of the island chain area, including geographical, military, political and other information. This comprehensive display enhances the visualization of intelligence information and can improve the understanding and cognition of decision-makers in the complex situation in the island chain region. In addition, in addition to the functions of general geographic information system such as ranging and area calculation, the system also integrates the functions of querying, modifying, and deleting knowledge graphs, so that users can find corresponding entities according to attributes, so as to achieve more intuitive, more convenient, and more interactive information acquisition and analysis, and provide stronger support for military decision-making (Fig. 5).

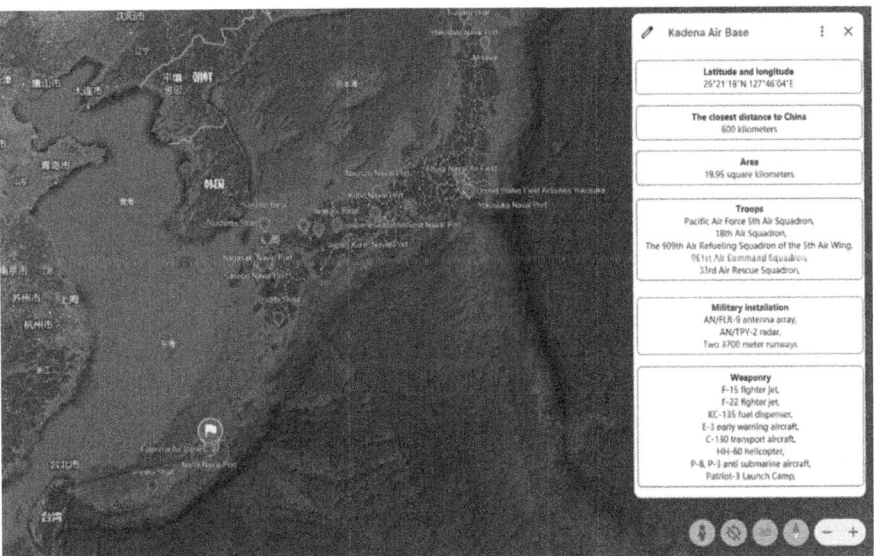

Fig. 5. Island Chain GIS

4.2 Optimize Machine Learning Algorithms as Prior Knowledge

Conventional machine learning algorithms, such as DQN [24] path planning algorithm, are difficult to comprehensively consider information such as troop deployment, weapons and equipment, and geographic layout when applied to maritime combat missions, so they cannot fully meet actual military needs. The island chain knowledge graph happens to contain this military information, which can improve this shortcoming. The specific strategy is as follows: firstly, in the state representation link, the information of the military base can be encoded into states, including the location of the base, the distribution of troops, the type and quantity of weapons, etc., as one of the inputs of the DQN algorithm; Then, in the design of the reward function, the weight of the reward function can be adjusted according to the strength, strength and quantity of the US military base in the island chain knowledge graph, and finally the optimal path obtained by the DQN algorithm is imported into the island chain geographic information system, displayed to the decision-maker, and further improved by the decision-maker according to the geographic intelligence information. The DQN algorithm framework optimized using the knowledge graph is shown in pseudocode 3.

---

pseudocode 3 : Improved DQN with island chain knowledge graph:

1. Initialization: Initialize the parameters of the DQN algorithm, including state space, action space, reward function, etc
2. Status representation: Encode the information of the military base into status, including the location of the base, the strength of the force, the number of weapon types, etc
3. Reward function design: Adjust the weight of the reward function according to the information in the island chain knowledge graph
4. Q network initialization: Initialize the parameters of the Q network, including weights and biases
5. Initialization of experience replay buffer: Initializes the experience replay buffer, which is used to store experience replay data
6. Training:
a. Select Action: Select an action based on the Q network
b. Perform Action: Perform the selected action
c. Get Rewards: Get rewards after performing actions
d. Storage Experience: Store experience replay data into the experience replay buffer
e. Update Q Network: Update the parameters of Q network
   7. Testing:
     a. Select Action: Select an action based on the Q network
     b. Perform Action: Perform the selected action
     c. Get Rewards: Get rewards after performing actions

Similarly, adding the island chain knowledge graph as prior knowledge to other machine learning algorithms such as resource allocation and target recognition can effectively improve the practicability and performance of these algorithms in the island chain field. By combining the detailed information about military bases, geographical topography, and marine environment in the island chain knowledge graph, these machine learning algorithms can more accurately perform tasks such as resource allocation optimization, target identification and tracking, and tactical deployment, and play a greater role in maritime operations. The introduction of the island chain knowledge graph can not only improve the decision-making ability and intelligence level of the algorithm, but also help the algorithm better understand the complex military environment, so as to provide more reliable support for military decision-making and action.

5 Concluding Remarks

Under the background of the serious blockade of the United States "Island Chain Strategy" against China, it is of great significance to master the geographical elements and military deployment of the island chain. Based on the demand of integrating complex information and fine-grained information in the field of island chain, this paper explores and applies the construction of island chain knowledge map. Based on the domain data of island chains such as military websites and professional books, the ontology layer is constructed under the guidance of domain experts. Then fine-tune CasRel model and use it to extract entity relationship and then fill in attributes to complete the construction of instance layer. At last, the intelligent search, intelligent recommendation and auxiliary strategic planning using island chain knowledge map are explored, which provides support for the next step of technology landing.

Due to the limitation of the number of public data we use, the number of entities and relationships in the island chain knowledge map is limited and it is difficult to cover all entity relationships well. Subsequent related research needs to dig more valuable data to build a more perfect island chain knowledge map. The island chain knowledge map constructed in this paper has not involved the marine environment and terrain elements in the expression of battlefield elements and it can be further explored in the future to increase the dimension of island chain knowledge map elements.

Acknowledgments. This work was supported by the National Natural Science Foundation of China (42071273), Fundamental Research Funds for the Central Universities (DUT24YG147)

References

1. Shi, C.L., Li, X.Y.: The impact of the US island chain blockage on China's access to sea. World Reg. Stud. **22**(02), 1–10 (2013)
2. Luo, R., Xiao, Y.J., Wang, L., Sheng, L.: Application of big data in command information system of naval battlefield. Ship Electron. Eng. **39**(3), 1–5+14 (2019)
3. Mohamed, A., Najafabadi, K.M., Wah, Y.B., Zaman, E.A.K., Maskat, R.: The state of the art and taxonomy of big data analytics: view from new big data framework. Artif. Intell. Rev. **53**(2), 989–1037 (2020)
4. Hoffmann, A., Kuwertz, A., Sander, J.: Towards information extraction and semantic world modelling to support information management and intelligence creation in defense coalitions. In: Judith, D. (ed.) Artificial Intelligence and Machine Learning in Defense Applications 2019. SPIE, vol. 11169, pp. 138–144. Strasbourg (2019)
5. Yu, Y.B., Wang, H.S., Fu, X.Q.: Research and judgment of sea battlefield situation based on multi-platform information fusion. Ship Electron. Eng. **43**(02), 34–37+152 (2023)

6. Wu, M.Y., Cai, H., Huang, Y.Y.: Multi-source Information Fusion Recognition Method for Sea Battlefield Situation Analysis. Nanjing University of Science & Technology, Nanjing (2023)
7. Chen, X.F., Liu, X., Gao, Y.B.: Structure for naval battlefield situation visualization platform based on data mining. Fire Control Command Control **40**(04), 144–147 (2015)
8. Zhang, H., Hou, D.W., Qu, Z.Y.: Probe into marine threat environment and technical development directions of unmanned underwater equipment. Digit. Ocean Underwater Warfare **5**(05), 448–452 (2022)
9. Wang, B.C.: Research on U. S Southwest Asian Military Base. Zhongnan University of Economics and Law, Wuhan (2021)
10. Shao, Z.F., Hu, B., Zhang, H.P.: Intelligent environment information assurance driven by knowledge graph. J. Geomatics **48**(01), 40–43 (2023)
11. Yang, Y.: Research on Battlefield Key Situation Analysis Based on Knowledge Graph. Xidian University, Xi'an (2022)
12. Xu, J.W., Gong, Y., Zhang, Y.Q.: Battlefield scenarios design based on knowledge graph. J. Command Control **9**(05), 573–579 (2023)
13. Li, X.: Research on Information Extraction Methods for Information Equipment Knowledge Graph Construction. National University of Defense Technology, Changsha (2021)
14. Fan, J.J., Ma, H.Q., Liu, X.L.: Research on intelligent question-answering services for military knowledge graphs based on open source intelligence in the era of digital wisdom. Data Anal. Knowl. Discov. **19**(05), 1–15 (2023)
15. Zhao, X.X., Li, S.L., Deng, K.B.: Search recommendation technology for multi-dimension military information. Command Inf. Syst. Technol. **15**(020), 70–75 (2021)
16. Chen, Y., Li, H., Li, H.: An overview of knowledge graph reasoning: key technologies and applications. J. Sens. Actuator Netw. **11**(4), 78 (2022)
17. Wu, Z.H., Liu, Q.H., Li, L.: Research on intelligent command and decision of electronic warfare based on knowledge graph technology. In: Chinese Institute of Command and Control. Proceedings of the 12th China Command and Control Conference, vol. 1, pp. 64–68, Ordnance Industry Press, Beijing (2024)
18. Hogan, A., et al.: Knowledge graphs. ACM Comput. Surv. **54**(4), 1–37 (2021)
19. Gruber, T.R.: A translation approach to portable ontology specifications. Knowl. Acquis. **5**(2), 199–220 (1993)
20. Mohamed, S., Nováček, V., Nounu, A.: Discovering protein drug targets using knowledge graph embeddings. Bioinformatics **36**(2), 603–610 (2020)
21. Van Der Vet, P.E., Mars, N.J.I.: Bottom-up construction of ontologies. IEEE Trans. Knowl. Data Eng. **10**(4), 513–526 (1998)
22. Wei, Z.P., Su, J., Wang, Y., Chang, Y.: A novel cascade binary tagging framework for relational triple extraction. In: Proceedings of the 58th Annual Meeting of the Association for Computational Linguistics, pp. 1476–1488. Association for Computational Linguistics. Seattle (2020)
23. Devlin, J., Chang, M., Lee, K., Toutanova, K.: BERT: pre-training of deep bidirectional transformers for language understanding. In: Proceedings of naacL-HLT, vol. 1, pp. 4171–4186. Google Incorporated. Minneapolis (2019)
24. Lv, L.H., Zhang, S.J., Ding, D., Wang, Y.: Path planning via an improved DQN-based learning policy. J. Sens. Actuator Netw. **11**(4), 78 (2022)

A Knowledge and Data Driven Method for Air Combat Intention Recognition

GuangFei Yang[1,2], Xuan Zheng[1(✉)], KeYun Wang[1], and Lian Liu[2]

[1] Dalian University of Technology, Liaoning, China
zx17422275572@163.com
[2] PLA Dalian Naval Academy, Liaoning, China
http://www.springer.com/gp/computer-science/lncs

Abstract. This paper introduces a novel approach to air combat intent recognition, emphasizing a model driven by the synergistic integration of expert knowledge and data. Departing from traditional knowledge base methods, our proposed framework leverages the interpretability of the Belief Rule Base (BRB) method and augments it with advancements in machine learning, particularly deep learning. The study explores the intricate interplay between BRB and neural networks, capitalizing on the strengths of each to create a robust decision support system. The belief rule base architecture is detailed, emphasizing fuzzy processing for handling information incompleteness. Additionally, the structure of the rule library, incorporating attribute importance, rule weights, and confidence levels, is outlined. Validation in the context of air combat target intention recognition demonstrates the model's efficacy in combining expert knowledge and data patterns, presenting a compelling advancement in decision-making systems.

Keywords: BRB · DNN · collaborative optimization · intent recognition

1 Introduction

The development of air combat intent recognition technology is essential for accurately and rapidly identifying the tactical intentions of aerial targets in the face of evolving air combat scenarios. Traditional methods relying solely on knowledge bases exhibit significant shortcomings in recognition accuracy and response speed, stemming from biased expert knowledge and underutilization of historical combat data. In response to these challenges, this study proposes a novel approach for air combat intent recognition, leveraging a combination of expert knowledge and data to address the complexities of future air battlefields.

Expert knowledge-based methods are indispensable when tackling intricate and dynamic problems in science and engineering. The Belief Rule Base (BRB) method, introduced by Yang et al. [8], is a representative prediction method based on expert knowledge. It organizes expert knowledge into a rule base using the IF-THEN structure and aggregates these rules through evidence reasoning. BRB has found applications in various fields due to its interpretability and effective utilization of expert knowledge. For instance, Guilan Kong et al. developed a clinical decision support system for risk stratification of patients with cardiac chest pain, demonstrating BRB's ability to handle uncertainties in both clinical domain knowledge and clinical data [2].

While BRB is robust, adopting the Dempster-Shafer (DS) evidence reasoning method for rule aggregation can lead to issues in high-conflict situations [1]. To address this, various scholars have proposed enhancements to evidence reasoning. The Yager combination rule [7] and Murphy combination rule [5] aim to resolve paradoxes arising from high conflicts. Martin et al. [4] optimized the robustness of combination rules, and Smarandache et al. [6] studied the reliability of these rules.

In recent years, machine learning has witnessed significant advancements, with algorithms automatically capturing patterns from data and transforming various fields. Notably, deep learning methods, simulating the human brain's information processing, have progressed rapidly [3]. Artificial neural networks, a key component of deep learning, excel at fitting mapping functions, offering a new avenue to enhance DS evidence reasoning. However, the inherent lack of interpretability in neural networks poses challenges, limiting their application in scenarios that demand transparency. BRB, with its strong interpretability, complements this drawback of neural networks.

Building on this analysis, this paper proposes a knowledge and data-driven model that integrates the strengths of BRB and neural networks. This approach not only harnesses the effectiveness of expert knowledge but also captures data patterns effectively. The proposed model is applied to air combat target intention recognition to validate its efficacy.

2   Model Structure

2.1   Belief Rule Base

The confidence rule library is a widely employed reasoning and decision-making technique in expert systems and decision support systems. It effectively incorporates both quantitative information and qualitative knowledge with uncertainty to model complex decision problems. This method enhances traditional rule representation by introducing rules, indicator weights, and output confidence. The implementation of rule base knowledge inference through evidence reasoning contributes significantly to improving data training efficiency.

Structure of Belief Rule Base. The belief rule base comprises the following components:

Fuzzy Processing: Utilizing membership functions to transform input data into fuzzy data with confidence, this component enables the system to handle information incompleteness or fuzziness effectively. This capability enhances the system's flexibility in making inferences and decisions. Commonly employed membership functions include:

- Gaussian functions,
- Triangular functions,
- Trapezoidal functions.

Rule Library: A rule library is a collection of rules, each specifying a particular context or condition along with its corresponding behavior or conclusion. The structure of the k-th rule is illustrated by Equation 1, where $A_i^k (i = 1, 2, ..., T)$ represents the reference value of the i-th prerequisite attribute $X_i (i = 1, 2, ..., T)$ in the k-th rule. Here, T denotes the number of prerequisite attributes. The logical relationship "AND" is denoted by $\wedge$, and θ_k signifies the importance of the k-th rule relative to other rules, referred to as the rule weight. Similarly, $\delta_i (i = 1, 2, ..., T)$ represents the importance of the i-th prerequisite attribute $X_i (i = 1, 2, ..., T)$ relative to other attributes, denoted as the attribute weight. Additionally, β_{kn} (n=1,2,...,n) represents the k-th rule's support for the n-th result, also known as the confidence level for the n-th result.

$$R_k : \text{IF } X_1 \text{ is } A_1^k \wedge X_2 \text{ is } A_2^k \wedge \cdots \wedge X_{T_k} \text{ is } A_{T_k}^k,$$

$$\text{THEN } \{(D_1, \beta_{1k}), \dots, (D_N, \beta_{Nk})\}, \left(\sum_{n=1}^{N} \beta_{nk} \leq 1 \right) \text{ with rule weight}$$

$$\theta_k (k = 1, \dots, L) \text{ and attribute weights } \delta_i (i = 1, \dots, T_k) \tag{1}$$

The completeness of confidence rules hinges on the presence of ignorance in the confidence distribution outlined by the rule's consequent. Specifically, it scrutinizes whether confidence has been allocated to every possible result within the identification framework. A k-th rule achieves completeness when $\sum_{i=1}^{N} \beta_{ik} = 1$, signifying the absence of unassigned confidence levels in the distribution articulated by the k-th rule. Conversely, if the sum deviates from 1, the k-th rule is classified as incomplete.

Reasoning Process of Belief Rule Base. In the inference process of the confidence rule base, the initial step involves calculating the activation weights for each rule in the rule base based on the input data. The activation degree w_k for the k-th rule is determined using Equation 2, where θ_k denotes the rule weight, representing the importance of the k-th rule relative to others. Similarly, δ_i (for $i = 1, 2, ..., T$) signifies the importance of the i-th prerequisite attribute relative to other attributes, and $\alpha_{i,j}^k$ represents the matching degree between the input data and the jth reference value of the i-th premise attribute in the k-th rule. Equation 2 converts $\alpha_{i,j}^k$ into the activation weights for each rule in the rule library.

Once the activation weights for each rule are obtained, the Dempster combination method is applied to aggregate the consequent of each rule, thereby deriving the inference results. Equation 9,10 illustrates the inference process of the confidence rule library, where β_j represents the confidence level of the j-th category in the resulting inference.

$$w_k = \frac{\theta_k \prod_{i=1}^{T_k} \left(\alpha_{i,j}^k\right)^{\bar{\delta}_i}}{\sum_{l=1}^{L} \left[\theta_l \prod_{i=1}^{T_l} \left(\alpha_{i,j}^l\right)^{\bar{\delta}_i}\right]} \tag{2}$$

$$\bar{\delta}_i = \frac{\delta_i}{\max_{i=1,\dots,T_k} \{\delta_i\}} \tag{3}$$

Optimization of Belief Rule Base. The output of the Belief Rule Base (BRB) is a confidence level vector for each category, denoted as $\beta = (\beta_1, \beta_2, \dots, \beta_m)$, where $\sum_{i=1}^{m} \beta_i = 1$. Assuming there are N feature-label pairs (X^n, y^n) for $n = 1, 2, \dots, N$, each pair can produce a corresponding inference result upon entering the BRB. The objective function for optimization is expressed in Formula4, utilizing the CrossEntropy function to calculate the cross entropy between (y^n, β^n) for each pair. The objective is to minimize the average cross entropy across all pairs, aiming for a closer alignment between the distribution of y^n and β^n, with lower cross entropy indicating a better fit.

$$\text{minimize } P = \frac{1}{N} \sum_{n=1}^{N} \text{CrossEntropy}(y^n, \beta^n) \tag{4}$$

Additionally, the optimization process of BRB parameters is subject to certain constraints:

- The confidence level in the consequent of each rule must be within the range $[0, 1]$:

$$0 \le \beta_{j,k} \le 1, \quad j = 1, \dots, N; \ k = 1, \dots, L. \tag{5}$$

- Each rule must be complete, meaning in the k-th rule, the sum of confidence levels for each category is 1:

$$\sum_{i=1}^{N} \beta_{ik} = 1, \quad k = 1, \dots, L. \tag{6}$$

- The weights of each rule should range from 0 to 1:

$$0 \le \theta_k \le 1, \quad k = 1, \dots, L. \tag{7}$$

- The weights of each feature should fall within the range $[0, 1]$:

$$0 \le \delta_t \le 1, \quad t = 1, \dots, T. \tag{8}$$

Drawing on the insights from Yang Jianbo et al. [8]., the optimization problem of BRB is formulated as a constrained optimization problem. The differential evolution algorithm is recommended for optimizing the parameters of BRB.

$$\mu = \left[\sum_{j=1}^{N} \prod_{k=1}^{L} \left(w_k \beta_{j,k} + 1 - w_k \sum_{i=1}^{N} \beta_{i,k}\right) - (N-1) \prod_{k=1}^{L} \left(1 - w_k \sum_{i=1}^{N} \beta_{i,k}\right)\right]^{-1} \tag{9}$$

$$\beta_j = \frac{\mu * \left[\prod_{k=1}^{L} \left(w_k \beta_{j,k} + 1 - w_k \sum_{i=1}^{N} \beta_{i,k}\right) - \prod_{k=1}^{L} \left(1 - w_k \sum_{i=1}^{N} \beta_{i,k}\right)\right]}{1 - \mu * \left[\prod_{k=1}^{L} (1 - w_k)\right]}, \quad j = 1, \dots, N \tag{10}$$

2.2 Deep Neural Network

The artificial neural network is a widely adopted machine learning method, designed to emulate the intricate connection relationships among neurons in the human brain for data learning and representation. Neural networks excel in automatically extracting features from extensive datasets, capturing complex patterns and relationships through layered abstraction and representation.

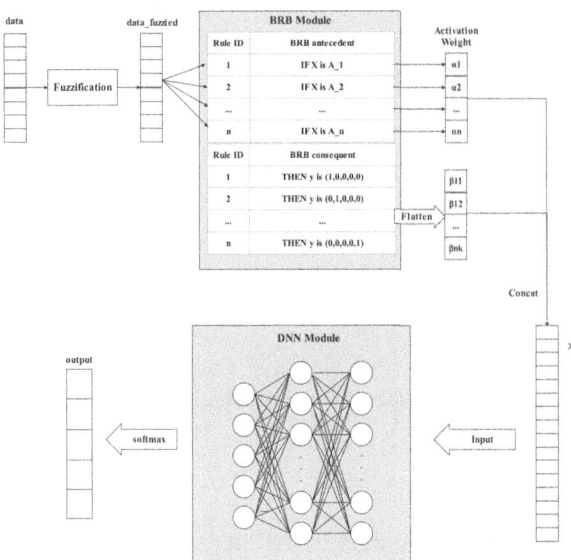

Fig. 1. Structure of BRB-DNN

Artificial neural networks typically comprise input layers, hidden layers, and output layers, facilitating data mapping between each pair of layers. Formula 11, 12 elucidate the operational mechanism of neural networks, involving the linear mapping of input X through Eq. 11 followed by application of a nonlinear activation function σ to map the result to y. In these formulas, w and b represent parameters that the model learns, with parameter optimization commonly achieved through the backpropagation mechanism.

$$z = w * x + b \tag{11}$$

$$y = \sigma(z) \tag{12}$$

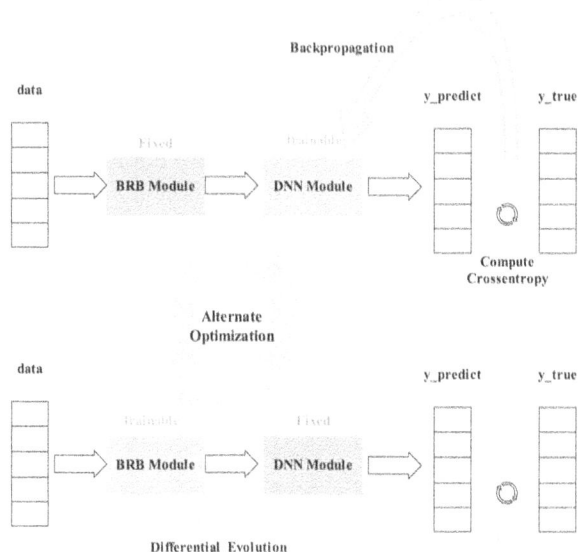

Fig. 2. Collaborative Learning Framework

2.3 BRB-DNN

Belief Rule Base (BRB) is a method rooted in expert knowledge, offering strong interpretability in the reasoning process. However, due to a limited number of internal parameters, BRB might struggle to fully capture historical data and may not be optimal in situations with high conflicts . On the other hand, Deep Neural Networks (DNN) boast a large number of internal parameters, providing robust fitting capabilities, but often suffer from poor interpretability. The integration of BRB and DNN allows for the creation of a model with both strong interpretability and fitting abilities, effectively leveraging expert knowledge and historical data to enhance overall performance.

The proposed BRB-DNN algorithm's framework, as illustrated in Figure 1, begins with the fuzzification module, converting input data into fuzzy data. This facilitates the calculation of the matching degree between input data and various reference values, leading to the computation of activation weights (w) for each rule. The activation weights (w) and the consequent information of the rule library (β) are concatenated into a vector (X). This vector captures both the original input information and the rule library information, completing the embedding of the rule library.

Subsequently, this vector X is fed into a fully connected DNN, where the DNN integrates, abstracts, and infers information in X, ultimately generating the prediction results.

2.4   Collaborative Optimization for BRB-DNN

The gradient descent method is a prominent optimization technique for Deep Neural Networks, relying on the availability of gradients for all trainable parameters. However, in the case of BRB-DNN, certain parameters lack gradient information, rendering the gradient descent method unsuitable for optimizing the BRB module of the model. Additionally, due to the extensive parameter count in the DNN module of the BRB-DNN model, using genetic algorithms for optimization confronts the challenge of the dimensionality disaster.

In response, this article advocates against employing a unified optimization algorithm for BRB-DNN. Instead, drawing inspiration from the collaborative evolution algorithm, different optimization algorithms are applied to distinct modules of BRB-DNN. Collaborative optimization is employed to alternately optimize the parameters of each module, mitigating issues associated with a unified algorithm.

The framework of collaborative optimization is depicted in Figure 2. Initially, the parameters of the confidence rule library module are fixed, and the backpropagation algorithm is employed to optimize the parameters of the DNN module. After multiple iterations in the DNN module, the parameters of the DNN are then fixed, and the differential evolution algorithm is utilized to optimize the parameters of the BRB module. This optimization process alternates and iterates repeatedly until convergence.

Table 1. Explanation of data

Feature Name	Meaning
β(mil)	Azimuth angle from my command ship to the direction of the aerial target
D(km)	Distance from my command ship position to the aerial target
V(m/s)	Speed of an aerial target on the horizontal plane
$\theta degree$	Direction of aerial target flight
H(km)	Vertical distance between an aerial target and sea level
$\delta(m^2)$	Size of the target's echo on the radar
$label$	Including 5 categories: 'reconnaissance', 'attack', 'cover', 'surveillance', 'other'

3   Experiments

(See Table 3).

Table 2. Rule Base

Rule ID	Rule Antecedent						Rule consequent				
	β	D	V	θ	H	δ	reconnaissance	attack	cover	surveillance	other
1	L	H	ML	M	any	any	1	0	0	0	0
2	M	any	MH	any	ML	any	0	1	0	0	0
3	MH	L	MH	L	L	ML	0	0	1	0	0
4	M	MH	L	MH	MH	H	0	0	0	1	0
5	any	any	any	any	any	any	0	0	0	0	1

Table 3. BRB Initial parameters

Rule ID	Feature Weight						Rule Weight
	β	D	V	θ	H	δ	
1	0.16	0.16	0.16	0.16	0.16	0.16	0.2
2	0.16	0.16	0.16	0.16	0.16	0.16	0.2
3	0.16	0.16	0.16	0.16	0.16	0.16	0.2
4	0.16	0.16	0.16	0.16	0.16	0.16	0.2
5	0.16	0.16	0.16	0.16	0.16	0.16	0.2

3.1 Source of Data and Prior Knowledge

The dataset used in this experiment originates from a specific air combat simulation platform, and a comprehensive explanation is presented in Table 1. The dataset consists of a total of 85 data points, with 80% (68 points) randomly selected for the training set and the remaining 20% (17 points) allocated to the testing set. Each data point encompasses 6 features and 1 label.

The ChatGPT model, embodying an extensive repository of human knowledge, assumes the role of a professional air combat commander. Leveraging its capacity to absorb recorded knowledge, ChatGPT generates rules for the Belief Rule Base (BRB), as outlined in Table 2.

3.2 Selection of Hyperparameters

The choice of hyperparameters significantly influences the performance of machine learning models. In the BRB-DNN model, key hyperparameters encompass the gradient descent learning rate (lr), batch size, iteration rounds of the differential evolution algorithm, and the hidden layer structure of the DNN. Below, we delve into a detailed exploration of the impact of various hyperparameters on the performance of the BRB-DNN model (Fig. 6).

Effect of Learning Rate. The learning rate plays a pivotal role in determining the magnitude of changes in model parameters and is among the most critical

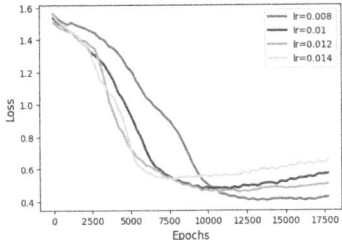

Fig. 3. Learning rate contrast (Loss)

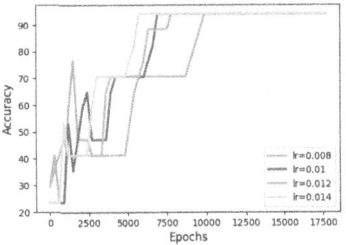

Fig. 4. Learning rate contrast (Accuracy)

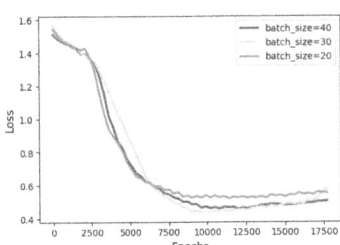

Fig. 5. Batchsize contrast (Loss)

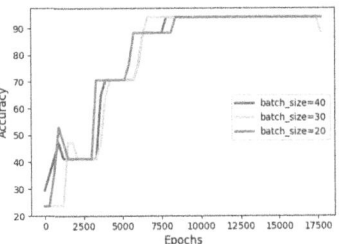

Fig. 6. Learning rate contrast (Accuracy)

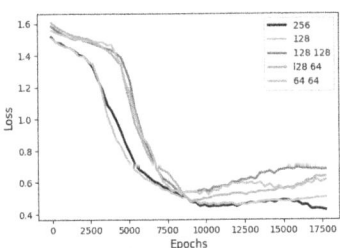

Fig. 7. Learning rate contrast (Loss)

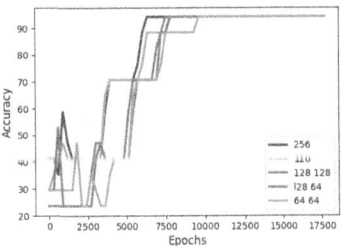

Fig. 8. Learning rate contrast (Accuracy)

hyperparameters influencing model quality. An excessively high learning rate can cause the model to overshoot the optimal solution, impeding convergence. Conversely, a low learning rate can limit the training speed, resulting in slow convergence. Figure 7, 8 illustrate the impact of different learning rates on BRB-DNN. The graph indicates that the least loss is achieved at a learning rate of 0.008, showcasing the best fitting effect. However, the learning process is slow, requiring approximately 12000 iterations to converge. Learning rates of 0.01 and 0.012 yield similar results. A learning rate of 0.014 accelerates convergence, reaching the lowest loss after around 6000 iterations. However, as iterations progress, the loss value begins to increase, suggesting potential overfitting or an excessively high learning rate, leading to departure from the optimal solution vicinity. Overall, the optimal performance is observed at a learning rate of 0.01, striking a balance between swift convergence and avoiding the risk of skipping the optimal solution.

Effect of Batch Size. Batch size, representing the length of the batch training dataset, significantly impacts the training process. Larger batch sizes involve more data in each iteration, providing more global information and stabilizing the convergence process. However, larger batches also result in time-consuming training. Conversely, smaller batch sizes use less data per iteration, leading to faster iteration rates but introducing stronger randomness and instability in the training process. Figure 3, 4 illustrate the impact of different batch sizes on BRB-DNN. A batch size of 20 results in quick convergence but with high loss values and poor fitting. With a batch size of 30, testing errors are less stable, exhibiting an upward trend post-convergence, accompanied by a decline in accuracy. A batch size of 40 offers stable model performance, making it a more suitable choice.

Effect of Differential Evolution Rounds. Table 4 illustrates the impact of different iteration rounds on model training error, testing error, training accuracy, and testing accuracy when using the differential evolution algorithm to optimize the BRB module. In theory, a larger number of rounds should result in smaller errors, higher accuracy, and potentially longer training times. However, achieving similar performance levels with fewer rounds is preferable. From Table 4, it is evident that the best fitting effect is achieved with 15 rounds, resulting in a testing error of only 0.477-ranking first-and both the training accuracy and testing accuracy reaching their highest levels.

Effect of DNN Hidden Layer Structure. The number of hidden layers in artificial neural networks can influence model performance. In theory, the more layers a neural network has, the stronger its ability to fit nonlinear functions. Similarly, more neurons in the same layer enhance the model's fitting capability. While increasing the number and layers of neurons improves fitting ability, it may lead to challenges such as an excessive number of parameters and difficulty in training. Figure 9 illustrate the impact of different hidden layer structures on BRB-DNN. Convergence is notably faster for a single-layer DNN compared to a double-layer DNN, with a smaller loss value post-convergence. The performance of the 256-neuron single hidden layer model closely matches that of the 128-neuron single hidden layer model. Following the Occam's razor principle, a DNN with a single hidden layer of 128 neurons should be considered the most suitable.

Table 4. Effect of Differential Evolution Rounds

Rounds	Training Loss	Training Accuracy	Testing Loss	Testing Accuracy
10	1.014	0.552	0.882	0.941
15	0.975	0.477	0.867	0.941
20	0.973	0.524	0.882	0.941

Following a comprehensive analysis of the aforementioned hyperparameters, we can ascertain the optimal configuration for BRB-DNN in the air combat target intention recognition dataset, as delineated in Table 5.

Table 5. Optimal Configuration For BRB-DNN

Learning Rate	Batch size	Rounds	DNN Structure
0.01	40	15	128(single layer)

3.3 Model Comparison

To thoroughly assess the performance of the BRB-DNN model proposed in this study, we conducted comprehensive comparative experiments involving various models. The models included in the horizontal comparison are BRB-DNN, BRB-DNN (train DNN only), BRB, and DNN. It is essential to highlight that BRB-DNN (train DNN only) maintains a structure identical to BRB-DNN, with fixed BRB modules to validate the collaborative optimization's effectiveness in preserving expert knowledge. Upon comparing BRB-DNN with BRB-DNN (train DNN only), notable distinctions emerge. BRB-DNN exhibits a significantly faster convergence rate, accompanied by a smaller loss value after convergence. Figure 10 further illustrates the accelerated increase in accuracy for BRB-DNN. These findings underscore the efficacy of the collaborative optimization algorithm in appropriately modifying prior knowledge within the confidence rule library. This modification enhances the accuracy of describing information in

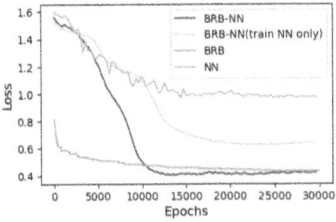

Fig. 9. Contrast by models (Loss)

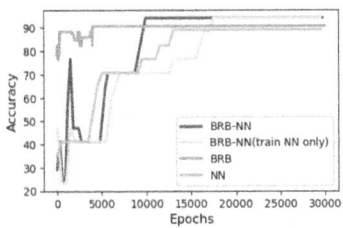

Fig. 10. Contrast by models (Accuracy)

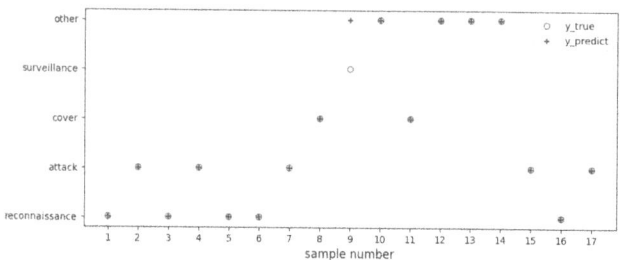

Fig. 11. Result on testing data

the training data, improving the quality of rules within the library and enriching the output of the BRB module with more comprehensive information.

In comparing DNN with BRB-DNN (train DNN only), a discernible enhancement in fitting accuracy is evident with the addition of the BRB module. While the DNN model relies solely on information derived from the training data, the BRB-DNN (train DNN only) model not only incorporates information from the training data but also integrates expert knowledge. This amalgamation empowers the model to exhibit superior fitting performance and testing accuracy.

The BRB model, primarily reliant on predefined prior knowledge, achieves commendable prediction accuracy and ralative low loss values without the need for extensive training. As depicted in Fig. 10, the BRB model attains a prediction accuracy of nearly 80% without training. However, owing to the limited number of internal parameters in BRB, its capacity to learn from data is constrained, resulting in marginal opportunities for both decreasing the loss value and increasing prediction accuracy.

Figure 11 showcases the performance of the trained BRB-DNN model on 17 test data points, highlighting an exceptional prediction accuracy of 94.12%, with just a single prediction error.

Further white box analysis was conducted, and we extracted the parameter information of the BRB module from the trained model, as shown in Figure 12. The observation results show that the weights of different rules undergo significant changes during the training process. Specifically, the weights of rules 2, 3, and 5 have increased, while the weights of rules 1 and 4 have significantly decreased. It is worth noting that the intention of Rule 4 is to "monitor", and the significant decrease in its weight directly leads to a decline in the performance of the model in identifying the "monitor" category. This phenomenon explains why all the samples with prediction errors in Fig. 5 belong to the "monitor" category. Figure 13 shows the distribution of feature weights in each rule. Rule 1 assigns a high weight to the flight speed of enemy aircraft, indicating that flight speed is a key indicator for determining whether the enemy aircraft has reconnaissance intent. Similarly, in Rule 2, the weights of flight altitude and reflection area are relatively low, indicating that these features have a relatively small impact on evaluating enemy aircraft attack intentions. Rule 3 has a higher weight for speed and distance, implying that the combination of "fast speed and short distance"

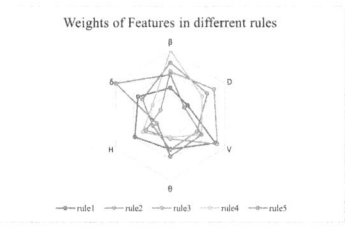

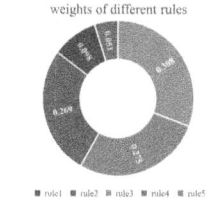

Fig. 12. Weights of different rules   **Fig. 13.** Weights of different features

features is an important basis for evaluating whether enemy aircraft have the intention of "covering up". Rule 4 demonstrates the importance placed on the azimuth of enemy aircraft, indicating that azimuth has a significant impact on determining the intention of enemy aircraft to "monitor". Rule 5 has a higher weight for radar reflection area and a lower weight for flight altitude and distance, revealing the importance of radar reflection area in identifying enemy aircraft's "other" intentions.

4 Conclusion

In conclusion, this study introduces a novel approach, BRB-DNN, for air combat target intention recognition. By combining the strengths of Belief Rule Base (BRB) and Deep Neural Network (DNN), our model leverages both expert knowledge and data-driven learning. The collaborative optimization algorithm enhances the model's ability to adapt expert knowledge during training, leading to improved fitting accuracy and testing performance.

Comparative experiments involving BRB-DNN, BRB-DNN (train DNN only), BRB, and DNN demonstrate the efficacy of our proposed model. BRB-DNN surpasses DNN by incorporating expert knowledge, showcasing accelerated convergence rates, and achieving higher fitting accuracy. The inclusion of the BRB module proves advantageous, enabling the model to embed richer information and enhancing its overall performance.

While the BRB model exhibits commendable prediction accuracy without extensive training, its limited capacity to learn from data highlights the importance of integrating data-driven approaches. The trained BRB-DNN model demonstrates outstanding performance on test data, achieving a remarkable prediction accuracy of 94.12.

In essence, our proposed BRB-DNN model emerges as a promising solution for air combat target intention recognition, effectively merging the interpretability of BRB with the learning capabilities of DNN. The collaborative optimization strategy proves instrumental in refining the model's performance, marking a significant stride in advancing the field of aerial target recognition.

References

1. Ding, Y., Wang, S., Zhao, X.: An improved method of ds evidence reasoning. In: 2009 IEEE International Conference on Communications Technology and Applications, pp. 761–765. IEEE (2009)
2. Kong, G., Xu, D.L., Body, R., Yang, J.B., Mackway-Jones, K., Carley, S.: A belief rule-based decision support system for clinical risk assessment of cardiac chest pain. Eur. J. Oper. Res. **219**(3), 564–573 (2012)
3. LeCun, Y., Bengio, Y., Hinton, G.: Deep learning. Nature **521**(7553), 436–444 (2015)
4. Martin, A.: Reliability and combination rule in the theory of belief functions. In: 2009 12th International Conference on Information Fusion, pp. 529–536. IEEE (2009)
5. Murphy, C.K.: Combining belief functions when evidence conflicts. Decis. Support Syst. **29**(1), 1–9 (2000)
6. Smarandache, F., Dezert, J.: Advances and applications of dsmt for information fusion. Collected works, volume 4 (2015)
7. Yager, R.R.: On the Dempster-Shafer framework and new combination rules. Inf. Sci. **41**(2), 93–137 (1987)
8. Yang, J.B., Liu, J., Wang, J., Sii, H.S., Wang, H.W.: Belief rule-base inference methodology using the evidential reasoning approach-RIMER. IEEE Trans. Syst. Man Cybern.-Part A Syst. Humans **36**(2), 266–285 (2006)

Model-Based Systems Engineering Supporting Architecture Modeling of Air Traffic Management System and Model Verifying Based on SMT

Tianning Liu[1], Xuesong Wang[2], Jinzhi Lu[1(✉)], Yao Tong[1], Yixiao Liu[1], and Xiaodu Hu[1]

[1] Beihang University, Beijing, China
jinzhl@buaa.edu.cn
[2] China Electronics Information Industry Group Co., Ltd. Sixth Research Institute, Beijing, China
99402006@sina.com

Abstract. The mission of air traffic management (ATM) is to effectively maintain and promote air traffic safety, maintain air traffic order and ensure smooth air traffic. For different environments and different aircraft types, ATM systems need to be designed differently. Now that the Concorde has been out of service for more than 20 years, there is a lack of supersonic aircraft in the civil aviation market. However, as a new type of aircraft, supersonic aircraft need to be introduced into the ATM system. Based on the above background, the Model-Based Systems Engineering (MBSE) approach in systems engineering approach is introduced. By using the multi-architecture modeling language KARMA and the multi-architecture modeling tool Airdraw, the metamodels are designed under the GOPPRR method, and the architectural models of ATM system are designed using these metamodels. After building the models, the comprehensiveness of the modeling design for the ATM system can be tested using the SMT checker in the modeling tool. Through this method, we found that this architecture model is functionally compliant with the ATM System requirements as defined by the ICAO through annexes and other documents.

Keywords: air traffic management System · MBSE · KARMA · GOPPRR · SMT

1 Introduction

As an important part of the aviation system, the ATM system ensures the safety and efficiency of air traffic. For different environments and aircraft types, ATM systems need to be designed differently. Now that the Concorde has been out of service for more than 20 years [1], there is a lack of supersonic airliners in the civil aviation market that can be officially put into commercial operation [2]. However, for the ATM system, since the supersonic passenger aircraft is a new type of aircraft involved in ATM services,

X. Tang et al. (Eds.): KSS 2024, CCIS 2269, pp. 183–197, 2025.
https://doi.org/10.1007/978-981-96-0178-3_13

the relevant indexes and properties of the existing air traffic control system need to be optimized and modified for the supersonic passenger aircraft.

In order to better integrate the system definitions and descriptions of the ATM system with the supersonic airliner, and to provide the design requirements under the ATM system, the MBSE approach [3] is introduced. Compared with the traditional document-based systems engineering approach, this approach realizes the modeling representation of the ATM system by establishing a system modeling of the ATM system to form a model library. The architecture model library obtained by this method can be updated for different ATM systems and different aircraft types by modifying the parameters of the corresponding views or adding relevant supplementary model views in the architecture model library, which greatly improves the compatibility and modifiability of the system definition.

The main purpose of the research is to focus on the construction of the architecture model of ATM system under the MBSE method under the platform of multi-architecture modeling tool Airdraw, which used the multi-architecture modeling language KARMA as its basic language. This paper mainly proposes an MBSE approach for the ATM system: (1) Metamodel designing based on GOPPRR method. (2) KARMA language-based architectural modeling, and UAF methodology for formalizing the model library. (3) Satisfiability Modulo Theories (SMT) is used to verify the obtained architectural models. By using the SMT checker in the multi-architecture modeling tool, the functional completeness of the model library comparing with the identification given by ICAO through files can be checked. First, the functions of ATM identified by ICAO through files are expressed through a graphical system model which is based on the KARMA language. The model captures all the functions mentioned on ICAO annex of the ATM system. Then the system models obtain parameter with the matrix and models in the architecture model library. After that, by coding and running the SMT checker in the tool, the parameters are approached and calculated, then the SMT checker then gives the output.

The rest of this article is organized as follows. In Sect. 2, the existing problems of ATM designing are introduced. In Sect. 3, descriptions of the architectural modeling and optimization of radar cabin layout are presented. Finally, we discuss our proposed approach in Sect. 4, and the conclusion is given in Sect. 5.

2 Problem Statement

To date, the traditional document-based system engineering method has been employed by CAAC to produce a substantial corpus of system engineering documents, including 296 civil aviation regulations and 1,227 normative documents. Additionally, the International Civil Aviation Organization (ICAO) has published 19 annexes to the International Civil Aviation Conventions pertaining to ATM systems.

As a system engineering research method that has gained considerable traction in the contemporary era, the development of system engineering models for ATM systems is also gradually becoming a priority. The Single European Sky Initiative was launched in the 1990s with the objective of advancing the modernization of ATM systems in Europe in order to meet future airspace capacity and safety requirements. In order to achieve harmonization of the structure of ATM systems across Europe and facilitate their management, the integration of disparate ATM systems across Europe using the MBSE approach has been included in the April 2019 report, "The Future of the Single European Sky," published by the European Commission.

In order to facilitate the integration of disparate system definitions and descriptions pertaining to ATM systems and supersonic airliners, this paper elucidates the process of incorporating the MBSE methodology into the design process of ATM systems. The metamodel, constructed according to the GOPPRR method, can be transformed into a set of model libraries following architectural modeling. These can be tailored to specific ATM systems and aircraft types by modifying the pertinent parameters of the corresponding views in the architectural model libraries or by incorporating supplementary model views. This significantly enhances the compatibility and modifiability of the definition of the ATM system.

3   Architecture Modeling Method Description

3.1   Designing Metamodels of ATM System Using GOPPRR

This paper uses the GOPPRR method [4] to build a metamodel library based on six meta-meta models (graph, object, point, property, relationship, role), and the definitions of the six meta-meta models are as follows:

Graph: The graph element contains several other elements, analyses a certain viewpoint of the system and describes the system in a block diagram.

Object: The main element in the model, used to represent an entity, which can exist alone or be linked to other objects.

Point: Usually attached to an object, used to represent the port connecting the object and the character.

Property: Cannot exist alone and is attached to other metamodels to represent its characteristics.

Role: At both ends of the relationship, it indicates the identity in which the object is connected.

Relationship: Connection between roles and objects, indicating the connection between objects.

After defining the meta–meta models, the metamodels can also be defined with the GOPPRR method. Considering the case environment of the ATM system, we finally give the definition of the metamodels as shown below:

(1) First is the metamodel of graph, which can be defined as:

$$Graph_i = \left\{ \sum Object_i, \sum Relationship_i, \sum Ob - RoBonding_i \right\} \tag{1}$$

where $\sum Object_i$ and $\sum Relationship_i$ refers to the summary of object metamodels and relationship metamodels that included in this graph, and $\sum Ob - RoBonding_i$ refers to the summary of the bonding that object metamodels and role metamodels made in this exact graph.

(2) To explain the definition of graph metamodels more deeply, below shows each of the three components included in the graph. As object metamodels can be defined like below:

$$Object_i = \left\{ \sum Property_i^{Ob}, \sum Point_i \right\} \tag{2}$$

where $\sum Property_i^{Ob}$ refers to the summary of all the properties that included in this object, and $\sum Point_i$ refers to the summary of all the points that included in this object.

(3) Below explains the definition of the point metamodels which can be seen in the defining method of the object metamodels:

$$Point_i = \left\{ \sum Property_i^{Po} \right\} \tag{3}$$

where $\sum Property_i^{Po}$ refers to the summary of all the properties that included in this point.

(4) Come back to the further explanation of the graph metamodels, below defines the relationship metamodels and the bonding of object metamodels and role metamodels:

$$Relationship_i = \left\{ \sum Property_i^{Re}, \sum Role_i \right\} \tag{4}$$

$$Ob - RoBonding_i = f_{Re_i} \left(\sum Role_i, \left(\sum Object_j, \sum Point_i \right) \right) \tag{5}$$

where $\sum Property_i^{Re}$ refers to the summary of all the properties that included in this relationship, and $\sum Role_i$ refers to the summary of all the roles that included in this relationship.

The definition of the bonding can be seen as $f_{Re_i}(Role_i, (Object_j, Point_i))$, which uses roles and suitable objects or points as input; after combing with the rules defined by the exact relationship metamodel, the bonding is defined as the output.

(5) Below explains the definition of the role metamodels which can be seen in the defining method of the bonding part:

$$Role_i = \left\{ \sum Property_i^{Ro} \right\} \tag{6}$$

where $\sum Property_i^{Ro}$ refers to the summary of all the properties that included in this role.

3.2   Architecture Modeling Framework: M0-M3

This paper uses the M0-M3 modeling framework [5] to implement multi-architecture modeling technology, which is divided into 4 levels, as shown in Fig. 1. Its specific meaning is as follows:

The M0 layer consists of 6 meta-meta models that refer to basic elements of the constructed model compositions and their interconnections, including graphs, objects, relationships, roles, points and property. The M1 layer is metamodels referring to the model compositions and connections needed to develop models, forming a metamodel library for a specific domain modelling language [6]. It is worth noting that precisely because the concept of metamodeling is derived from meta-metamodeling, metamodeling itself is divided into the six GOPPRR concepts introduced above, namely, graph metamodel, object metamodel, etc. M2 is the model layer, which is designed based on the metamodel to support system development. It is an abstract expression of a certain viewpoint in the real world [7]. For example, the requirements diagram is used to represent certain design requirements of the system during modelling. M3 represents a certain viewpoint in the real world, that is, expressing the system's concerns from a certain system perspective.

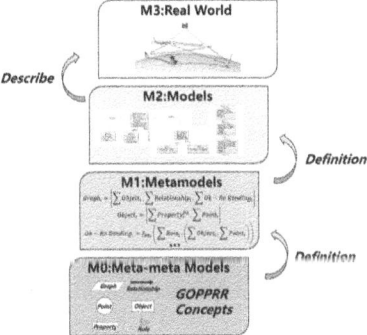

Fig. 1. M0-M3 Modelling Framework

4   Modeling Study and Verification

4.1   The Architectural Modeling of the ATM System

The main purpose of the research is the construction of the architecture model of ATM system. Based on the relevant documents provided by ICAO, the specifications and functionalities of the ATM System were extracted and the architectural modeling was completed with the following specific missions:

1. The ATM system is described through different viewpoints to create a methodological model view.
2. The nature of the ATM System, as defined in the relevant ICAO documents, is completed to be universal, i.e., it conforms to the standard ATM architecture model regardless of whether or not the ATM supports the operation of supersonic airliners.
3. A model of the architecture of part of the ATM system specially designed for supersonic passenger aircraft was completed.

With these missions, the modeling work can be finished through the progress below:

Modeling of the UAF Methodology
To finish the modeling work, we finally choose to adopt the methodology model based on UAF [8]. In the multi-architecture modeling tool, the methodology tool module is used to complete the modeling of the methodology model, and the final methodology model structure is shown in Fig. 2.

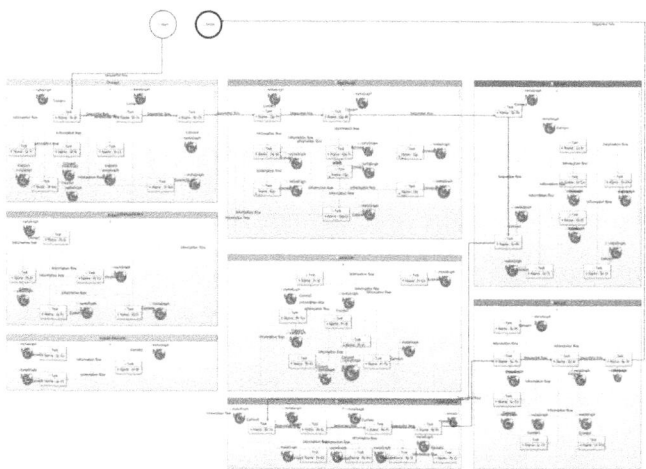

Fig. 2. Structural schematic the UAF methodology model

Settings of the Metamodels
Using the definitions show in Sect. 3.1 and the basic logic of UAF methodology, we can define the metamodels on the work of ATM system. This defining process can be finished on the modeling tool, taking the definition of object metamodel Actual Consisting Mission as an example, which is shown in Fig. 3.

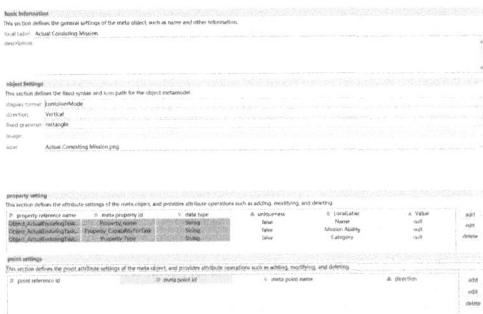

Fig. 3. Definition of object metamodel in Airdraw

The metamodel defining process ends up with 102 object metamodels, 8 point meta-models, 98 property metamodels, 41 relationship metamodels, 36 role metamodels, and 65 graph metamodels [9]. The basic information and settings about these metamodels are shown in Table 1.

Table 1. Interrelationship Among GOPPRRE Meta-Meta Models

Metamodel Category	Quantity	Included Property Types	Typical Classification	Typical Metamodel	Average Number of Metamodel/Graph
Graph Metamodel	65	–	Classified by different domains	Defined by UAF Methodology	1
Object Metamodel	102	Name, Provider, Usage, etc.	Entity, Concept, etc.	System, Requirement, Ability, etc.	1.57
Point Metamodel	8	Name, Status, etc.	Input Port, Output Port	Operation Port Input, Operation Port Output, etc.	0.12
Property Metamodel	98	–	String Property, Mathematical Property, etc.	Name, Provider, Latitude, Altitude, etc.	1.51
Relationship Metamodel	41	Name	Logical relationship, Connector, etc.	Make up, Coupling Connector, etc.	0.48
Role Metamodel	36	–	Relationship Provider & Receiver, Neutral role, etc.	From Ability Demander, From Ability Provider, etc.	0.8

• **Inspection of Metamodels: Construction of Model Library**

As shown in Fig. 2, the UAF methodology is used to build the KARMA models for the ATM system [10].

According to the guidance document of UAF methodology, the model covers 8 domains [11]. The Actuals Resources (Fig. 4 a) analysis evaluates different options and assumptions and weighs actual resource allocations to illustrate desired or realized actual resource allocations and relationships between actual resources. For example, the Real Resource Structure view (Fig. 5 a) shows that ATM system is mainly divided into two practical organizations in the actual operation of ATM organization and air service operation organization.

(a)   (b)

(c)   (d)

(e)   (f)

Fig. 4. Domain architecture model based on UAF methodology

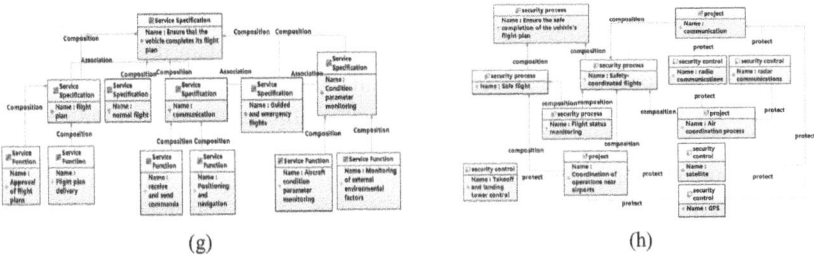

(g)   (h)

Fig. 4. (*continued*)

The Resource domain (Fig. 4 b) defines the architecture of the resources required for the solution to implement the operational requirements of the system, including the organization, software, artifacts, capability configuration, natural resources, and so on. For example, the Resource Structure view (Fig. 5 b) describes the target information resources needed to ensure the flight plan execution of the aircraft in the system, including aircraft state parameters, external environmental factors, flight plans and other resources, which are related to the emergency and communication system.

In the Strategic domain (Fig. 4 c), the architect can model the definition and logic of the system at the strategic level, and the model in the strategic domain describes the top-level strategic design of the system. For example, the strategic structure view (Fig. 5 c) defines that ensuring aircraft to complete the flight plan is one of the most important capabilities in the ATM system, which is composed of five capabilities, including reviewing flight plans, avoiding collisions and making air plans.

The Project domain (Fig. 4 d) shows the relationships and structure between different projects. It describes the dependencies between organizations and projects that contribute to the project. For example, the Project Connection view (Fig. 5 d), for example, shows the project of ensuring aircraft to complete the flight plan in the ATM system consists of three sub-projects: flight plan approval, communication and condition parameter monitoring.

The Operational domain (Fig. 4 e) describes the logical architecture. It describes the requirements, operational behavior, and structure required to support the functionality. The operational domain independently defines all the operational elements in the solution. For example, the Operational Structure view (Fig. 5 e) determines the composition relationship between different operational architectures and the resources required for each level of operational structures.

The Personnel domain (Fig. 4 f) describes human factors and aims to clarify the role of human factors in the creation of an architecture to facilitate human factor integration and systems engineering. For example, the People Structure view (Fig. 5 f) illustrates the analysis of staffing describes which specific positions are made up of each level of organization and gives a partial description of the responsibilities of these organizations.

The Service domain (Fig. 4 g) shows the services that can be provided in the system and the types of services that can be provided or carried out by each business activity. For example, the service structure view (Fig. 5 g) shows the structural relationship between the different services provided by the system. This view can intuitively show the hierarchy and parallel relationship between the services provided by the system.

The Security domain (Fig. 4 h) shows all the contents related to security and risk of the system, and the model in the security domain will explain the security control process and risk impact of the system. For example, the security structure (Fig. 5 h) shows the radio communication security in the air traffic relations system can protect the communication security and the air coordination process, and the air coordination process is an integral part of the safety process of flight under the safety coordination. Different from the traditional aircraft, the introduction of the new supersonic passenger

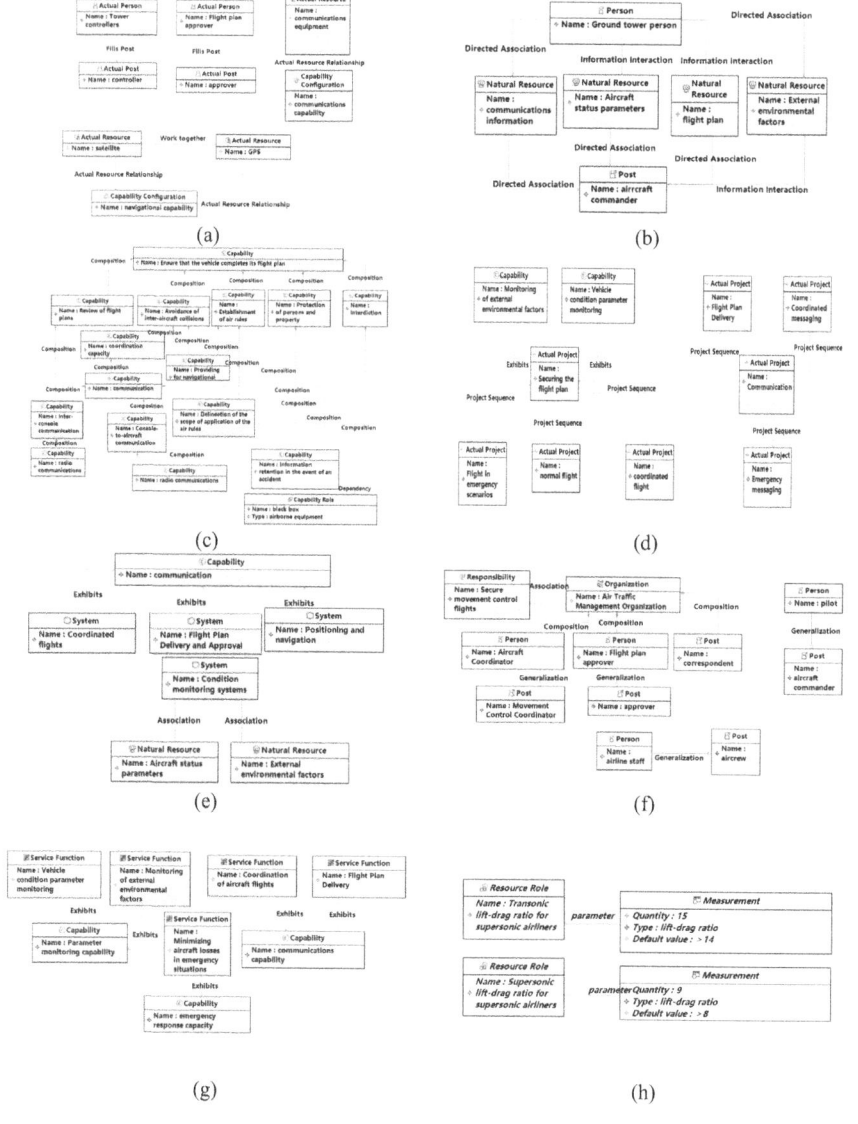

Fig. 5. Supersonic aircraft ATM system models

aircraft in the ATM system adds the protection means of the supersonic passenger aircraft visual enhancement system to the safety control of the communication system in the model view.

After completing the modeling work for each viewpoint model gradually according to the UAF methodology, the model library for the architecture of the ATM system is obtained, consisting of 65 main models, and 19 of them are customized for the supersonic aircraft, allowing the completed ATM architecture model library to meet the requirements for the introduction of this type of aircraft.

From this result, it can be reflected that the designed metamodel has good richness and can meet the actual needs of architecture modeling in the case.

4.2 Model Verifying Based on Satisfiability Modulo Theories (SMT)

In this research, the comprehensiveness of the modeling design for the ATM system can be tested [12] using the SMT checker in the modeling tool. Firstly, the KARMA model is constructed according to the functions of air traffic services as specified in the ICAO annex file as shown in Fig. 6.

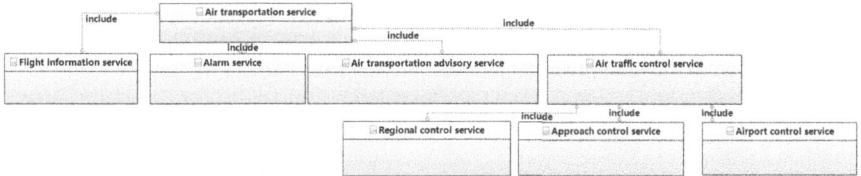

Fig. 6. Model diagram of the ICAO definition of air traffic services

Since we have already finished the Project Activity to Capability Mapping Matrix in the real case model library as shown in Fig. 7, it's available to give each functional module in Fig. 6 a capability parameter as Fig. 8 shows.

For each functional module in Fig. 8, the parameters of the requirement function are given [13] as shown in the Table 2. (In the table, "1" means that the requirement function is realized by this function in the ATM system defined by the model library, and "0" means that it is not realized.)

For the calculation of the parameters of the demanded and provided functions in this case, there are the following formulas:

$$Requirement = \sum_{i=1}^{7} Req_i \tag{7}$$

$$Real = \sum_{i=1}^{7} Re_i \tag{8}$$

in which Req_i is each specific functional requirement in the annex file issued by ICAO; and Re_i is the actual implementation capability of each function given in the ATM System Model Library completed by the modeling tool.

	Guiding Flight	Normal flight	Emergency flight
Flight Plan Transmission		1	
Aircraft condition parameter monitoring	1	1	1
External factors monitoring	1	1	1
Guiding Flight	1		
Normal flight		1	
Emergency flight			1
Communication	1	1	1

Fig. 7. Project Activity to Capacity Mapping Matrix

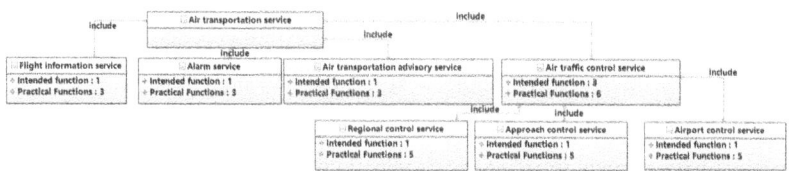

Fig. 8. Model of ICAO's definition of air traffic services after assignment

After the calculation, the result is shown in Fig. 9. SMT Parameter Resultsand Fig. 10. The metrics validation script calculations conclude that the ATM System model library, based on the KARMA language, is functionally compliant [14] with the ATM System requirements as defined by the ICAO through annexes and other documents.

Table 2. Required Functions-Display Function Parameters Table

	flight plan	Vehicle condition parameter monitoring	Monitoring of external factors	Guided Flight	Flight under normal conditions	Flying in emergency situations	correspond (by letter etc.)
Flight information services	1	0	1	0	0	0	1
alarm service	0	1	0	0	0	1	1
Air transportation advisory services	0	1	1	0	0	0	1
Air traffic control services	1	1	1	1	1	0	1
Regional control services	1	0	1	1	1	0	1
Approach control services	0	1	1	1	1	0	1
Airport control services	1	0	1	1	1	0	1

name	type	value
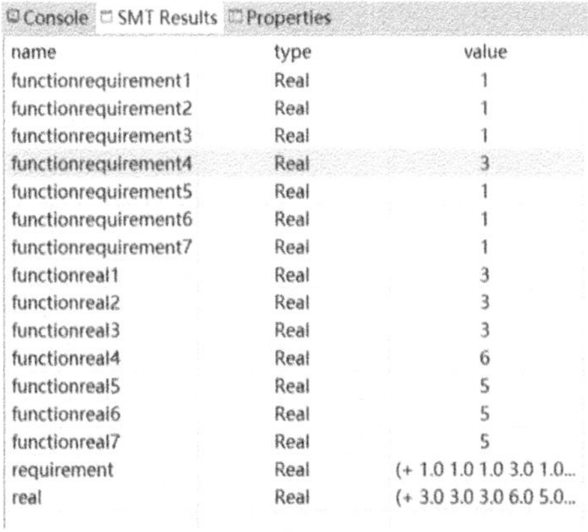 ⊡ Console ⊡ SMT Results ⊡ Properties		
functionrequirement1	Real	1
functionrequirement2	Real	1
functionrequirement3	Real	1
functionrequirement4	Real	3
functionrequirement5	Real	1
functionrequirement6	Real	1
functionrequirement7	Real	1
functionreal1	Real	3
functionreal2	Real	3
functionreal3	Real	3
functionreal4	Real	6
functionreal5	Real	5
functionreal6	Real	5
functionreal7	Real	5
requirement	Real	(+ 1.0 1.0 1.0 3.0 1.0...
real	Real	(+ 3.0 3.0 3.0 6.0 5.0...

Fig. 9. SMT Parameter Results

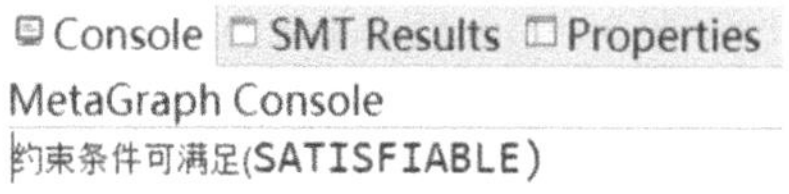

⊑ Console ⊡ SMT Results ⊡ Properties

MetaGraph Console

约束条件可满足(SATISFIABLE)

Fig. 10. SMT Checker Output Results

5 Conclusion

In this paper, through the study of the ATM system under the ICAO regulations, combined with the introduction of the MBSE methodology and the KARMA language which based on GOPPRR method, a set of metamodels was proposed, from which the M0-M3 modeling framework based on the GOPPRR methodology was derived and refined. After that, a model library of architecture models for the ATM system is developed. The architectural model library of the ATM system was constructed in modeling tool, and the constructed model library contains 8 viewpoints and 65 main models. Finally, in order to verify the integrity of the model, we wrote a set of SMT checker scripts based on the KARMA language. By using the SMT checker, the verification of the model is finished, in which we found that this architecture model is functionally compliant with the ATM System requirements as defined by the ICAO through annexes and other documents.

This paper mainly proposes a system engineering description of ATM system using MBSE method and GOPPRR method, and completes the whole process of modeling according to this method. The architectural model library obtained by this method can be modified according to the different parameters of the ATM system, which greatly improves the scalability compared with the traditional file-based description method.

Another major work in this paper is to write a KARMA-based SMT calculation script, and according to this script, the functional integrity check of the obtained architectural models is completed. The computational results show that the library of architectural models of ATM systems for supersonic airliners constructed by the methodology mentioned above is functionally complete.

References

1. Candel, S.: Concorde and the future of supersonic transport. J. Propul. Power **20**(1), 59–68 (2004)
2. Sun, Y., Smith, H.: Review and prospect of supersonic business jet design. Progress Aerosp. Sci. **90**, 12–38 (2017)
3. Wymore, A.: Model-based Systems Engineering. CRC Press, Boca Raton, FL, USA (2018)
4. Wang, H., Wang, G., Lu, J., Ma, C.: Ontology supporting model-based systems engineering based on a GOPPRR approach. In: Rocha, Á., Adeli, H., Reis, L., Costanzo, S. (eds.) New Knowledge in Information Systems and Technologies, vol.1, pp.426–436. Springer, Cham (2019). https://doi.org/10.1007/978-3-030-16181-1_40
5. Lu, J., Wang, G., Ma, J., Kiritsis, D., Zhang, H., Törngren, M.: General modeling language to support model-based systems engineering formalisms. In: INCOSE International Symposium, vol. 30, no. 1, pp. 323–338 (2020)
6. Yang, P., Lu, J., Feng, L., Wu, S., Wang, G., Kiritsis, D.: A knowledge management approach supporting model-based systems engineering. In: Rocha, Á., Adeli, H., Dzemyda, G.,

Moreira, F., Ramalho Correia, A.M. (eds.) Trends and Applications in Information Systems and Technologies, vol. 2, pp. 581–590. Springer, Cham (2021). https://doi.org/10.1007/978-3-030-72651-5_55

7. Lu, J., Wang, G., Yan, Y.: System engineering modeling methodology based on multi-architectural modeling language. J. Syst. Eng. **38**(2), 146–160 (2023) (in Chinese)

8. Hause, M., Bleakley, G., Morkevicius, A.: Technology update on the unified architecture framework (UAF). Insight **20**(2), 71–78 (2017)

9. Forman, B.: The political process and systems architecting. In: The Art of Systems Architecting, 2nd ed. CRC Press LLC, Boca Raton, FL, USA (2000)

10. Li, S., Deng, K.: UAF-based air transportation cooperative information SoS modeling under 'belt and road initiative'. In: 2022 IEEE 6th Information Technology and Mechatronics Engineering Conference (ITOEC), vol.6, pp.746–754. IEEE (2022)

11. Forman,B.: The political process in systems architecture design. In: INCOSE International Symposium, vol. 3, no. 1, pp. 92-97 (1993)

12. De Moura, L., Bjørner, N.: Satisfiability modulo theories: introduction and applications. Commun. ACM **54**(9), 69–77 (2011)

13. Barrett, C., Tinelli, C.: Satisfiability modulo theories. Handbook of Model Checking, pp. 305–343 (2018)

14. Barrett, C., Sebastiani, R., Seshia, S., Tinelli, C.: Satisfiability modulo theories. Handbook of Satisfiability, pp. 1267–1329 (2021)

Mission Modeling for the Perseverance Rover Based on KARMA Language

Zhiqing Liu[1], Guoxin Wang[1], Junda Ma[1,2(✉)], Jinzhi Lu[3], and Mengru Dong[1]

[1] Industrial and Systems Engineering Laboratory, Beijing Institute of Technology,
Beijing 100081, China
3120220369@bit.edu.cn
[2] Yangtze Delta Region Academy of Beijing Institute of Technology,
Jiaxing 314019, Zhejiang, China
mjd2015@sina.cn
[3] Beihang University, Beijing 100191, China
jinzhl@buaa.edu.cn

Abstract. Over the last two decades, Model-Based System Engineering (MBSE) has been progressively introduced into the design of complex systems as an essential method, aiming to address the complexity management challenges in traditional document-based approaches. However, limited by a lack of diversity in project and constraints on knowledge, the application of MBSE faces several issues. These include insufficient hierarchical decomposition of missions, overly abstract meta-models, and inadequate adaptability of methods. When MBSE is applied to the rover, the complexity of the Mars environment and the multidisciplinary nature of the Mars rover's design increase the learning curve for researchers due to the high-level abstraction of the meta-model. This paper proposes a new multi-architecture modeling method based on KARMA language, particularly focusing on the mission and operation stages. Including how to start from client objectives to precisely decompose stakeholder needs and successfully build a conceptual model library, the method supports more accurate and adaptable missions analysis in complex systems. Finally, we use the Perseverance rover design as an example to validate the effectiveness of the proposed method.

Keywords: MBSE · Mission modeling · Meta model · KARMA Language · Mars rover

1 Introduction

MBSE is an advanced theory of systems engineering, which has greatly facilitated the progress of SoS over past two decades. Including modeling method, modeling tool and modeling language as three pillars [1], MBSE significantly enhances design efficiency and management quality [2]. Methods have proven to be powerful ways for addressing complex architectural design challenges. Researchers continuously develop and refine methods to enhance the design, management, and efficiency of complex projects.

X. Tang et al. (Eds.): KSS 2024, CCIS 2269, pp. 198–212, 2025.
https://doi.org/10.1007/978-981-96-0178-3_14

However, existing modeling methods lacks a multi-perspective hierarchical analysis of mission, which often remains at a superficial lever without exploring the underlying business goals, user expectations, and environment factors. The relationship and dependencies are not fully revealed, such as cost overruns or design products not meeting user objectives. Moreover, the high abstract level modeling methods present a learning challenge for personnel from diverse backgrounds. Current high-abstraction meta-models complicate understanding and application. Additionally, existing methods lack of flexibility and adaptability limits their effective response to the unknown variables. As technology rapidly develops, method should also be flexible enough to adapt to new knowledge to support long-term innovation of projects.

The aerospace industry is the first to successfully apply MBSE. However, as we continue to use MBSE, we are finding some challenges in the field of Mars rover development. First, manufacturing a Mars rover is a high-tech process that involves many disciplines and complex technological innovations. This makes the modeling process highly specialized, leading to high learning costs and long learning cycles for non-experts. Second, the Martian environment is highly changeable and largely unknown, with variations in light, radiation, and other cosmic factors. Mars is also much farther from Earth than the moon, making rover development very difficult. Therefore, the design and modeling of the rover need to adapt to the changing Martian environment and must be flexible and robust.

To deal with these issues, this paper first customizes a meta-model library for Mars rover design, then proposes a multi-architecture modeling method based on KARMA language, supporting multi-perspective architectural design of mission and operation, and finally validates the proposed method with the Perseverance rover project.

The rest of the paper is organized as follows: Sect. 2 discusses related work in MBSE modeling methods. Section 3 proposes a multi-architecture mission modeling method for "mission analysis" and "operation analysis". Section 4 presents a modeling example based on the Perseverance rover and discusses it. Finally, Sect. 5 concludes the paper.

2   Related Work

In this chapter, we first summarize the current mainstream modeling methods in MBSE. Then some meta-models approach used in the methods are introduced. Research status of Mars rover design in the field of MBSE is also analyzed later. Finally, based on the related work, we summarize the objectives of this study.

2.1   Current MBSE Methods

Current mainstream modeling methods, such as OPM, STRATA, RFLP, and DoDAF, each have their strengths in mission or requirements analysis but can benefit from enhanced multi-perspective analysis. In OPM [3], requirements are typically modeled as states of objects and connected to components through labels such as "satisfies," "realizes," or "derives." While effective, this approach may benefit from clearer representation of relationships and multi-perspective analysis. Vitech's MBSE modeling methodology STRATA [4] breaks down requirements analysis into four system domains and focuses on the hierarchical decomposition of high-level requirements. However, incorporating a more comprehensive multi-perspective view could further enhance its utility. RFLP [5] (Requirement, Function, Logic, Physical) integrates mission and stakeholder needs at the requirement stage but could more fully reveal the interconnectivity and complexity of different perspectives. These observations suggest that there is room for improvement in multi-perspective analysis in complex system engineering projects.

2.2 Meta-models Used in Methods

Most mainstream MBSE modeling methods are designed for general domains, such as DoDAF and SysML. While powerful in many respects, these methods exhibit a high level of abstraction when applied to the Mars rover field, posing significant learning challenges for engineers. For example, DoDAF 2.0 is based on the UPDM [6] meta-model library, and MagicGrid refers to the SysML 1.0 [7] meta-model library. These general frameworks often do not fit well with the specific complexities and unique missions. Additionally, modeling tools like Capella, which supports the Arcadia method based on DSML [8] meta-models and has been successfully applied in specific domains like avionics and rail transportation, are not suitable for direct application in other field.

2.3 Mars Rover Design Using MBSE

In the field of Mars exploration, the flexibility and adaptability of MBSE modeling methods are crucial. Although methods like DoDAF and SysML demonstrate strong capabilities in multiple domains, they show certain limitations in the specific complex environment of Mars exploration. For instance, the NASA Perseverance rover project [9], despite widespread application of MBSE theory throughout the project's management process, the specialized nature of its modeling processes has resulted in high learning costs, particularly for non-specialists. This not only limits the widespread adoption of the methodology but also slows its practical application efficiency in Mars rover projects. Similarly, the ExoMars [2] mission collaboration between the European Space Agency (ESA) and Roscosmos shows that SysML-based modeling work failed to fully cover the system model, limiting comprehensive system analysis of complex missions. Additionally, Chinese scholars, such as Gao Jinyan [10], have achieved some success in modeling the maintenance and management systems of Mars rovers, but still lack a complete architectural design for the entire lifecycle of Mars rover projects, reflecting the existing methodologies' insufficiencies in supporting comprehensive system engineering. These examples illustrate that existing

MBSE modeling methods need to enhance their flexibility and adaptability to better address the specific challenges of Mars exploration, ensuring the effectiveness of design solutions and supporting the long-term success and innovation of projects.

2.4 Summary and Motivation

From the above literature review, we summarize the main motivations and contributions of this study: We provide a comprehensive MBSE method that includes multi-level analysis for Mars rover missions, addressing the gaps in existing methods. We make the MBSE method more accessible for engineers by reducing the abstraction levels of meta-models, thus lowering the learning difficulty. We enhance the flexibility and adaptability of MBSE methods to better handle the uncertainties and technological changes specific to the Martian environment.

This study is motivated by the need to overcome these limitations. We aim to develop a more accessible and flexible MBSE approach that better suits the complexities of Mars exploration, ultimately enhancing mission success and adaptability in an ever-evolving technological landscape.

3   A Mission Modeling Method Based on KARMA Language

The following is the framework of research in this article. From top to bottom, it is Meta-metamodel, Metamodel, System model, Physical world. In Meta-metamodel hierarchy, GOPPRR-E(Graph, Object, Property, Point, Relationship, Role and Extension) is a kind of meta modeling language. In Metamodel hierarchy, based on GOPPRR-E, we use modeling language KARMA to create the domain metamodel in order to express propriety concepts, such as mission task, operation task, mission phase, operation phase. In System model, we develope a set of viewpoints, divided into mission phase and operation phase, and based on the meta-model, according to each step of the method to build the model. In Physical world, MBSE models map to real-world objects, such as the Perseverance rover, the Zhu Rong rover (see Fig. 1).

3.1 GOPPRR-E Meta Modeling

Kelly [11] synthesized several widely used meta-meta models from existing meta-modeling languages, such as ARIS, Ecore, GME, GOPPRR, and MS DSL Tools, among which GOPPRR stands out for its strong descriptive capabilities supporting more complex model concepts. In systems engineering practice, Ding [12] et al. extended the GOPPRR meta-meta model to GOPPRR-E, where 'E' stands for Extension. GOPPRR-E consists of Graph, Object, Point, Property, Relationship, Role, and Extension, each representing fundamental elements and extended derived concepts, detailed in paper [12]. This chapter utilizes the GOPPRR-E meta-modeling method, customized through the KARMA language, to develop a meta-model library specifically for the Mars rover domain.

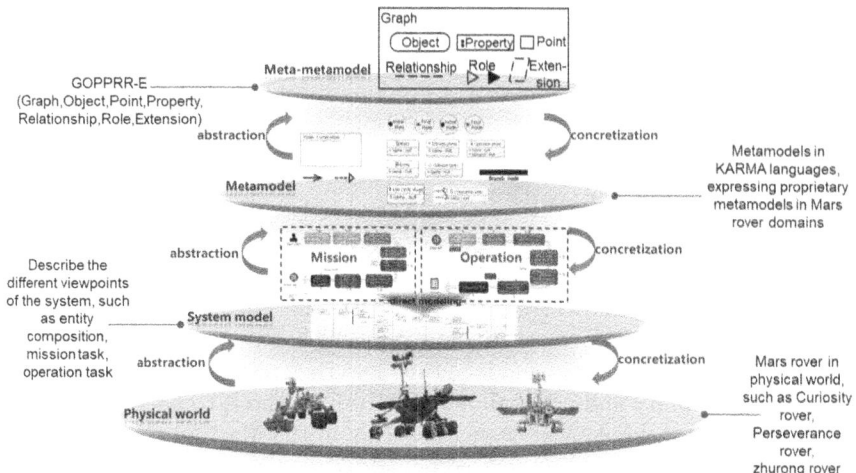

Fig. 1. Technical architecture

3.2   KARMA Language

Based on GOPPRR-E meta modeling, we use KARMA language to create meta-model in specific domain. KARMA [13], a multi-architecture unified modeling language for model-based systems engineering, supports architecture-driven design, code generation, dynamic behavior description, performance analysis, verification, and numerical analysis. It visualizes data as tables and Gantt charts while offering 2D and 3D modeling capabilities. KARMA's nature focus on "text-readable formalized language" to model requirements, functions, logic, and architectures. It enables to verify performance metrics and requirements via simulation and testing, enabling early formalization and validation of system views and reducing development costs and risks during conceptual design.

3.3   Mission Modeling Using KARMA

Metamodel in Specific Domain. Based on GOPPRR-E, We use KARMA language to create metamodel. Table 1 gives a summary of metamodel.

Methods. The method points out the modeling sequence, with each box in the method representing the viewpoints. Each viewpoint is instantiated with a metamodel or other, such as requirements forms(ReqIF), mTable and tables. The design modeling method based on KARMA language is divided into mission phase and operation phase.

Mission Phase. In the mission analysis phase, the objective is to define the problem to be solved, describing the business and problem space. This phase

Table 1. Metamodel in mission phase and operation phase.

Metamodel type	Metamodel example	Quantity
MetaGraph	Mission concept diagram,Task diagram, Life cycle diagram,Entity diagram, Operation concept diagram	5
MetaObject	Entity,Block,Mission phase,Mission task, Operation phase,Operation task,Life cycle stage	7
MetaProperty	Name,Behavior	2
MetaPoint	Input,Output	2
MetaRelationship	Directed aggregation,Object flow,Control flow, Sequential flow,Message flow	5
MetaRole	Beginning with no style,Beginning with dot diamond, Beginning with solid diamond,End with arrow	4

begins with "mission-client-objective" as inputs, where the mission is defined as a set of objectives to be achieved, stakeholders are entities related to but not directly involved in achieving the mission, and the client represents a specific category of stakeholders. The first step in the analysis is to define stakeholders, entities, and operational systems to determine the participants and relevant parties for subsequent analysis. Following this, the mission is decomposed to identify mission phases based on a time sequence, to understand the components of the mission thoroughly. This process includes defining transition events between mission phases, which are necessary conditions for moving from one phase to another, thereby building a conceptual framework for the mission. Further, entities, operational systems, and clients involved in each mission phase are analyzed to clarify their respective tasks, refining the mission phases into specific mission tasks and analyzing the capabilities and expected outcomes for each task. Finally, the operational system versions are determined, and their capability coverage assessed to provide a basis for defining the operation system in the next phase. Mission phase analysis is a system-level approach designed to move from an abstract mission to precisely identify the specific operational systems involved in the mission implementation (Fig. 2).

Operation Phase. The core purpose of the operation analysis phase is to explore in detail the interactions between external entities and the system of interest (SOI), defined by the mission phase as the operational system. This stage begins with a comprehensive consideration of the SOI's lifecycle activities, covering all aspects from design and development to application and eventual decommissioning. The SOI is then positioned within the entire mission task, analyzing its operational activities and the conditions and triggering events between these activities, thus constructing the operational concept of the SOI. The in-depth analysis during this phase associates operational activities with mission tasks to establish the specific contributions of the SOI to the mission. These operational activities are traced back to lifecycle stages of the SOI, providing guidance for implementation activities. Moreover, identifying the external entities involved

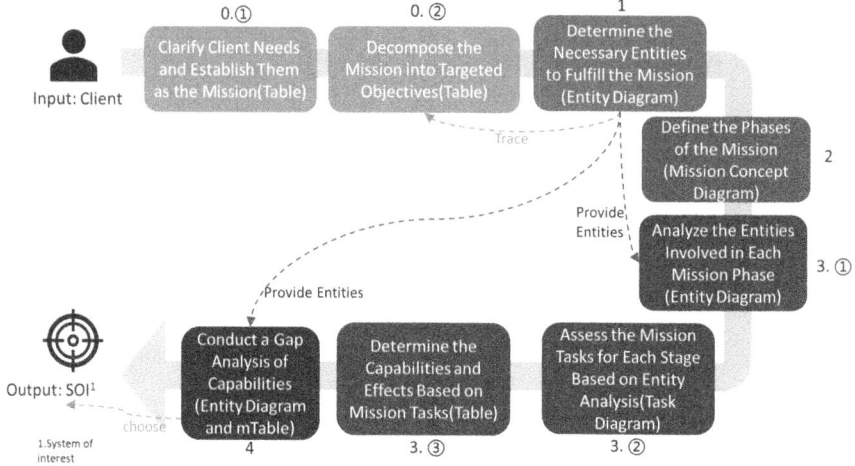

Fig. 2. Mission phase

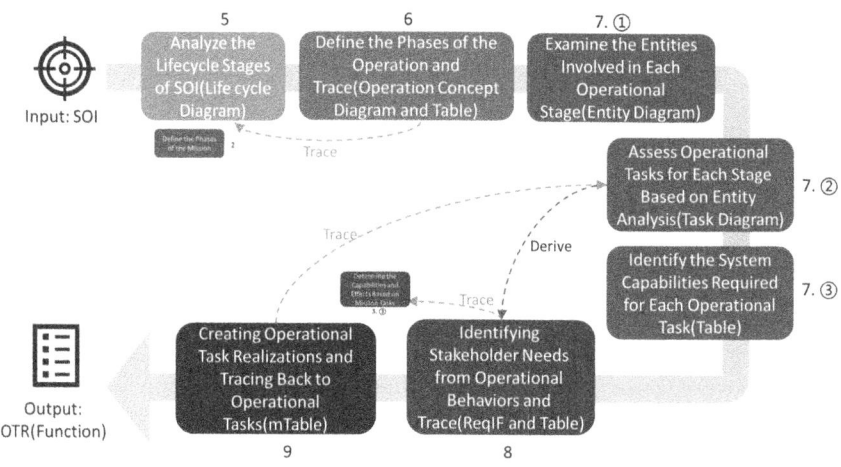

Fig. 3. Operation phase

in each operational stage provides clear definitions for the implements of operational tasks. Then, defining the operational tasks executed by the operational system and external entities in each operational phase lays the foundation for deriving stakeholder requirements. As operational tasks give rise to stakeholder requirements, each requirement corresponds to one or more operational tasks, ensuring that the requirements can be traced back to the capabilities and effects of the mission. Ultimately, creating Operational Task Realization (OTR) and tracing it back to related operational tasks ensures that an operational task may be associated with one or more OTRs, while ensuring that OTRs provide accurate information to the functional architecture phase (Fig. 3).

4   Case Study

4.1   Overview

The book "Perseverance and the Mars 2020 Mission" by Manfred [14] provides a detailed account of the Perseverance Rover's construction background, mission tasks, architectural design, landing site selection, and implementation process. Using the rich information from this book, we conduct a requirement design modeling of the Perseverance Rover. Initially, referencing the SysML meta-model library, a domain-specific meta-model library is constructed. Subsequently, based on the "mission and operation" requirement modeling methodology proposed in this paper, a model library focuses on the Perseverance Rover was developed. This model library not only includes the design, manufacturing, assembly, testing, and transportation processes of the Perseverance Rover but also encompasses the functional requirements of the Perseverance Rover in the Mars exploration mission.

4.2   Perseverance Rover's Mission Modeling

Step 0 (Client Input Mission). Using the Metagraph tool's table, customer needs for the "Exploring Mars" mission are analyzed based on government documentation. The mission decomposition targets four objectives: determining Mars' ability to support life, understanding the development and history of Mars' climate, comprehending the origins and evolution of the Martian geological system, and preparing for human exploration. These four objectives are specific breakdowns of the top-level mission "Exploring Mars," meaning that achieving all four objectives would signify the mission's completion (see Fig. 4(a)).

Step 1 (Determine the Necessary Entities to Fulfill the Mission). Considering the entities required to accomplish the "Exploring Mars" mission includes the "Perseverance Rover," "Atlas V-541 rocket," "Mars 2020 mission team," "JPL team," "Kennedy Space Center," "Mars Reconnaissance Orbiter," and "launch site operators" among others-totaling 22 entities. Steps 3.① and 7.①, based on Step 1, select entities that must not exceed the range defined(see Fig. 4(b)).

Step 2 (Define the Phases of the Mission). Initially, the mission "Exploring Mars" is analyzed to determine specific implementation stages, dividing it into six mission phases along with the initial and final states. The transitions between each phase, such as from the initial state to the early stages of the project being triggered by the project's commencement and from early preparation to the Atlas V-541 rocket launch triggered by the launch window, are defined(see Fig. 4(c)).

Step 3. ① (Analyze the Entities Involved in Each Mission Phase). This step involves analyzing the entities associated with the six mission phases using the cloning feature in the tool (where all cloned entities share all properties at the base level), selecting relevant entities from step 2. For instance, the mission phase 2 "Atlas V-541 Rocket Launch" involves entities such as Launch complex 41 at cape canaveral air force station, launch operators, and Atlas V-541 rocket(see Fig. 4(d)).

Step 3. ② (Assess the Mission Tasks for Each Stage Based on Entity Analysis). Based on the entities from Step 3.①, the mission tasks for phase 2 "Atlas V-541 Rocket Launch" are decomposed, starting with Launch complex 41 at cape canaveral air force station performing task T.2.1 Provide launch site for Atlas V-541 rocket, followed by the Launch site operators performing task T.2.2 Give launch order, and finally the Atlas V-541 rocket performing task T.2.3 Launch, concluding the task decomposition for mission phase 2(see Fig. 4(e)).

Step 3. ③ (Determine the Capabilities and Effects Based on Mission Tasks). All tasks of the Atlas V-541 rocket launch are summarized in a table, individually analyzing the capabilities and effects corresponding to each mission task, such as "T.2.1 Provide launch site for Atlas V-541 rocket" corresponds to the capability "launch Complex 41 provides the facilities," with the effect "launch -Complex 41 provides the well-established facilities" (see Fig. 4(f)).

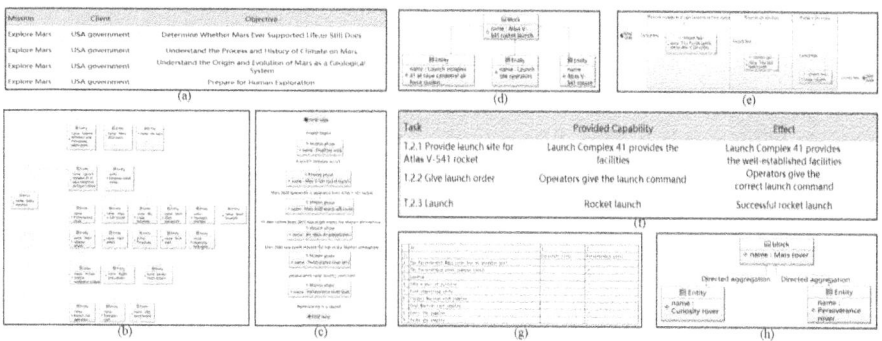

Fig. 4. Perseverance rover models (a) Mission, Client, Objectives (b) The relevant entities involved in the Mars mission (c) Mission concept (d) Atlas V-541 launch mission phase involves entities (e) Analyze mission task (f) Mission capability and capability effect (g) Identify potential operational systems (h) Curiosity versus Perseverance models

Step 4 (Conduct a Gap Analysis of Capabilities). Initially, based on the design history of the Mars rovers, the historical and latest versions of the rovers are analyzed, such as the latest version being the Perseverance Rover, with historical versions including Curiosity, Opportunity, and Spirit. Then, using the mTable relational mapping function provided by Metagraph, the mapping relationship between Curiosity and Perseverance in terms of mission tasks is established, demonstrating the capability gap between them, indicating Perseverance's full coverage of execution capabilities, based on which Perseverance is identified as the SOI for this analysis, being the focus and center of the analysis(see Fig. 4(g) and Fig. 4(h)).

Step 5 (Analyze the Lifecycle Stages of SOI). This step marks the beginning of the operation phase, starting with the Perseverance Rover identified in the mission phase as the System of Interest (SOI). The life cycle of the Perseverance Rover is decomposed, including stages such as the design, production, and manufacturing of the rover, assembly and testing, arrival at the launch site, successful landing on the Martian surface, and participation in Mars exploration missions. Transition conditions are added at each stage, specifying that the Perseverance Rover progresses to the next life cycle stage only when conditions are met(see Fig. 5(a)).

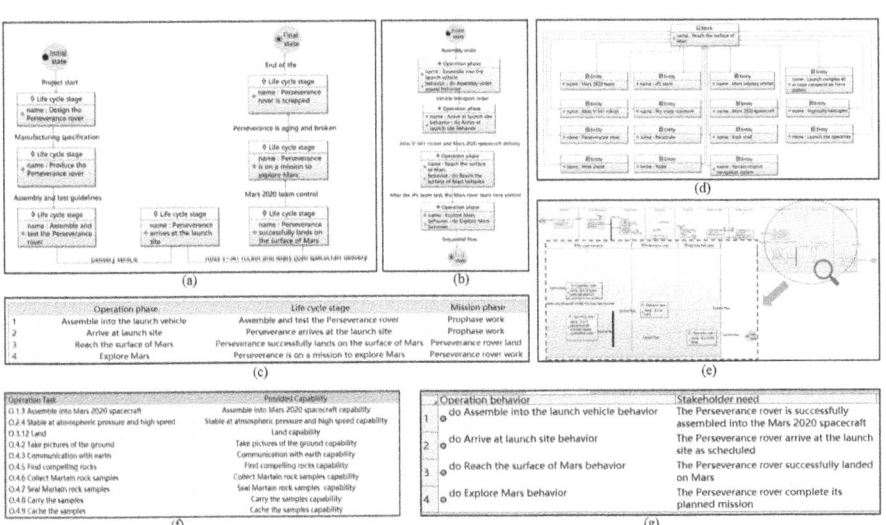

Fig. 5. Perseverance rover models (a) The perseverance lifecycle (b) The perseverance operation concept (c) Operation phase traceability to Lifecycle stage and mission phase (d) Operation phase 3 involved entities (e) Operation phase 3 operation tasks

Step 6 (Define the Phases of the Operation and Trace). This step involves defining the main operational phases of the Perseverance Rover in the context of exploring Mars, including assembling into the launch vehicle, arriving at launch site, reach the surface of Mars, and exploring Mars. The operational phases are traced back to both life cycle stages and mission phases, illustrating the relationship between the Perseverance Rover's operational and mission stages. Each operational phase includes phase names and operational behaviors, with operational behaviors providing a verb-oriented description for inputs into Step 8 (see Fig. 5(b) and Fig. 5(c)).

Step 7. ① (Examine the Entities Involved in Each Operational Stage). Taking operational phase 3 as an example, the entities involved include the Mars 2020 mission team, JPL team, Mars Odyssey Orbiter, parachutes, and the Ingenuity helicopter, among others(see Fig. 5(d)).

Step 7. ② (Assess Operational Tasks for Each Stage Based on Entity Analysis). Based on the entities involved in operational phase 3, operational tasks are analyzed. For instance, at the end of the task sequence, a sky crane performs "O.3.10 Nylon Cords spooled out," followed by the trigger event "When the spacecraft sensed the rover had touched," after which the sky crane performs the next task "O.3.11 Pyrotechnically activated blades severed at cords," and subsequently the Perseverance Rover and Ingenuity perform "O.3.12 Land" and "O.3.13 Fly away" respectively (see Fig. 5(e)).

Step 7. ③ (Identify the System Capabilities Required for Each Operational Task). As shown in Fig. 16, this step involves first identifying the operational tasks related to the Perseverance Rover, then analyzing the system capabilities required for these tasks(see Fig. 5(f)).

Step 8 (Identifying Stakeholder Needs from Operational Behaviors and Tracing Back to Mission Capabilities and Effects). From the operational behaviors identified in Step 6, stakeholder needs are derived, typically describing the specific execution effects of the operational behaviors. These stakeholder needs are then traced back to the mission capabilities and effects identified in Step 3.③, ensuring that the Perseverance Rover's capabilities satisfy the stakeholder needs, and verifying the completion of implementing stakeholder needs(see Fig. 5(g) and Fig. 6).

Step 9 (Creating Operational Task Realizations and Tracing Back to Operational Tasks). Operational Task Realization (OTR) describes the completion effects of operational tasks and also serves as the output of the operational phase, providing inputs to the architectural design of the Mars rover as a basis for fundamental functionalities. This step ensures that each operational task may be associated with one or more OTRs, simultaneously ensuring that OTRs provide accurate information to the functional architecture phase (see Fig. 7).

	Stakeholder need	1	2	3	4
Mission capability effect		The Perseverance rover is successfully assembled into the Mars 2020 spacecraft	The Perseverance rover arrive at the launch site as scheduled	The Perseverance rover successfully landed on Mars	The Perseverance rover complete its planned mission
1	Perseverance was successfully assembled into the Mars 2020 spacecraft	✓			
2	The Perseverance rover remains steady		✓		
3	Successfully land			✓	
4	Take a pair of picture with the engineering imagers known as Hazard Camera				✓
5	Look for rocks that formed in, or were altered by				✓
6	Drill a core sample				✓
7	Break off the core sample from the rock and cap and hermetically seal it inside the tube				✓
8	Place each sealed tube in a storage rack onboard and transport it until the mission team decides to deposit it on the Martian surface				✓
9	Put the Martian samples in the same place on the Martian surface so that a future mission could potentially retrieve and return them all together				✓

Fig. 6. Stakeholder need trace to mission capability effect

	Operation task	1	2	3	4	5	6	7	8	9	10
OTR		O.1.3 Assemble into Mars 2020 spacecraft	O.2.4 Stable at atmospheric pressure and high speed	O.3.12 Land	O.4.2 Take pictures of the ground	O.4.3 Communication with earth	O.4.5 Fing compelling rocks	O.4.6 Collect Martian rock sample	O.4.7 Seal Martian rock samples	O.4.8 Carry the samples	O.4.9 Cache the samples
1	Assembly into Mars 2020 spacecraft successfully	↗									
2	Stable at atmospheric pressure and high speed		↗								
3	Land successfully			↗							
4	Take a pair of picture with the engineering imagers known as Hazard Camera				↗						
5	Communicate with earth successfully					↗					
6	Look for rocks that formed in, or were altered by						↗				
7	Drill a core sample							↗			
8	Break off the core sample from the rock and cap and hermetically seal it inside the tube								↗		
9	Place each sealed tube in a storage rack onboard and transport it until the mission team decides to deposit it on the Martian surface									↗	
10	Put the Martian samples in the same place on the Martian surface so that a future mission could potentially retrieve and return them all together										↗

Fig. 7. OTR trace back to operation task

4.3   Discussion

In the case study of the rover system's mission design, the Perseverance model is built using the mission and operation analysis methods proposed in this paper. Compared to previous methods, the proposed approach offers the following advantages:

1 The modeling method proposed in this paper divides requirement modeling into "mission" and "operation" phases, significantly enhancing the depth of mission analysis and the ability to decompose missions on multiple levels. By integrating auto-generated Sankey diagrams, this paper effectively maps the capability needs from mission phase analysis to the final stakeholder needs, demonstrating how the design of the Perseverance covers and meets all stakeholder needs. This approach not only exemplifies the multi-angle analysis of requirements but also highlights the deep exploration of multi-level need decomposition (see Fig. 8(a)).

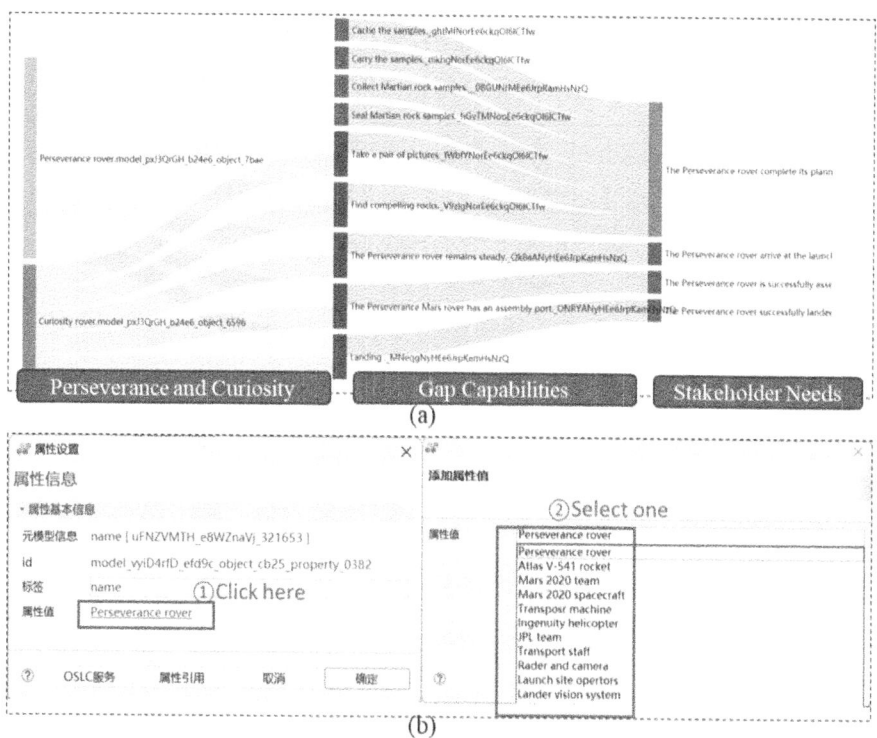

Fig. 8. (a) Stakeholder needs traceability (b)Meta model for entity

2 Facing the high abstraction and generality issues presented by traditional meta-models in Mars exploration projects, this paper adopts innovative steps to reduce these levels of abstraction using KARMA language. KARMA is a language that integrates text and graphics, providing a rigorous mapping between them for enhanced readability and intuitive visualization in meta-model design. Specifically, the meta-model states are detailed into "mission phases," "operational phases," and "life cycle stages," with invocation actions specified as "mission tasks" and "operational tasks." This detailing not only provides a clearer and more operable framework for projects but also makes the mission analysis and design processes more consistent with the actual operational modes of the projects. Moreover, by defining modules as "entities" and pre-recording attributes including names for these entities, such as the Perseverance Rover, Atlas V-541 rocket, and Mars 2020 spacecraft, the abstraction level of the model is further reduced (see Fig. 8(a)).

3 This paper introduces a more flexible modeling framework and dynamic adjustment mechanisms, significantly enhancing the adaptability of the modeling method. Firstly, by dividing the modeling process into multiple stages and defining clear goals and tasks for each stage, it allows for adjustments and optimizations during project implementation based on new information and conditions. Secondly, the use of modular design enables the replacement or updating of modeling components as needed, thereby increasing the model's flexibility and extensibility. Additionally, by specifying the components of the meta-model as entities and pre-recording their attributes, the model becomes more closely aligned with actual application scenarios, further enhancing the adaptability of the modeling method. These innovative measures address the deficiencies in flexibility and adaptability present in existing modeling methods and provide a modeling approach capable of effectively handling unknown environmental variables and technological advancements in Mars exploration projects.

5 Conclusion

In this paper, we propose a new mission modeling method tailored to the Mars rover system engineering field, utilizing a multi-level modeling approach based on the KARMA language. This method includes modeling both the mission and operation phases and involves the development of a domain-specific meta-model library customized for the Mars rover sector. The feasibility and logical soundness of the proposed method and meta-model are validated using the case study of the Perseverance Rover. This method not only ensures that our modeling framework is directly applicable to the unique missions of Mars rover projects but also demonstrates its effectiveness in handling the complexities associated with such ambitious engineering endeavors.

Acknowledgement. This study was funded by JCKY2021203A001 and D020101.

References

1. Delligatti, L.: SysML Distilled: A Brief Guide to the Systems Modeling Language. Pearson Education, London (2013)

2. Gregory, J.R.: A Model-Based Framework for Early-Stage Analysis of Spacecraft. Ph.D. thesis, University of Bristol (2022)
3. Di Maio, M., Weilkiens, T., Hussein, O., et al.: Evaluating MBSE methodologies using the FEMMP framework. In: 2021 IEEE International Symposium on Systems Engineering (ISSE), pp. 1–8. IEEE, Vienna (2021)
4. Booth, S.T.: Network centric architectures: are we up to the task? In: INCOSE International Symposium, vol. 15, no. 1, pp. 600–609 (2005)
5. Omoarebun, E.N., Cimtalay, S., Mavris, D.N.: Formalizing the decomposition process between elements in the RFLP framework using axiomatic design theory. In: AIAA AVIATION 2023 Forum, p. 3771. AIAA (2023)
6. Hause, M.: The unified profile for DoDAF/MODAF (UPDM) enabling systems of systems on many levels. In: 2010 IEEE International Systems Conference, pp. 426–431. IEEE (2010)
7. Kapos, G.-D., Dalakas, V., Tsadimas, A., et al.: Model-based System Engineering Using SysML: deriving executable simulation models with QVT. In: 2014 IEEE International Systems Conference Proceedings, pp. 531–538. IEEE (2014)
8. Bouba, K., Ait Wakrime, A., Ouhammou, Y., Benaini, R.: A transformation methodology for Capella to Event-B models with DSL verification. J. Comput. Lang. **77**, 101241 (2023)
9. Hill, T.: Small Steps, Giant Leaps: Episode 56, Model-based Systems Engineering. https://www.nasa.gov/podcasts/small-steps-giant-leaps/small-steps-giant-leaps-episode-56-model-based-systems-engineering/. Accessed 24 Mar 2021
10. Gao, J.Y., Wang, L.Y., Pan, Z.S., et al.: MBSE architecture modeling of mars maintenance and management devices. Syst. Eng. Electron. **45**(5), 1441–1450 (2023)
11. Kelly, S., Tolvanen, J.-P.: Domain-Specific Modeling: Enabling Full Code Generation. John Wiley and Sons, Hoboken (2008)
12. Ding, J., Reniers, M., Lu, J., et al.: Integration of modeling and verification for system model based on KARMA language. In: Proceedings of the 18th ACM SIGPLAN International Workshop on Domain-Specific Modeling, pp. 41–50. ACM (2021)
13. Lu, J., Wang, G., Ma, J., et al.: General modeling language to support model? Based systems engineering formalisms (Part 1). In: INCOSE International Symposium, vol. 30, no. 1, pp. 323–338 (2020)
14. Von Ehrenfried, M.: Dutch: Perseverance and the Mars 2020 Mission: Follow the Science to Jezero Crater. Springer International Publishing, Cham (2022). https://doi.org/10.1007/978-3-030-92118-7

Social Media Oriented Fake News Detection Based on Social Context and Cascade Graph

Zhihua Yan[1][(✉)] ⓘ, Xijin Tang[2,3] ⓘ, Zhenpeng Li[4], and Xuxian Yan[1]

[1] School of Management Science and Engineering, Shanxi University of Finance and Economics, Taiyuan 030006, China
zhyan@amss.ac.cn
[2] Academy of Mathematics and Systems Science, Chinese Academy of Sciences, Beijing, China
xjtang@iss.ac.cn
[3] University of Chinese Academy of Sciences, Beijing 100039, China
[4] School of Electronics and Information Engineering, Taizhou University, Taizhou 318000, Zhejiang, China
lizhenpeng@amss.ac.cn

Abstract. Fake news detection has received great attention in recent years due to the proliferation of disinformation. Existing researches focus on modeling the news content and user comments on social media. However, these researches ignore the cascades of news propagation. Accordingly, we propose a Social Content and Cascade Neural Network (SCCN) that integrates cascade information into the model. Specifically, we adopt Bi-LSTM, graph attention mechanism and average pooling to represent the social content as a vector, encoding the cascade structures by graph neural network. We compare our model with representative baselines to validate the performance and perform an ablation analysis to verify the effectiveness of social context and cascade structures. The results show that SCCN model performs better and yields improvement over the baselines according to precision, recall, and F1-score based on two real-world datasets.

Keywords: Fake news detection · Social media · Social context · Cascade graph

1 Introduction

Due to the widely used of the social media, the ways by which people access various types of news have become increasingly convenient. Social media have facilitated the public to access and share recent news, and increase the speed and impact of news dissemination. However, false news disseminated across social media have posed significant challenges in providing readers with accurate and valuable information, leaded to anxiety and panic among individuals, as well as social unrest. Hence, fake news detection helps to bolster the credibility of social media networks, preserve a stable cyberspace and safeguard societal stability.

Heavy human efforts have been invested in fake news detection. Fact-checking sites, such as Snopes[1], annielab[2] and factcheck[3], determine the authenticity of controversial news manually. However, it is difficult to deal with massive amounts of fake news generated by individuals and organizations every day. As the rapid development of machine learning techniques in recent years, machine learning-based fact-checking has gradually replaced manual fact-checking and receives wide attention by researchers [1]. Existing researches on fake news detection mainly focus on rumor detection [2] and social spam detection [3].

Regarding the deceitful content of news, a large number of researchers try to leverage text content of news to mitigate the detrimental consequences. These content-based language methods typically focus on the textual features associated with the news articles, including the linguistic features, syntactic features, writing styles, and emotional signals [4,5]. When dealing with well-crafted fake news that closely mimics real news in terms of textual features, these methods suffer from the low discriminativeness between the fake news and genuine information, leading to compromised performance. However, these studies do not take the propagation characteristics of fake news into considerations. Cascade denotes the paths of news propagation, and helps to characterize the dynamics of fake news [6].

In this paper, a framework named Social Content and Cascade Neural Network (SCCN) is proposed to detect fake news. The primary goal of the SCCN model is to improve the performance and realize automation of fake news detection by combining news text data and propagation structures: representing words, pos-tags and word positions as vectors; using Bi-LSTM, graph attention mechanism and average pooling to represent the social content as a vector; encoding the cascade by graph neural network to gain dynamic features of news; classifying news articles using a multi-layer perception network. The main contributions of this research are as follows:

(1) We introduce the cascade structures to reveal the propogation features of news, and represent the cascade graph by GNN to capture the dynamics of fake news.
(2) We propose SCCN model which effectively leverages the social content and the cascade structures of news to improve the accuracy of model, realize automatic fake news detection.

2   Related Works

2.1   Graph Neural Networks

Graph neural network (GNN) represents the data as graphs to extract and uncover features and patterns from neighborhood propagation and aggregation.

---

[1] https://www.snopes.com/.
[2] https://annielab.org/.
[3] https://factcheck.afp.com/.

Generally, the methods of GNN parameter updating are categorized into spatial approaches [7] and spectral approaches [8]. GNN gains a better representation of dataset by fusing node features and the relations between the nodes together. GNN achieves better performance in the areas of recommender system [9], entity extraction, event extraction, text classification [10] and sentiment analysis [11].

Graph structure learning (GSL) helps to alleviate the effects of noisy data by node embeddings and graph structure [12]. Existing GSL contains the metric-learning-based methods, the probabilistic methods and the direct-optimized methods. The metric-learning-based methods combine kernel function [13] with node embeddings to build the adjacency matrix and optimize the graph topology. In the probabilistic methods, a specific probabilistic distribution, such as gaussian distribution, is used to construct the adjacency matrix [14]. The direct-optimized methods learn parameters of GNN based on task-specific parameters without considering node embeddings and distributions [15]. Graph pooling methods that can be considered as GSL algorithms capture the graph structural information by merging or dropping nodes [16].

2.2 Fake News Detection Based on Content

False news refers to news whose content is false, intentionally misleading or fabricated information [17]. Those news always fail to report the fact fully, correctly and appropriately, resulting in deception. Fake news detection is an important NLP task, and deep learning based mothods have attracted a large number of researchers because of the good performance.

Content-based methods capture the text features of news, such as sentiment, contextual semantic, and Part of Speech taging, to determine the authenticity of news. These methods mine context features of fake news from news articles by NLP techniques, such as syntactic, sentiment, linguistic and so on. Kim utilized convolutional neural networks to gain local linguistic features from news articles [18]. Similarly, differences in wording and writing style have been widely used in recent researches [19]. Kaliyar et al. conducted a news content analysis based on BERT and three parallel CNN blocks [20]. Yang et al. improved performance by a dual-attention model which extracts hierarchical features of news articles [21]. Similarly, a great number of researches devote to how to integrate auxiliary textual information associated with news articles, such as comments [22], stance, attitude, and sentiment [23], to enhance the performance of models.

In content-based methods, textual features of each news article are extracted and utilized to verify authenticity of news. However, an increasing number of AI tools can assist in fabricating fake news by mimicking the semantic and linguistic style of real news, and their detection performance is not satisfactory. Such defective results typically come from a failure to consider the features of fake news your spread.

2.3 Fake News Detection Based on Graph

In contrast to content-based fake news detection, approaches based on underlying structures, such as word relations, news propagation and social structures, are referred to as graph-based fake news detection. Yao et al. used GCN to constructe a weighted graph using the word-word relations contained within the news article [24]. Similarly, the relations of the interior and exterior of news article are represented by heterogeneous graph attention networks [25]. However, these approaches only consider the syntactic structure and ignore the relations between news articles. Besides, Ma et al. [26] employed RNN and Bi-GCN to capture news propagation features on social media. Other researches also model the relations between news and users, or even news and external knowledge graph [27].

3 Method

As illustrated in Fig. 1, SCCN model contains four distinct modules: word embedding, graph-based content encoder, graph-based cascade encoder and fake news classifier. In contrast to the traditional approach that utilizes textual information for fake news detection, SCCN model adopts social content and cascade graph to improve performance.

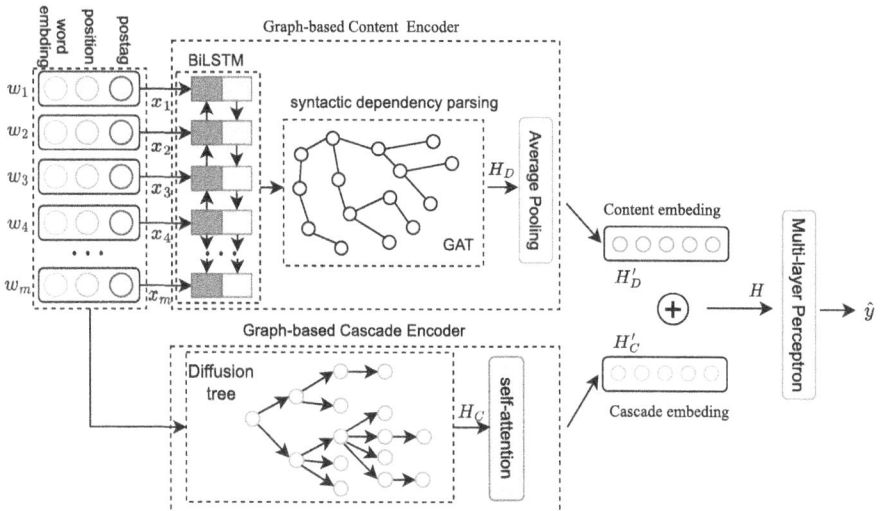

Fig. 1. SCCN model consists of four modules: (a) Representing words, pos-tags and word positions as vectors; (b) Using Bi-LSTM, graph attention mechanism and average pooling to represent the social content as a vector; (c) Encoding the cascade by graph neural network and self-attention; (d) Classifying news articles using a multi-layer perception layer.

3.1   Problem Formulation

Fake news detection can be implemented by either manual methods or machine learning methods. In a general way, it can be treated as a text classification problem. Given news from social media, social content refers to the content of news, and the behaviors of replying and forwarding constitute cascade graph. We use D for social content and C for cascade structure, respectively. $\widehat{y} = f(D, C, \theta)$ is the predicted value, f is the predicted model, θ is the training parameter and $\widehat{y} \in \{0, 1\}$.

3.2   Word Embedding

As semantic context, part of speech of words and position of words may help to reveal the authenticity of news, we conduct a presentation learning based news article. First, the LTP tool[4] is used to perform Part of Speech tagging. Each news document contains m words, w_i denotes the ith word and is represented as a vector using word2vec. Considering the influence of pos-tag and word position on the truthiness of news, the pos-tag and word position are represented as random vectors, respectively. The word w_i is denoted as vector x_i by concatenating word vector, pos-tag vector and word position vector together and the document is represented as $D = [x_1, x_2, \cdots, x_m]$.

3.3   Graph-Based Content Encoder

The content of news contains semantic and structural information that are the key features of fake news. Bi-LSTM is employed to capture the context semantic information across the news article. Given ducument $D = [x_1, x_2, \cdots, x_m]$, $\overrightarrow{p}_t$ and $\overleftarrow{p}_t$ are the word vectors trained by Bi-LSTM:

$$\overrightarrow{p}_t = \overrightarrow{LSTM}(\overrightarrow{p}_{t-1}, x_t), \tag{1}$$

$$\overleftarrow{p}_t = \overleftarrow{LSTM}(\overleftarrow{p}_{t-1}, x_t) \tag{2}$$

Then x_t is represented as $h_t = [\overrightarrow{p}_t, \overleftarrow{p}_t]$, and $[\ ,\ ]$ denotes concatenation of vectors.

As fake news usually has special structural features, we use dependent syntactic analysis to identify hierarchical syntactic relations between phrases in each sentence, which provides important features for fake news detection. CoreNLP tool is utilized to perform dependency syntactic analysis on sentences to obtain dependency syntactic trees. In this paper, undirected graph $G = (V, E)$ denotes dependency syntactic tree, where V denotes nodes, E denotes edges, and $(v_i, v_j) \in E$ denotes the relation between word w_i and word w_j. Then the root nodes of the syntactic dependency tree in each news article are connected together. In the end, the news article is represented as an undirected graph.

---

[4] https://www.ltp-cloud.com/.

To alleviate the effect of noisy information, we obtain a weighted sum of neighboring nodes by graph attention mechanism. In the lth layer graph attention network, the similarity coefficient between word w_i and neighbor word w_j is $e_{ij}^{(l)} = f([W^{(l)}h_i^{(l)}, W^{(l)}h_j^{(l)}])$, then the graph attention coefficient of w_i is:

$$\alpha_{ij}^{(l)} = \frac{exp\left(LeakyReLU(e_{ij}^{(l)})\right)}{\sum_{k \in N_i} exp\left(LeakyReLU(e_{ij}^{(l)})\right)} \tag{3}$$

The word w_j is computed as follows:

$$h_j^{(l+1)} = \sigma\left(\sum_{k \in N_i} \alpha_{ij}^{(l)} W^l h_j^{(l)}\right) \tag{4}$$

where N_i denotes the neighbor nodes of the word w_i. Eventually, the content features of document D are represented as matrix H_D. Then, We utilize average pooling on the content matrix H_D to gain the feature vector $H'_D = mean(h_1^{(l+1)}, h_2^{(l+1)}, \cdots, h_m^{(l+1)})$.

3.4   Graph-Based Cascade Encoder

The cascade structures of news reveal the behaviors of posters and the paths of news spreading, and help to detect fake news. Generally speaking, the cascade structure contains cascade sequence and cascade graph. The cascade graph reflects the news diffusion structure and interactions between posters. We use multilayer GCN to represent the cascade graph which is aggregated from neighboring nodes:

$$H^{(l+1)} = \sigma\left(\tilde{D}^{-\frac{1}{2}}\tilde{A}\tilde{D}^{-\frac{1}{2}}H^{(l)}W^{(l)}\right) \tag{5}$$

Here, $H^{(l)} \in R^{n \times d^{(l)}}$ denotes the lth layer, $W^{(l)} \in R^{d^{(l)} \times d^{(l+1)}}$ is the corresponding weight matrix. D denotes the degree of the diagonal nodes of $\tilde{A}$, $\tilde{A} = A + I$. Besides, $\sigma(\cdot)$ denotes the activation function, such as sigmoid function and RELU function.

To improve the effectiveness of representation learning of cascade graph, this paper introduces the attention mechanism. Given a cascade graph $G_i = (V_i, E_i)$, A_i denotes the adjacency matrix of G_i. $\overline{X}$ is the input, Q_i, K_i and V_i are the queries, keys and values vectors generated by attention mechanism. $Q_i = \overline{X}W_q$, $K_i = \overline{X}W_k$, $V_i = \overline{X}W_v$, where W_q, W_k and W_v are the corresponding weight matrixes. Based on the attention mechanism, cascade graph can be computed as follow:

$$H'_C = Attention(Q_i, K_i, V_i) = softmax\left(\frac{Q_i K_i^T}{\sqrt{d_k}}\right)V_i \tag{6}$$

3.5 Fake News Classifier

After obtaining the content embeding $H_D^{'}$ and cascade embeding $H_C^{'}$, we concatenate them together and feed into a 2-layer perceptron to predict the authenticity of news. The result of prediction $\widehat{y}$ is calculated as follows:

$$\widehat{y} = softmax(\text{MLPs}([H_D^{'}, H_C^{'}])) \tag{7}$$

The standard cross-entropy loss function is calculted as follow:

$$L = -(y\log\widehat{y} + (1 - y)\log(1 - \widehat{y})) \tag{8}$$

where $y, \widehat{y} \in \{0, 1\}$ are the true and predicted values, respectively.

4 Experiments

4.1 Datasets

Social media have become the major channel of fake news production and dissemination due to its fast-spreading and huge influence. Weibo is widely used in China, and twitter is used across the world. Hence, a large number of scholars carry out fake news researches based on Weibo and Twitter. In this paper, we choose Weibo dataset [28], Twitter15 [29] and Twitter16 dataset [30] to evaluate the performance of SCCN model. The Weibo dataset contains news articles and user attributes, while Twitter15 and Twitter16 dataset only have news articles and propagation trees. Since the volume of twitter dataset is small, the twitter15 and twitter16 datasets are merged together to construct the new bigger Twitter dataset.

As shown in Table 1, we use the number of real news (#Real News), the number of fake news (#Fake News) and the average number of tweets of each news (Avg # of tweets) to describe the datasets. Each dataset contains social contents and cascade structures. We carry out preprocessing to remove noise information and duplicate posters, and utilize tweets as nodes and replies or retweets as edges to form a cascade graph. In the end, the Weibo dataset contains 2,351 real news, 2,312 fake news and the average tweets of each news is 378, while the Twitter dataset contains 579 real news, 578 fake news and the average tweets of each news is 30. Compared with Twitter, Weibo has more active users, and the scale of replies and retweets of fake news on Weibo is greater. Hence, the average number of tweets of Weibo is larger than Twitter.

4.2 Experimental Settings

We choose Python 3.8 and Pytorch to build deep learning model. We choose Adam optimizer to train our model. Pre-trained Word2vec is use for word embedding and the dimensions is 100. The learning rate is 1e-3, the training epochs is 200, and the batch size is 64. 5-fold cross validation is used to gain experimental results with high robustness. Besides, we choose the precision, recall and F1-score as the evaluation metrics.

Table 1. The Statistics of Weibo dataset and Twitter dataset.

Platform	Weibo dataset	Twitter dataset
#Real News	2351	579
#Fake News	2312	578
#Total	4338	1147
Avg # of tweets	378	30

4.3 BaseLines

We choose typical fake news detection algorithms proposed in recent years, and compare them with SCCN model:

(1) BERT [31]. The context and user attribute of news article are represented as features based on BERT model to judge the authenticity of news.
(2) PPC [32]. News propagation paths which are represented as a series of user behaviors through RNN and CNN models are used for fake news detection at an early stage.
(3) FNED [33]. Multi-region mean-pooling mechanism is used to aggregate text features, while user features and position-aware attention mechanism are employed to capture important user responses to obtain better performance of fake news detection.
(4) Bi-GCN [34]. A GCN with a top-down and an opposite directed graph is used to gain the features of news propagation.
(5) HGNNR4FD [35]. News content and relationship embeddings is generated through knowledge graph to provide a factual basis, and the knowledge graph and the heterogeneous graph is aggregated by Transformer.
(6) HeteroSGT [36]. Word-level and sentence-level semantics are captured by BERT model, and subgraphs is extracted by Restart Random Walk. Then, semantic featrues and subgraphs are fed into the Transformer to determine the authenticity of news.

4.4 Performance Comparison

The comparing of baselines model and SCCN model on Weibo dataset and Twitter dataset are shown in Table 3. Compared with baselines, SCCN model performs better. In Weibo dataset, the precision, recall and F1-score of SCCN model are 0.912, 0.921 and 0.916, respectively. In Twitter dataset, the precision, recall and F1-score of SCCN model are 0.891, 0.895 and 0.893, respectively. Baseline models only consider some typical features of the news. For example, the BERT model only considers the textual features of the news, while the PPC model use the behavioral features of the posters. In FNED model, the textual features of the news and the replying behaviors of the posters are merged together. HGNNR4FD improves the performance by aggregating the knowledge and the heterogeneous graph. HeteroSGT conbines semantic featrues and subgraphs by

Transformer to determine the authenticity. In contrast, SCCN model combines the social content and cascade graph of news, leveraging both of the content features and cascade features of fake news for better performance (Table 2).

Table 2. Comparison of baselines model and SCCN model on Weibo dataset and Twitter dataset.

Method	Weibo dataset			Twitter dataset		
	Precision	Recall	F1-score	Precision	Recall	F1-score
BERT	0.833	0.825	0.829	0.785	0.801	0.793
PPC	0.812	0.807	0.809	0.748	0.755	0.751
FNED	0.859	0.863	0.861	0.773	0.765	0.769
Bi-GCN	0.871	0.852	0.861	0.852	0.859	0.855
HGNNR4FD	0.891	0.887	0.889	0.878	0.873	0.875
HeteroSGT	0.903	0.912	0.907	0.883	0.885	0.884
SCCN (our model)	0.912	0.921	0.916	0.891	0.895	0.893

4.5   Ablation Analysis

In this paper, we carry out an ablation study to assess the contribution of the main modules of the SCCN model, as shown in Table 4. It is evident that the performance is less than satisfactory when we implement SCCN model without utilizing the social content and cascade graph. In Weibo dataset, the performance of SCCN model can be improved 0.085 by social content and 0.047 by cascade graph module according to F1-score.

Table 3. Ablation analysis of SCCN model on Weibo and Twitter Datasets.

Method	Weibo dataset			Twitter dataset		
	Precision	Recall	F1-score	Precision	Recall	F1-score
w/o social content	0.825	0.838	0.831	0.812	0.803	0.807
w/o cascade	0.862	0.877	0.869	0.855	0.874	0.864
SCCN (our model)	0.912	0.921	0.916	0.891	0.895	0.893

5   Conclusions

The spreading of fake news creates public opinion confusion and disagreement, destroys public trust of the mass media, enterprise and government, triggers social panic and disrupts the order of society. With the development of generative artificial intelligence, fake news become easier to be generated and more difficult to be detected. Nowadays, fake news regulation has become a huge challenge for the government. Fake news can be identified by content and dissemination features. In this paper, a novel SCCN model combining social content and cascade information is proposed. The main goal of this model is to determine the authenticity of news by combining news context and propagation structures: represent words, pos-tags and word positions as vectors; use Bi-LSTM, graph attention mechanism and average pooling to represent the social content as a vector; encode the cascade based on GNN and self-attention; classify news articles using a multi-layer perception layer. Compared with baseline models, SCCN model gain better performances and yields improvement over the baselines according to precision, recall, and F1-score based on Weibo dataset and Twitter dataset.

Acknowledgement. This paper is supported by the National Social Science Fund of China (No. 23 & ZD331).

References

1. LeCun, Y., Bengio, Y., Hinton, G.: Deep learning. Nature **521**(7553), 436–444 (2015)
2. Rani, N., Das, P., Bhardwaj, A.K.: Rumor, misinformation among web: a contemporary review of rumor detection techniques during different web waves. Concurr. Comput. Pract. Exp. **34**(1), e6479 (2022)
3. Rao, S., Verma, A.K., Bhatia, T.: A review on social spam detection: challenges, open issues, and future directions. Expert Syst. Appl. **186**, 115742 (2021)
4. Horne, B., Adali, S.: This just in: fake news packs a lot in title, uses simpler, repetitive content in text body, more similar to satire than real news. In: Proceedings of the 11th International AAAI Conference on Web and Social Media, pp. 759–766 (2017)
5. Pelrine, K., Danovitch, J., Rabbany, R.: The surprising performance of simple baselines for misinformation detection. In: Proceedings of the Web Conference 2021, pp. 3432–3441 (2021)
6. Vosoughi, S., Roy, D., Aral, S.: The spread of true and false news online. Science **359**(6380), 1146–1151 (2018)
7. Hamilton, W.L., Ying, R., Leskovec, J.: Inductive representation learning on large graphs. In: Proceedings of the 31st International Conference on Neural Information Processing Systems, pp. 1025–1035 (2017)
8. Defferrard, M., Bresson, X., Vandergheynst, P.: Convolutional neural networks on graphs with fast localized spectral filtering. In: Proceedings of the 30th International Conference on Neural Information Processing Systems, pp. 3844–3852 (2016)
9. Chen, T., Wong, R.C.W.: Handling information loss of graph neural networks for session-based recommendation. In: Proceedings of the 26th ACM SIGKDD International Conference on Knowledge Discovery & Data Mining, pp. 1172–1180 (2020)
10. Yao, L., Mao, C., Luo, Y.: Graph convolutional networks for text classification. In: Proceedings of the 33th AAAI Conference on Artificial Intelligence, pp. 7370–7377 (2019)

11. Li, R., Chen, H., Feng, F., Ma, Z., Wang, X., et al.: Dual graph convolutional networks for aspect-based sentiment analysis. In: Proceedings of the 59th Annual Meeting of the Association for Computational Linguistics and the 11th International Joint Conference on Natural Language Processing, pp. 6319–6329 (2021)
12. Jin, W., Ma, Y., Liu, X., Tang, X., Wang, S., et al.: Graph structure learning for robust graph neural networks. In: Proceedings of the 26th ACM SIGKDD International Conference on Knowledge Discovery & Data Mining, pp. 66–74 (2020)
13. Wu, X., Zhao, L., Akoglu, L.: A quest for structure: Jointly learning the graph structure and semi-supervised classification. In: Proceedings of the 27th ACM International Conference on Information and Knowledge Management, pp. 87–96 (2018)
14. Franceschi, L., Frasconi, P., Salzo, S., Grazzi, R., Pontil, M.: Bilevel programming for hyperparameter optimization and meta-learning. In: The 35th International Conference on Machine Learning, pp. 1568–1577 (2018)
15. Yang, L., Kang, Z., Cao, X., Di, J., Yang, B., et al.: Topology optimization based graph convolutional network. In: Proceedings of the 28th International Joint Conference on Artificial Intelligence, pp. 4054–4061 (2019)
16. Lee, J., Lee, I., Kang, J.: Self-attention graph pooling. In: Proceedings of the 36th International Conference on Machine Learning, p. 3734 (2019)
17. Shu, K., Sliva, A., Wang, S., Tang, J., Liu, H.: Fake news detection on social media: a data mining perspective. ACM SIGKDD Explor. Newsl. 19(1), 22–36 (2017)
18. Kim, S., Kang, I., Kwak, N.: Semantic sentence matching with densely-connected recurrent and co-attentive information. In: Proceedings of the 33th AAAI Conference on Artificial Intelligence, vol. 33, no. 1, pp. 6586–6593 (2019)
19. Horne, B., Adali, S.: This just in: fake news packs a lot in title, uses simpler, repetitive content in text body, more similar to satire than real news. In: Proceedings of the 31th International AAAI Conference on Web and Social Media, vol. 11, no. 1, pp. 759–766 (2017)
20. Kaliyar, R.K., Goswami, A., Narang, P.: FakeBERT: fake news detection in social media with a BERT-based deep learning approach. Multimed. Tools Appl. 80(8), 11765–11788 (2021)
21. Yang, Z., Yang, D., Dyer, C., He, X., Smola, A., et al.: Hierarchical attention networks for document classification. In: Proceedings of the 2016 Conference of the North American Chapter of the Association for Computational Linguistics: Human Language Technologies, pp. 1480–1489 (2016)
22. Rao, D., Miao, X., Jiang, Z., Li, R.: STANKER: stacking network based on level-grained attention-masked BERT for rumor detection on social media. In: Proceedings of the 2021 Conference on Empirical Methods in Natural Language Processing, pp. 3347–3363 (2021)
23. Zhang, X., Cao, J., Li, X., Sheng, Q., Zhong, L., et al.: Mining dual emotion for fake news detection. In: Proceedings of the Web Conference 2021, pp. 3465–3476 (2021)
24. Yao, L., Mao, C., Luo, Y.: Graph convolutional networks for text classification. In: Proceedings of the 33th AAAI Conference on Artificial Intelligence, vol. 33, no. 1, pp. 7370–7377 (2019)
25. Linmei, H., Yang, T., Shi, C., Ji, H., Li, X.: Heterogeneous graph attention networks for semi-supervised short text classification. In: Proceedings of the 2019 Conference on Empirical Methods in Natural Language Processing, pp. 4821–4830 (2019)

26. Ma, J., Gao, W., Wong, K.F.: Rumor detection on Twitter with tree-structured recursive neural networks. In: Proceedings of the 56th Annual Meeting of the Association for Computational Linguistics, pp. 1980–1989 (2018)

27. Dun, Y., Tu, K., Chen, C., Hou, C., Yuan, X.: KAN: knowledge-aware attention network for fake news detection. In: Proceedings of the 35th AAAI Conference on Artificial Intelligence, vol. 35, no. 1, pp. 81–89 (2021)

28. Ma, J., Gao, W., Mitra, P., Kwon, S., Jansen, B.J., et al.: Detecting rumors from microblogs with recurrent neural networks. In: Proceedings of the Twenty-Fifth International Joint Conference on Artificial Intelligence, pp. 3818–3824 (2016)

29. Liu, X., Nourbakhsh, A., Li, Q., Fang, R., Shah, S.: Real-time rumor debunking on Twitter. In: Proceedings of the 24th ACM International on Conference on Information and Knowledge Management, pp. 1867–1870 (2015)

30. Ma, J., Gao, W., Wong, K. F.: Detect rumors in microblog posts using propagation structure via kernel learning. In: Proceedings of the 55th Annual Meeting of the Association for Computational Linguistics, pp. 708–717 (2017)

31. Devlin, J.: 2018. Bert: pre-training of deep bidirectional transformers for language understanding. 2018, arXiv preprint arXiv:1810.04805

32. Liu, Y., Wu, Y. F.: Early detection of fake news on social media through propagation path classification with recurrent and convolutional networks. In: Proceedings of the 32th AAAI Conference on Artificial Intelligence, vol. 32, pp. 354–361 (2019)

33. Liu, Y., Wu, Y.F.B.: FNED: a deep network for fake news early detection on social media. ACM Trans. Inf. Syst. **38**(3), 1–33 (2020)

34. Bian, T., Xiao, X., Xu, T., Zhao, P., Huang, W.: Rumor detection on social media with bi-directional graph convolutional networks. In: Proceedings of the AAAI Conference on Artificial Intelligence, vol. 34, no. 1, pp. 549–556 (2020)

35. Xie, B., Ma, X., Wu, J., Yang, J., Xue, S., et al.: Heterogeneous graph neural network via knowledge relations for fake news detection. In: Proceedings of the 35th International Conference on Scientific and Statistical Database Management, pp. 1–11 (2023)

36. Zhang, Y., Ma, X., Wu, J., Yang, J., Fan, H.: Heterogeneous subgraph transformer for fake news detection. In: Proceedings of the ACM on Web Conference 2024, pp. 1272–1282 (2024)

Data Augmentation Using Large Language Model for Fake Review Identification

Qingxu Li[1], Jindong Chen[1,2(✉)], and Wen Zhang[3]

[1] Beijing Information Science and Technology University, Beijing 100192, China
j.chen@bistu.edu.cn
[2] Intelligent Decision Making and Big Data Application Beijing International Science and Technology Cooperation Base, Beijing, China
[3] Beijing University of Technology, Beijing, China
zhangwen@bjut.edu.cn

Abstract. With the development of e-commerce, fake review identification is crucial for both platform users and merchants. Most datasets of fake reviews are imbalance, and the number of fake reviews is far less than that of real reviews. Due to the excellent performance of Large Language Modeling (LLM) in text generation, this paper proposes a fake review recognition method based on data augmentation by LLM. Firstly, several typical LLMs are selected to generate fake reviews, and the quality of the generated text is evaluated by the computational novelty and diversity. Subsequently, based on the data generated by different LLMs, and the performances of the different classification methods are compared. Finally, all the reviews generated by different LLMs are added to the original dataset, and the classification performances are compared with the original dataset. The experimental results of fake review identification show that the augmented dataset has about 10% improvement in classification accuracy compared to the original dataset only. It can be concluded that data augmentation through LLM is indeed significant for fake review identification.

Keywords: false review identification · LLM · text generation · data augmentation

1 Introduction

Online review systems are an integral part of modern e-commerce, helping consumers make purchasing decisions and providing feedback to merchants. However, the reliability of online review systems is often threatened by the presence of fake reviews, especially when attacked by trolls or dishonest marketing practices [1]. With the rapid growth of online reviews and comments on various platforms, fake review identification has become an important issue. The number of fake reviews in existing fake review datasets is often much less than that of real reviews, and there is a serious data imbalance problem. Classification models trained on imbalanced datasets may not learn features of the small classes well, and tend to predict categories that occur more frequently. This causes the classification model to perform poorly on minority categories and affects the model's generalization ability.

© The Author(s), under exclusive license to Springer Nature Singapore Pte Ltd. 2025
X. Tang et al. (Eds.): KSS 2024, CCIS 2269, pp. 225–238, 2025.
https://doi.org/10.1007/978-981-96-0178-3_16

At present, the most commonly used data augmentation methods in natural language processing (NLP) include synonym replacement, random insertion, random deletion, context generation, and back translation [2]. These methods improve the generalization ability of the classification model by increasing the diversity of the training data without changing the original meaning of the sentence. In the field of deep learning, data augmentation methods mainly include variational autoencoders (VAEs) [3] and generative adversarial networks (GANs) [4]. VAE is able to generate diverse and realistic data by encoding input data into latent space and generating new samples from it [5]. GAN uses the generator to generate fake data and the discriminator to distinguish the real data through adversarial training, so as to generate high-quality new samples [6]. These methods are widely used in various tasks such as image and text generation.

With the improvement of computility, large language models (LLMs) based on Transformer architecture have made breakthroughs in the field of text generation. LLMs typically have billions to hundreds of billions of parameters and are capable of generating high-quality and diverse texts. By learning massive amounts of text data, LLMs are not only able to understand and generate text that is very close to natural human language, but also drive major breakthroughs in a number of application areas such as machine translation, content creation, and fake review detection [7].

Therefore, we generate fake review based on LLMs to improve the performance of fake review identification. The limited data of fake reviews in the original dataset are expanded by the LLMs. Prompts are constructed using the fake review in the original dataset and passed to the large model for text generation. The quality of the generated text is evaluated using diversity and novelty metrics. The data generated by the LLMs can be added to the original dataset for data augmentation. Experiments are conducted on multiple classification models to demonstrate that the augmented dataset improves the classification performance.

2 Relevant Studies

In recent years, the development of deep learning models in the field of Natural Language Processing (NLP) has greatly contributed to the advancement of text generation technology. Impressive achievements have been made in terms of text generation. Such as Recurrent Neural Networks (RNN), Generative Adversarial Networks (GAN), variational autoencoder(VAE).

Recurrent neural network (RNN)is a kind of neural network suitable for processing sequence data, which can effectively capture the temporal information in the sequence. Hochreiter proposed Long Short Term Memory (LSTM) [8] in 1997 to solve the problem of gradient vanishing and gradient explosion in the traditional RNN model. Graves [9] et al. proposed the Bidirectional Long and Short-Term Memory Networks (BiLSTM) model which is able to capture richer contextual information in the input sequences by combining two LSTM networks, the forward and the backward. Bahdanau [10] et al. introduced the attention mechanism into the LSTM structure, which significantly improved the performance and interpret ability of the model.

Generative Adversarial Networks (GAN) can also be applied to the field of text generation. Several GAN variants based on GAN sequences have emerged, such as Seq-GAN [11] or LeakGAN [12] for text generation. TextGAN [13] has also been proposed to solve the problem of discrete outputs in GAN models by using LSTM as the generator of the GAN and CNN as the discriminator of the GAN.GANs require a large amount of text data for training, and if the training data is not sufficiently rich or diversified, the generated text may be subject to the sparsity of the data impact.

The variational autoencoder (VAE) is a generative model that can be used for text generation tasks. Models such as Skip-VAE [14] model Seq2Seq VAE [15] have been applied to text generation and conversion, especially for tasks such as translation, dialog generation, and so on. However, since the VAE generation process is based on random sampling of latent spaces, the generated text content may appear blurry or incoherent, which may lead to a lack of clarity and semantic coherence in the generated text.

In recent years, LLMs have made significant progress in the field of text generation. Ming [16] et al. used the GPT2 model in their study to augment the dataset for false comment identification with data through data generation, Junjie [17] et al. proposed a new enhancement technique LLM-DA based on large language modeling, With the continuous progress and development of large models, the GPT family has also evolved with the emergence of GPT3, such as Suhaeni [18] et al. used GPT3 to achieve good results for text generation on sentiment class imbalanced dataset. Zhang et al. fine-tuned chatglm to generate advertising text. Zhipu AI also released the latest GLM4 model, which has a 60% performance improvement and excels in understanding evaluation, mathematics and programming capabilities. Fake Review Identification is a challenging task, and LLM have demonstrated strong capabilities in this area. By pre-training on large-scale datasets, LLM are able to capture the complex semantic and contextual information in the comments to generate more diverse data, which significantly improves the accuracy and robustness of fake review recognition.

3 Models and Experiments

3.1 LLMs Selection

The purpose is to resolve the data imbalance problem through LLMs that generate data from a small number of scarce categories in the dataset. Depending on the training corpus as well as the modeling details, the following three large language models are chosen for this study.

GPT-3.5. The Generative Pre-trained. Transformer (GPT) family is a very powerful family of LLMs proposed by OpenAI, which can achieve amazing results in very complex NLP tasks (e.g., article generation, code generation, machine translation, Q&A, etc.), and no supervised learning is required for model fine-tuning to accomplish these tasks. GPT-3.5-turbo [20] is a natural language processing model from OpenAI that is a streamlined and high-performance variant of GPT-3.The model uses a similar architecture and training methodology with powerful language generation and comprehension capabilities. It learns patterns and structures of language through large-scale data and self-supervised learning, and can be widely used in natural language processing tasks such as text generation, text summarization, translation and question answering.

GLM4. GLM is a pre-trained grand model in the field of natural language processing released by the Knowledge Engineering Group at Tsinghua University. The GLM pre-trained grand model is a generalized language model based on autoregressive gap filling. The gap-filling pre-training is improved by adding 2D positional encoding and allowing arbitrary order prediction of spans, resulting in improved performance over BERT and T5 in natural language understanding (NLU) tasks.GLM4 [21] is the latest generation of the basic bigram model released by Wisdom Spectrum Artificial Intelligence, which employs grouped query attention for its attention mechanism [22], and the overall performance of the model has been improved over the previous generation, supports longer contexts, and greatly reduces the inference cost.

Starfire. Xinghuo V3.5 is a big model of cognitive intelligence built by iFLYTEK and released in January, 2024. It is designed to provide high-quality text generation, dialog understanding, semantic analysis and other natural language processing services. Trained on large-scale data, the model has excellent text generation capabilities, generating coherent and varied text and showing strong performance in multiple NLP tasks. With good language understanding and generation capabilities, it can handle complex language tasks.

3.2 Prompt Building

The prompt construct is shown in Fig. 1. In order to enhance the scarce fake reviews in the original dataset, the fake reviews in the dataset are used to construct prompts and feed them to the LLM for text generation. We use the same template to build prompts for each fake review, and in each prompt, we tell the LLM, as a fake review author, to emulate a similar sentence based on the statements provided below. After the prompt is provided to the LLM, we can get the fake comment statement it generates.

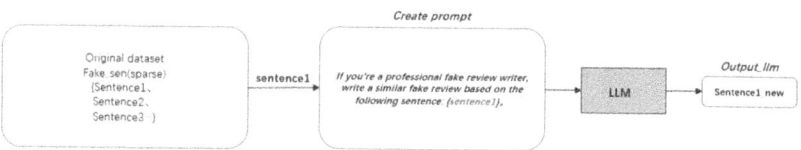

Fig. 1. The Prompt build method

3.3 Experimental Process

The Performance Comparison of FRI of Different LLM-Generated Data. As shown in Fig. 2, we add each LLM-generated data to the original data separately for data augmentation to obtain three augmented data, and then compare the classification performance of the classification models trained on these three augmented datasets. In order to ensure that the amount of training data does not affect the training effect of the model, we use the same prompt constructor and select the same amount of data from the data generated by the three large models and add it to the original data to construct the training dataset.

Machine learning and deep learning models are used in the experimental classification models, where the machine model uses SVM and random forest model, the embedding part of the deep learning model uses pre-trained models such as BERT, RoBERTa, while the classifier uses neural network models such as RNN. Through such a design, the effectiveness and applicability of the method can be more comprehensively verified.

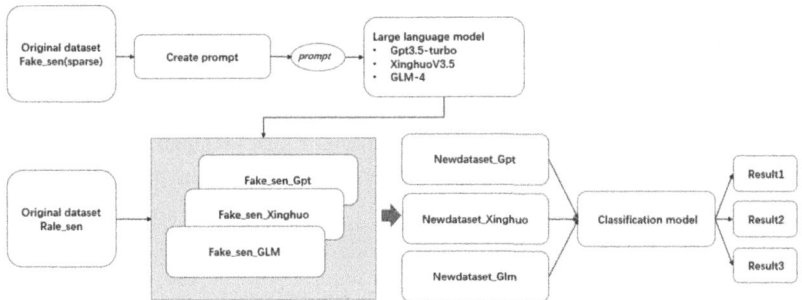

Fig. 2. The performance comparison of different LLM-generated data

The Performances of Fake Review Identification Based on Data Fusion of Different LLMs. The purpose of the experiment is to verify the feasibility of data augmentation through LLMs. Specifically, this experiment adds all three LLMs generated fake comments to the original data for data augmentation. The augmented dataset is then used to train a classifier, which is compared with the classifier trained using the original dataset.

In order to verify the effectiveness of data enhancement using LLMs, the experiments are tested on multiple classification models. The number of classification models is the same as in the previous experiment. The specific flow of the experiment is shown in Fig. 3. The fake reviews generated by several LLMs are fused as a whole, and the original dataset is added to the original dataset to enhance the original dataset, and the enhanced dataset is classified by the classification model, and the classification results are compared with the classification results of the original dataset to evaluate the effectiveness of data augmentation.

The Performance of Different Data Augmentation Methods In order to verify the effectiveness of the data augmentation method we proposed, we also compared it with other data augmentation techniques. In the traditional data augmentation method, Easy Data Augmentation (EDA) method was used to randomly perform synonym replacement, random insertion, random deletion and random replacement operations on the original text to obtain new text. The Back Translations method was also used to translate the original text into other languages, and then translate the translated language back to the original language to obtain new text. In the deep learning model, we used the original data to train the VAE (Variational Auto Encoder) and GAN (Generative Adversarial Networks) models, and used the trained models to generate new text for data augmentation.

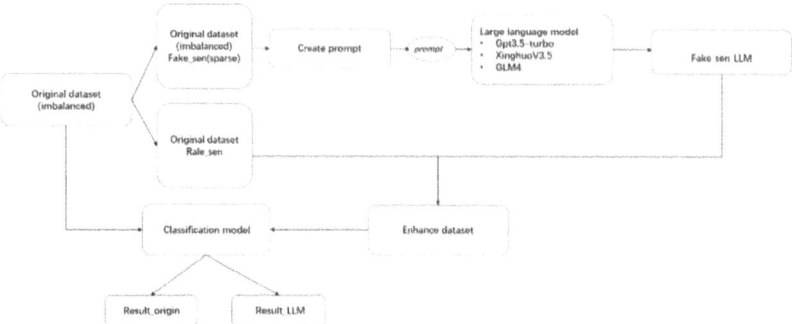

Fig. 3. The performances of fake review identification based on data fusion of different LLMs

We used four methods to generate the same amount of fake text, and then we compared the effects of several different text augmentation methods with the use of LLM for text augmentation.

4   Experimental Results

4.1   Data Sets

The experiments in this paper use the Yelp-restaurant dataset. The Yelp dataset is provided by Yelp company [23], and contains information from user reviews and ratings of merchants. This data can be used for tasks such as analyzing user perceptions of merchants, sentiment tendencies, and modeling of recommender systems. The dataset has a significant imbalance between the different categories. Of the total number of reviews, 53,397 (87.8%) were genuine reviews. In contrast, the number of fake reviews is significantly lower at 8141 (13.2%). These unbalanced data pose a significant challenge in categorizing fake reviews [24].

4.2   Evaluation Guidelines

Guidelines for Generating Quality Assessments.   Based on the work of Suhaeni [18] we assessed the quality of the generated data mainly by evaluating the novelty as well as the diversity of the fake comments generated by each large model.

Novelty. Novelty is the level of uniqueness of the comment data generated by a large model compared to the original comment data. It evaluates whether the model generates new reviews or just copies the reviews from the original data. We use the following formula to calculate the novelty of each generated review.

$$Novelty(R_i) = 1 - avg\{\varphi(R_i, C_j)\}_{j=1}^{j=|C|} \tag{1}$$

where C is the original false comment dataset, and φ is the cosine similarity function. The degree of similarity between the generated comments and the original dataset is reflected by calculating the similarity between the generated comments and the original comments taking the mean value. The novelty score tends to 0 indicates that the generated data is not novel, which indicates that the model has a low ability to generate new sentences, but can only simply replace the original comments with simple modifications. And close to 1 indicates that the generated data is very different from the original data, away from the original target dataset.

Diversity. Diversity is the level of diversity in the data generated by the big models compared to each other. It evaluates whether the fake reviews generated by each big model are too similar to each other and determines whether the generated review data are too similar. We use the following formula to calculate the diversity of each generated review.

$$Diversity(R_i) = 1 - avg\{\varphi(R_i, R_j)\}_{j=1}^{j=|R|, j \neq i} \tag{2}$$

where R is the generated dataset, and φ is the cosine similarity function. The diversity of the generated data is reflected by calculating the similarity of the generated comment data with other comment data to take the mean value. The diversity score tends to 0 means that the text is similar to other generated texts and the generated statements are relatively homogeneous. While the score tends to 1 means that the text is more different from other generated texts, indicating that the comments generated by the large model are more diverse.

Classification Model Evaluation Indicators. The evaluation metrics of the classification model used Accuracy, recall, and F1 metrics to assess the classification effectiveness of the classification model. The formula for calculating the indicator. In formula 5, R stands for Recall and P stands for Precision.

$$Accuracy = \frac{TP + TN}{TP + TN + FP + FN} \tag{3}$$

$$Recall = \frac{TP}{TP + FN} \tag{4}$$

$$F1 = \frac{2 * P * R}{R + P} \tag{5}$$

4.3 Text Generation Quality Comparison

In the one-shot text generation phase, we first pre-processed the raw dataset, which included removing non-English sentences and invalid characters, etc., to ensure that the raw data used as input for the data generation process was clean and relevant. After filtering the generated utterances, the final result was 6,120 generated spurious comments. The raw statements and the statements generated by the larger model are shown in Table 1.

Table 1. Samples of LLM generation

Sentence of Yelp	Sentence Generation	LLM
I liked the services offered by (Alinea) restaurant and will definitely come again and again whenever I get a chance or ceremony to come in	I enjoyed the services provided by (Alinea) restaurant and plan to return whenever the opportunity arises	Sentence-gpt3.5-turbo
	I so enjoyed the services provided by (Bistro Paris) restaurant and will surely return whenever I have the opportunity or a special occasion	Sentence-GLM4
	I loved the services offered by (Alinea) restaurant and will definitely come back whenever I get the chance or have a special occasion	Sentence-xinghuo V3.5
I wanted to go to a Mexican Restaurant, but instead we came here. It was very good	I had planned to visit an Mexican Restaurant, but we ended up coming here instead. It turned out to be fantastic	Sentence-gpt3.5-turbo
	I wanted to go to a Mexican Restaurant, but instead, we came here. It was a disappointing alternative	Sentence-GLM4
	I had my heart set on a Mexican Restaurant, but we ended up here. It was quite disappointing	Sentence-xinghuo V3.5

We evaluate the effectiveness of several LLMs in generating fake reviews through the previously mentioned novelty and diversity metrics. The scoring results obtained are shown in the Tables 2 and 3:

Table 2. Comparison of text novelties of different LLMs

	Mean	Std	25%	50%	75%
GPT3.5-turbo	0.516	0.105	0.484	0.546	0.591
GLM4	0.534	0.075	0.515	0.550	0.624
Xinghuov3.5	0.526	0.087	0.485	0.550	0.592

Table 3. Comparison of text Diversities of different LLMs

	Mean	Std	25%	50%	75%
GPT3.5-turbo	0.624	0.111	0.606	0.666	0.690
GLM4	0.672	0.106	0.664	0.716	0.732
Xinghuov3.5	0.571	0.121	0.535	0.609	0.643

The novelty represents the similarity of the generated reviews with the original reviews. The analysis of the obtained results shows that the GLM4 model scored the highest, indicating that the text it generated was more innovative than the other two models. The GPT model has poor performance and large variance, indicating that some of the text generated by it is highly reproducible with the original data. The novelty scores of the data generated by several big models are around 0.5, which means that the fake comment data generated by these big models have some degree of novelty compared with the original fake comment data, while keeping the relevance to the original data, rather than simply replacing and modifying the original comment data, and do not deviate excessively from the original data, but still retain the relevant data information. This suggests that the large model has successfully generated new comments based on the original comment data that are different from the original data.

The diversity represents how different a generated comment is from the other generated comments. The diversity scores reveal a gap in the comments generated by various large models, with the highest diversity score of the comments generated by the GPT, which indicates that the GPT-generated comments are less similar to each other and the outputs are more diverse, followed by the Starfire large model with the GLM, which generates comments with less diversity and a certain degree of similarity between multiple comments. The results of the diversity assessment indicate that there is some similarity between the comments generated by the models.

4.4   Results of the Performance Comparison of Different LLM-Generated Data

The corresponding datasets of the three LLMs are constructed by adding the fake comments generated by the three LLMs to the original dataset respectively, and the classification models were trained using these three datasets as the training set, and the classification models used the pre-trained language model combined with the structure of the neural model. In machine learning, the C parameter of the SVM model is set to 1, the number of trees in the random forest is set to 100, the maximum depth of the tree is selected to 10, the learning rate parameter of the deep learning model is uniformly set to 1e-7, and the weight attenuation parameter is set to 0.001. Training 20 epochs. The following table shows the classification results after augmentation of different LLMs (Table 4).

Through the overall analysis of the classification effects of different classification models, it can be seen that from machine learning models to deep learning models, the classification effects of classification models have been greatly improved, and the pre-trained language model uses the Roberta model to achieve a more comprehensive classification effect. Better than the model using Bert. By comprehensively comparing the classification performance of the classification models trained by several large-scale models, the GLM4 model achieved the optimal effect on the deep learning model, with the highest accuracy rate of 81.4. Followed by the GPT-3.5 Turbo model, the highest accuracy reached 80.58, while the highest accuracy of the Spark V3.5 model only reached 77.63. This result also shows from the side that the model transformation of GLM4 has made special optimizations in improving the performance of generation tasks, and may be unique in enhancing the natural fluency and coherence of text. In comparison, GPT-3.5 Turbo focuses more on maintaining high efficiency while expanding its wide range of application capabilities. On the other hand, Spark V3.5 seems to be more inclined to optimize application performance in the Chinese context, so it appears to be more accurate in processing Chinese content. As a result, its performance in the field of English text generation is slightly insufficient. Compared to simple machine

Table 4. Comparison of the effect of data generated by different LLMS

Methods	Xinghuov3.5	Gpt3.5-turbo	GLM4
SVM	Acc: 66.85 f1: 0.66 recall:0.67	**Acc: 67.89** **f1: 0.67** **recall:0.68**	Acc: 66.33 f1: 0.65 recall:0.77
Random Forest	Acc: 60.69 f1: 0.54 recall:0.61	**Acc: 61.63** **f1: 0.57** **recall:0.63**	Acc: 61.29 f1: 0.56 recall:0.62
Bert+LSTM	Acc: 72.33 f1: 0.66 recall:0.71	**Acc:76.84** **f1: 0.72** **recall:0.79**	Acc: 76.19 f1: 0.73 recall:0.77
Bert+RNN	Acc: 71.20 f1: 0.67 recall:0.66	Acc:75.84 f1:0.67 recall:0.81	**Acc:78.53** **f1:0.78** **recall:0.83**
Roberta+RNN	Acc: 77.63 f1: 0.71 recall:0.75	Acc:79.42 f1: 0.78 recall:0.79	**Acc:81.40** **f1: 0.83** **recall:0.87**
Roberta+FNN	Acc: 76.21 f1: 0.74 recall:0.76	Acc: 80.58 f1: 0.82 recall:0.81	**Acc: 81.32** **f1: 0.81** **recall:0.79**

learning methods, deep learning models are able to learn more data features through a multi-layer neural network structure, and RoBERTa outperforms BERT mainly because it employs a larger dataset and longer training time during the training process, and there are improvements in both model details, which make RoBERTa perform These improvements make RoBERTa perform better in a number of natural language processing tasks.

4.5 Results of the Performances of Fake Review Identification Based on Data Fusion of Different LLMs

According to the model classification effect, it can be seen that on multiple models of machine learning and deep learning, the classification model trained with the enhanced dataset has a better classification effect than the classification model trained with the original dataset. The accuracy index of the classification model increased by about 10%. The new dataset enhanced by LLMS achieves the optimal classification effect on the Roberta + RNN classification model. The model's classification accuracy reaches 84.46%. In comparison, the classification model using the original dataset achieved an accuracy of 73.62%. This validates our approach. Data augmentation of the dataset through LLMS can indeed improve the classification effect of the final classification model.

We believe that the fundamental reason for the improvement of the classification performance of the classification model is that the large model generates new samples, which alleviates the problem of insufficient number of minority samples in the dataset, so that the model can better learn the characteristics of the minority class, and then improve the discrimination ability of the classification model. In addition, the diversity of the enhanced data is improved, which helps to reduce the overfitting of the model to minority classes, improves the generalization of the model, and improves the overall learning effect of the model (Table 5).

Table 5. Comparison of classification performances of different classification models

Methods	Origin Dataset	New Dataset
SVM	Acc: 64.18 f1: 0.63 recall:0.66	Acc: 77.15 f1: 0.75 recall:0.76
Random Forest	Acc: 58.21 f1: 0.56 recall:0.56	Acc: 69.80 f1: 0.68 recall:0.65
Bert+rnn	Acc: 69.74 f1: 0.65 recall:0.64	Acc: 78.89 f1: 0.77 recall:0.79
Bert+LSTM	Acc: 67.19 f1: 0.65 recall:0.66	Acc: 79.55 f1: 0.76 recall:0.78
Roberta+fnn	Acc: 73.59 f1: 0.72 recall:0.69	Acc: 84.05 **f1: 0.84** **recall:0.82**
Roberta+rnn	Acc: 73.62 f1: 0.74 recall:0.72	**Acc: 84.46** f1: 0.82 recall:0.81

4.6 Results of the Performance of Different Data Augmentation Methods.

By comparing the augmentation effects of different data augmentation methods, we can see that the data augmentation effect based on EDA is the worst. Compared with the original data, the ACC is only improved by about 5, followed by the Back translation method. Compared with them, the data augmentation effect of using a deep learning model is better. Using LLM for data augmentation has achieved the best augmentation effect under all indicators.

This is because the EDA and Back translation methods only make simple modifications based on the original sentences. Although the methods are simple and easy to implement, such methods cannot capture the complexity of the language. The generated sentences are highly similar to the original data, and the data is relatively simple, which can easily lead to overfitting of the classification model. The VAE method can generate new texts with similar features by learning the potential distribution of the data. GAN learns the features of the original sentence and generates similar sentences through an adversarial mechanism. However, the training process of the two models is usually complicated and requires a lot of computing resources. There is a risk of instability in training. Moreover, these two methods have high requirements on the amount of data.

In the case of insufficient data, overfitting or poor generation effect may occur. LLM has been pre-trained on a large-scale dataset and can understand and generate the complexity of language. The generated text is usually more grammatically correct and fluent. In addition, by constructing prompts, the generation of LLM can be made more diverse, and new expressions and information can be introduced to make the generated data richer and more diverse, thereby achieving excellent data augmentation effects (Table 6).

Table 6. Comparison of different data augmentation methods

Enhance Methods	ACC	recall	F1	AUC-ROC
Origin	72.51	0.70	0.74	76.23
EDA	77.67	0.76	0.77	82.84
Back translation	78.69	0.76	0.78	85.35
VAE	80.85	0.79	0.79	83.21
GAN	83.34	0.81	0.82	85.42
LLM	**84.56**	**0.83**	**0.84**	**89.26**

5 Conclusions

In order to effectively solve the current impact of unbalanced data sets on false review identification, our experiment uses the large model to generate data for the unbalanced review data set, builds prompts based on the false reviews in the original data set and sends them to the large model generation. The original dataset is augmented by generating fake reviews, and the quality of the generated text is evaluated using novelty and diversity metrics. The final experimental results on the classification model show that the classification effect of multiple classification models trained based on the enhanced dataset is better than that of the original dataset, the accuracy of each model has been improved by about 10%, and the classification model based on Roberta + RNN achieved the highest accuracy after training with the enhanced data set, proving that the classification effect of the enhanced dataset in the classification model has been improved to a certain extent.

This article also compared the enhancement effects of text generated by three large models: GPT3.5turbo, GLM4, and Xinghuo v3.5. It is hoped that in the future, large models will be applied to imbalanced data sets in other directions, and LLMs can be used to augment data for scarce categories in datasets, thereby improving the performance of various downstream tasks such as text classification and entity recognition.

Acknowledgments. The work is supported by The Project of Cultivation for Young Topmotch Talents of Beijing Municipal Institutions "Research on the comprehensive quality intelligent service and optimized technology for small medium and micro enterprises" (Grant No. BPHR202203233). National Natural Science Foundation of China "Research on the influence and governance strategy of online review manipulation with the perspective of E-commerce ecosystem" (Grant No. 72174018). National Natural Science Foundation of China "Future scenario generation method and empirical research based on multi-source data fusion" (Grant No. L2324224).

References

1. Zhang, W., Wang, Q., Li, J., Ma, Z., Bhandari, G., Peng, R.: What makes deceptive online reviews? A linguistic analysis perspective. Humanit. Soc. Sci. Commun. **10**(1), 1–14 (2023)
2. Wei, J., Zou, K.: EDA: easy data augmentation techniques for boosting performance on text classification tasks. arxiv preprint arXiv:1901.11196 (2019)
3. Li, J., Luong, M.T., Jurafsky, D.: A hierarchical neural autoencoder for paragraphs and documents. arxiv preprint arXiv:1506.01057 (2015)
4. Goodfellow, I., et al. Generative adversarial nets. In: Advances in neural information processing systems, vol. (2014), 27
5. Kingma, D.P., Welling, M.: An introduction to variational autoencoders. Found. Trends. Mach. Learn. **12**(4), 307–392 (2019)
6. Salehi, P., Chalechale, A., Taghizadeh, M.: Generative adversarial networks (GANs): an overview of theoretical model, evaluation metrics, and recent developments. arXiv preprint arXiv:2005.13178 (2020)
7. Zhang, W., Li, R., Quan, P., Chang, J., Bai, Y., Su, B.: Lightweight deep learning for missing data imputation in wastewater treatment with variational residual auto-encoder. IEEE Internet Things J. https://doi.org/10.1109/JIOT.2024.3445965
8. Zhang, W., Zhao, J., Quan, P., Wang, J., Meng, X., Li, Q.: Prediction of influent wastewater quality based on wavelet transform and residual LSTM. Appl. Soft Comput. **148**, 110858 (2023)
9. Graves, A., Jaitly, N., Mohamed, A.: Hybrid speech recognition with deep bidirectional LSTM. In: 2013 IEEE Workshop on Automatic Speech Recognition and Understanding, pp. 273–278. IEEE (2013)
10. Bahdanau, D., Cho, K., Bengio, Y.: Neural machine translation by jointly learning to align and translate. arXiv preprint arXiv:1409.0473 (2014)
11. Lantao, Z.W.: Sequence generative adversarial nets with policy gradient. In: Proceedings of the Thirty First AAAI Conference on Artificial Intelligence (AAAI), p. 2852. AAAI, San Francisco (2017)
12. Sun, Z., Li, X., Zhu, G.: LeakGAN-based causality extraction in the financial field. In: Abawajy, J.H., Xu, Z., Atiquzzaman, M., Zhang, X. (eds.) International Conference on Applications and Techniques in Cyber Intelligence, pp. 247–255. Springer, Cham (2022). https://doi.org/10.1007/978-3-031-28893-7_30
13. Zhang, Y., et al.: Adversarial feature matching for text generation. In: International Conference on Machine Learning, pp. 4006–4015. PMLR (2017)
14. Dieng, A.B., Kim, Y., Rush, A.M, Blei, D.M.: Avoiding latent variable collapse with generative skip models. In: The 22nd International Conference on Artificial Intelligence and Statistics, pp. 2397–2405. PMLR (2019)
15. Bowman, S.R., Vilnis, L., Vinyals, O., Dai, A.M., Jozefowicz, R., Bengio, S.: Generating sentences from a continuous space. arXiv preprint arXiv:1511.06349 (2015)
16. Liu, M., Poesio, M.: Data augmentation for fake reviews detection. In: Proceedings of the 14th International Conference on Recent Advances in Natural Language Processing, pp. 673–680 (2023)
17. Ye, J., et al.: LLM-DA: data augmentation via large language models for few-shot named entity recognition. arxiv preprint arXiv:2402.14568 (2024)

18. Suhaeni, C., Yong, H.S.: Mitigating class imbalance in sentiment analysis through GPT-3-generated synthetic sentences. Appl. Sci. **13**(17), 9766 (2023)
19. Zhang, X., Zhang, X., Yu, Y.: ChatGLM-6B fine-tuning for cultural and creative products advertising words. In: 2023 International Conference on Culture-Oriented Science and Technology (CoST), pp. 291–295. IEEE (2023)
20. Abramski, K., Citraro, S., Lombardi L, Rossetti, G., Stella, M.: Cognitive network science reveals bias in GPT-3, GPT-3.5 turbo, and GPT-4 mirroring math anxiety in high-school students. Big Data Cogn. Comput. **7**(3), 124 (2023)
21. Yang, A., Li, Z., Li, J.: Advancing GenAI assisted programming--a comparative study on prompt efficiency and code quality between GPT-4 and GLM-4. arxiv preprint arXiv:2402.12782 (2024)
22. Team, G.L.M., et al. ChatGLM: a family of large language models from GLM-130B to GLM-4 all tools. arXiv e-prints arXiv:2406.12793 (2024)
23. Dai, W., Jin, G., Lee, J., Luca, M.: Aggregation of consumer ratings: an application to yelp.com. Quant. Mark. Econ. **16**, 289–339 (2018)
24. Wang, Q., Zhang, W., Li, J., Ma, Z., Chen, J.: Benefits or harms? The effect of online review manipulation on sales. Electron. Commer. Res. Appl. **57**, 101224 (2023)

Knowledge Management

Digital Transformation Mechanisms for Emergency Management in Chemical Enterprises: An Industrial Agglomeration Perspective

Yue Feng[1], Meiqi Niu[2], Yingyi Zhang[1], and Lili Rong[1(✉)]

[1] Dalian University of Technology, Dalian 116024, China
`llrong@dlut.edu.cn`
[2] Shanxi University of Finance and Economics, Shanxi 030006, China

Abstract. The digital transformation of emergency management in chemical enterprises within industrial clusters is significantly influenced by neighboring enterprises and third-party regulatory agencies. The decision-making process is complex due to varying interests and behavioral motives among different stakeholders. We investigate the digital transformation mechanisms in emergency management for chemical enterprises from the perspective of industrial agglomeration leveraging evolutionary game theory, considering the impact of decisions made by neighboring chemical enterprises and third-party management departments in industrial park. The results highlight how direct potential benefits, potential benefits discount factor, relative benefits, and input costs differentially impact decision-making processes. Two stable evolutionary states are identified between neighboring enterprises. However, when park management departments implement regulatory interventions, the evolutionary system lacks a stable strategy, resulting in cyclical strategy choices over time and leading to the "regulatory dilemma". These insights offer valuable theoretical support for management strategies aimed at enhancing digital emergency management in industrial agglomerations.

Keywords: Industrial Agglomeration · Chemical Enterprises · Emergency Management · Digital Transformation · Evolutionary Game

1 Introduction

The digital transformation of emergency management is a strategic initiative focused on digital innovation to enhance the efficiency and effectiveness of emergency response activities. It involves applying digital technologies and concepts across all aspects of emergency management, including the digital innovation of organizational structures and actions [1]. The digital emergency management system utilizes advanced technologies and platforms to transmit and process relevant data in real time, enabling decision-makers to conduct accurate accident assessments and implement more effective response measures. As the nation transitions to an era of ecological civilization characterized by

X. Tang et al. (Eds.): KSS 2024, CCIS 2269, pp. 241–255, 2025.
https://doi.org/10.1007/978-981-96-0178-3_17

green, low-carbon, and sustainable development, the chemical industry has emerged as a key sector where the benefits of industrial agglomeration are increasingly apparent. Industrial agglomeration is the process by which enterprises in the same industry concentrate in a specific geographic area, facilitating the convergence of industrial capital and resources [2]. Chemical parks, resulting from this agglomeration in the chemical industry, serve as key platforms for regional chemical enterprises. These parks provide infrastructure, physical space, and technical cooperation, enhancing competitive advantages and reducing management costs. However, as specialized physical areas housing numerous hazardous chemicals, chemical parks are susceptible to catastrophic accidents, where the consequences can have a domino effect. Thus, digital transformation in emergency management for chemical enterprises is both necessary and urgent. However, the Deloitte report "Digital Transformation: Are Chemical Enterprises Ready?" in collaboration with the China Petroleum and Chemical Industry Federation (CPCIF), indicates that 52% of surveyed chemical enterprises lack a comprehensive digital strategy or transformation roadmap. Compared to sectors like finance, retail, and automotive manufacturing, the chemical industry lags significantly in digital transformation efforts, including emergency management practices. On-site investigations reveal that the digital transformation of emergency management within chemical parks faces numerous challenges, hindering progress.

In the context of industrial agglomeration, the decision-making process for the digital transformation of emergency management in chemical enterprises is complex and features several distinct characteristics. First, it involves multiple stakeholders with intricate and dynamic relationships. Interactions among neighboring chemical enterprises, as well as between enterprises and park management, can be either cooperative or competitive, evolving over time and in response to external conditions. Second, there is a significant spillover effect; when one chemical enterprise adopts digital emergency response measures, it not only reduces its own accident risks but also enhances the emergency response capabilities of other enterprises in the park. Third, the process is influenced by third-party park regulators, who are responsible for ensuring production safety and environmental monitoring. These regulators may disseminate information on macro policies, market trends, and technical standards to promote digital transformation in emergency management, which can also affect enterprises' motivations to transform. In this decision-making landscape, it is crucial to identify the factors influencing digital transformation strategies for emergency management in chemical enterprises within industrial agglomeration and how these factors manifest. Addressing these questions is essential for helping chemical enterprises achieve effective digital transformation in emergency management, which is of significant practical importance.

This study makes three main contributions: (1) It constructs an evolutionary game model under industrial agglomeration, incorporating neighboring chemical enterprises and third-party park management departments, to illustrate the evolutionary process of promoting digital transformation in emergency management and elucidate the underlying mechanisms. (2) By simulating the evolutionary game under various conditions, the research identifies key factors and interaction mechanisms that influence digital transformation decisions in emergency management, emphasizing the differential impacts of direct potential benefits, potential benefits discount factors, relative benefits, and input costs. (3) The study highlights how regulatory interventions lead to cyclical strategy choices and create a "regulatory dilemma", thereby enhancing our understanding of regulatory challenges.

2 Related Work

Many countries began researching digital emergency management in the 20th century, leading to the widespread application of digital technology in emergency management systems across various fields. Digital emergency management integrates the Internet, computer technology, and other information technologies with traditional emergency management systems. Scholars increasingly focus on the theoretical aspects of this integration, emphasizing modeling and simulation analysis to enhance emergency management models through digital means [3]. Researchers have also studied the application of information technology in emergency management extensively. For instance, some scholars proposed a novel algorithmic structure for wide-area rescue systems, which supports the development of hazardous event detection systems for smartphones [4]. Others addressed the shortcomings of insufficient informatization and low intelligence in emergency plans, offering intelligent response strategies for various scenarios using wireless sensors and near-field communication technology [5]. IoT technology has proven significant in optimizing emergency evacuation systems [6]. Additionally, scholars have explored ex-ante prediction in disaster response, arguing that digital twin technologies provide an interactive testbed for understanding and predicting changes in disaster environments [7]. These technologies can distill and encapsulate specialized emergency response knowledge, facilitating precise planning and optimization of emergency management programs.

The development of big data, cloud computing, artificial intelligence, blockchain, and 5G advanced significantly after the COVID-19 pandemic [8]. In the era of big data, information now comes from a diverse range of sources, including businesses, communities, and citizens, not just the government [9]. Digital technology is increasingly integrated into daily life and production. While it presents new opportunities for building digital emergency management systems, it also reveals challenges such as inadequate institutional mechanisms, slow digital empowerment, and insufficient synergy between information technology and business processes [10, 11]. To tackle these issues, scholars have explored digital transformation paths and mechanisms for emergency management at macro-society, meso-industry, and micro-enterprise levels. At the macro-social level, the focus is on government policies, public resource allocation, social collaborative governance, and the construction of information infrastructure, aiming to enhance the efficiency of digital transformation and improve disaster response capacity [12, 13]. At the meso-industry level, researchers examine digital practices across sectors like military, energy, public health, industrial manufacturing, and retail, analyzing how industry characteristics shape digital transformation in emergency management [14]. At the micro-enterprise level, scholars investigate the use of digital technologies in enterprise emergency management, discussing how big data analytics, artificial intelligence, and the Internet of Things (IoT) can enhance the speed and quality of emergency response and decision-making [15].

Existing research has yielded valuable insights into digital emergency management. However, most studies primarily focus on technology application and system optimization, with limited exploration of multi-stakeholder interactions in digital decision-making. Furthermore, research on the digital transformation of emergency management in chemical enterprises, especially within industrial agglomerations, is sparse. A comprehensive understanding of these transformation mechanisms remains elusive. Integrating the characteristics of industrial agglomeration with the influences of multi-stakeholder decision-making is essential.

3 Construction of Evolutionary Game Model Between Neighboring Chemical Enterprises

3.1 Problem Description

When chemical enterprises pursue the digital transformation of emergency management within industrial agglomerations, their decision-making is influenced by neighboring enterprises and third-party regulators, making it no longer solely independent. This multi-stakeholder process involves dynamic mutual influence and interaction, compounded by differing interests and motivations. Evolutionary game theory is well-suited for optimizing strategies in such complex and uncertain systems [16].

3.2 Basic Hypotheses

In the game process, if a chemical enterprise finds the benefits of not implementing digital transformation lower than those of implementing it, it will choose to implement in the next round. Conversely, if the benefits of not implementing are higher, it will opt not to. This decision-making continues until equilibrium is reached. The following hypotheses are proposed:

(1) Chemical enterprises exhibit finite rationality. They face two options: "implement" and "not implement", with probabilities x and y, respectively.
(2) If both enterprises choose "not implement", they receive only basic benefits.
(3) When an enterprise implements digital transformation of emergency management, it gains direct benefits by reducing its own safety risks and losses, while also indirectly benefiting neighboring enterprises by minimizing their potential losses.
(4) The potential benefits discount factor reflects the balance between immediate and long-term interests. A larger factor indicates a focus on long-term benefits, while a smaller value prioritizes immediate benefits.
(5) Implementing digital transformation generally results in profit gains. Enhanced decision-making improves customer service and market responsiveness, while forgoing transformation leads to inefficiencies and reduced competitiveness. Thus, those choosing "implement" see increased benefits, while "not Implement" see a relative decrease.
(6) If both enterprises choose "implement", the collaborative effect reduces transformation costs, making investment lower than if only one enterprise proceeds.

3.3 Model Construction

Based on the above hypotheses, variables have been defined, as detailed in Table 1.

Enterprises are labeled as enterprise A and B. Their benefits are analyzed and the benefit payoff matrix is constructed, as shown in Table 2.

Table 1. Variables involved in the evolutionary game

Variables	Explanation
R_i	Basic benefits of the enterprise
D_i	Direct potential benefits
I_i	Indirect potential benefits
$\eta_i(0 < \eta_i < 1)$	Potential benefits discount factor of enterprises. The importance that enterprises place on their own future potential benefits
η_1	Potential benefits discount factor of park management department. The importance that park management department places on potential benefits
L_i	Relative benefits arise when one enterprise implements digital transformation in emergency management while the other does not
C_i	Input costs of the digital transformation in emergency management
$C_i'(C_i' < C_i)$	Collaboration costs. If both neighboring companies choose to promote digital emergency management, collaboration effect will occur
C_1	Costs incurred by park management department in enforcing regulations
ε	Subsidy coefficients granted by the park management to enterprises
P	Penalties are imposed by the park management
k	Responsibility cost transfer coefficient. The proportion of responsibility cost borne by the chemical enterprise is transferred to the regulator

Table 2. Payoff matrix of neighboring enterprises in evolutionary game

		Enterprise B	
		Implement (y)	Not Implement ($1-y$)
Enterprise A	Implement (x)	$R_A + \eta_A(D_A + I_A) - C_{A'}$ $R_B + \eta_B(D_B + I_B) - C_{B'}$	$R_A + \eta_A D_A + L_A - C_A$ $R_B + \eta_B I_B - L_B$
	Not Implement ($1-x$)	$R_A + \eta_A I_A - L_A$ $R_B + \eta_B D_B + L_B - C_B$	R_A R_B

For enterprise A and B, the average expected benefits are represented in formulas 1 and 2.

$$\overline{E_A} = x\eta_A D_A + x L_A - x C_A + xy(C_A - C_{A'}) + y\eta_A I_A - y L_A + R_A \tag{1}$$

$$\overline{E_B} = y\eta_B D_B + y L_B - y C_B + xy(C_B - C_{B'}) + x\eta_B I_B - x L_B + R_B \tag{2}$$

We obtain the replicated dynamic equations for the evolutionary game of neighboring enterprises in the chemical park, as shown in formula 3.

$$\begin{cases} F(x) = \dfrac{dx}{dt} = x(E_{A1} - \overline{E_A}) = x(1-x)[\eta_A D_A + L_A - C_A + y(C_A - C_{A'})] \\ F(y) = \dfrac{dy}{dt} = y(E_{B1} - \overline{E_B}) = y(1-y)[\eta_B D_B + L_B - C_B + x(C_B - C_{B'})] \end{cases} \tag{3}$$

3.4 Evolutionary Stability Analysis

The game subjects eventually reach the stable state, known as an effective Nash equilibrium, through continuous trial and error and learning in the evolutionary game process [17]. To find the possible equilibrium points in the game, we set the equations in the set of replicated dynamic equations to zero to solve for $O(0, 0)$, $P(0, 1)$, $M(1, 0)$ and $N(1, 1)$. And the possible equilibrium point $E(x^*, y^*)$ is obtained, where $x^* = \frac{C_B - \eta_B D_B - L_B}{C_B - C_{B'}}$, $y^* = \frac{C_A - \eta_A D_A - L_A}{C_A - C_{A'}}$.

According to the equilibrium state determination method proposed by Friedman [18], the stability of equilibrium point can be assessed using local stability analysis of the Jacobian matrix. The Jacobian matrix of the evolutionary system is derived from replicated dynamic equations, as shown in formula 4.

$$J = \begin{bmatrix} \frac{\partial F(x)}{\partial x} & \frac{\partial F(x)}{\partial y} \\ \frac{\partial F(y)}{\partial x} & \frac{\partial F(y)}{\partial y} \end{bmatrix}$$
$$= \begin{bmatrix} (1-2x)[\eta_A D_A + L_A - C_A + y(C_A - C_{A'})] & x(1-x)(C_A - C'_A) \\ y(1-y)(C_B - C_{B'}) & (1-2y)[\eta_B D_B + L_B - C_B + x(C_B - C_{B'})] \end{bmatrix} \tag{4}$$

Using the local stability analysis method of Jacobian matrix, when the equilibrium point satisfies $Det(J) > 0$ and $Tr(J) < 0$, it indicates that the system is in a locally asymptotic stable state. This point is considered an evolutionarily stable strategy (ESS) [19]. The stability analysis of five possible equilibrium points is detailed in Table 3.

As shown in Table 3, $O(0, 0)$ and $N(1, 1)$ are equilibrium points. The (not implement, not implement) and (implement, implement) as final evolutionary stable strategies. $P(0, 1)$ and $M(1, 0)$ are unstable points. $E(x^*, y^*)$ serves as a center point. Based on the stability analysis results, the evolutionary phase diagram can be obtained, which visually represents the dynamic decision-making process for digital transformation in neighboring chemical enterprises, as shown in Fig. 1.

Table 3. Stability analysis of possible equilibrium points in evolutionary game systems

Possible equilibrium points	$Det(J)$	$Tr(J)$	Stability
$O(0,0)$	$+$	$-$	ESS
$P(0,1)$	$+$	$+$	Non-ESS
$M(1,0)$	$+$	$+$	Non-ESS
$N(1,1)$	$+$	$-$	ESS
$E(x^*, y^*)$	$-$	0	Center point

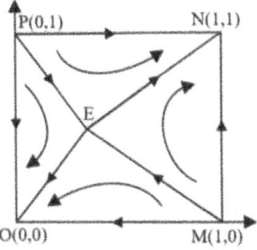

Fig. 1. Evolutionary phase diagram of neighboring chemical enterprises

As shown in Fig. 1, the square OMEN is divided into left and right parts by the broken line MEP. When the transformation decisions of both subjects fall in the right part NMEP, the game evolves to $N(1,1)$, implying that both neighboring chemical enterprises implement the digital transformation of emergency management. Consequently, both enterprises A and B obtain direct and indirect benefits. When the transformation decisions fall in the left part OMEP, neither enterprise will proceed with transformation. The probability of system evolution is determined by the area of these two sections, which depends on the coordinates of $E(x^*, y^*)$. Smaller values of x^* and y^*, the larger the area of NMEP, increasing the likelihood that both subjects will implement the transformation. The area of NMEP is calculated in formula 5.

$$S_{NMEP} = \frac{1}{2}\left[\left(1 - x^*\right) + \left(1 - y^*\right)\right] \tag{5}$$

The outcome of the evolutionary game is closely related to the coordinates of $E(x^*, y^*)$. The coordinates of $E(x^*, y^*)$ depend on direct potential benefits D_i, potential benefits discount factor η_i, relative benefits L_i, and input costs (C_i or C_i').

Next, we analyze these parameters to understand their influence on the decision-making of both subjects. The following propositions can be obtained.

Proposition 1. As direct potential benefits increase, the likelihood that enterprises will undergo digital transformation in emergency management also increases.

The larger the value of direct potential benefits D_i, the larger the value of S_{NMEP}. When D_i increases, the values of x^* and y^* become smaller, which means the possibility of system converging to $N(1,1)$ increases.

Proposition 2. The greater the utility of potential future benefits, the more motivated both enterprises will be to choose the "implement" strategy.

A larger potential benefits discount factor η_i results in a higher value of S_{NMEP}. As η_i increases, the values of x^* and y^* decrease, increasing the probability of convergence to $N(1, 1)$.

Proposition 3. The greater the relative benefits, the stronger the impetus for chemical enterprises to implement digital emergency management.

Relative benefits L_i have the same effect as D_i, η_i on S_{NMEP}.

Proposition 4. The higher the input costs of digital emergency management, the greater the probability that enterprises will choose not to implement it.

As the input cost $C_i(C_i')$ increases, the NMEP area shrinks, boosting the likelihood of system converging to $O(0, 0)$. This means the likelihood of chemical enterprises A and B choosing the "not implement" strategy will increase.

4 Construction of Evolutionary Game Model Between Park Management Departments and Chemical Enterprises

4.1 Basic Hypotheses

The park management department guides the digital transformation behaviors of enterprises. Different regulatory measures are tailored to the varying decision-making behaviors of chemical enterprises. The following hypotheses are proposed.

(1) The probability that park management departments choose to "supervise" is x_1. The probability that chemical enterprises choose to "implement" is y_1.
(2) The park management department incurs costs when carrying out supervision.
(3) To promote the digital transformation of emergency management, park management departments offer subsidies. They also impose penalties on enterprises with safety accidents, which can be effectively avoided by adopting digital emergency management.
(4) Safety risks are shared between park management departments and enterprises. Meaning that liability costs from safety accidents are partially transferred from chemical enterprises to management departments.

4.2 Model Construction

The benefits of park management departments and chemical companies are analyzed based on above hypotheses. The resulting benefit payoff matrix is shown in Table 4.

Table 4. Payoff matrix between park management department and chemical enterprises in the game

		Chemical enterprises	
		Implement (y_1)	Not Implement ($1-y_1$)
Park management departments	Supervise (x_1)	$\eta_1 kD - C_1 - \varepsilon C$	$P - C_1$
		$R + \eta D - (1 - \varepsilon)C$	$R - P$
	Not Supervise ($1-x_1$)	$\eta_1 kD$	0
		$R + \eta D - C$	R

The average expected benefits for the park management departments and chemical enterprises are shown in formulas 6 and 7.

$$E_g = x_1 P - x_1 C_1 - x_1 y_1 (\varepsilon C + P) + y_1 \eta_1 kD \tag{6}$$

$$E_m = y_1 \eta D - y_1 C + x_1 y_1 (\varepsilon C + P) + R - x_1 P \tag{7}$$

The replicated dynamic equations of the evolutionary game between the park management departments and chemical enterprises are shown in formula 8.

$$\begin{cases} F(x_1) = \dfrac{dx_1}{dt} = x_1(E_{x_1} - E_g) = x_1(1 - x_1)[P - C_1 - y_1(\varepsilon C + P)] \\ F(y_1) = \dfrac{dy_1}{dt} = y_1(E_{y_1} - E_m) = y_1(1 - y_1)[\eta D - C + x_1(\varepsilon C + P)] \end{cases} \tag{8}$$

4.3 Evolutionary Stability Analysis

The possible equilibrium points $O_1(0, 0)$, $P_1(0, 1)$, $M_1(1, 0)$, $N_1(1, 1)$, $E_1(x_1^*, y_1^*)$ can be obtained, where $x_1^* = -\frac{\eta D - C}{\varepsilon C + P}$ and $y_1^* = \frac{P - C_1}{\varepsilon C + P}$. These five possible equilibrium points constitute the solution domain.

The Jacobi matrix of the evolutionary system can be derived, as shown in formula 9. The stability results for potential equilibrium points are presented in Table 5.

$$\begin{aligned} J &= \begin{bmatrix} \frac{\partial F(x_1)}{\partial x_1} & \frac{\partial F(x_1)}{\partial y_1} \\ \frac{\partial F(y_1)}{\partial x_1} & \frac{\partial F(y_1)}{\partial y_1} \end{bmatrix} \\ &= \begin{bmatrix} (1 - 2x_1)[P - C_1 - y_1(\varepsilon C + P)] & x_1(1 - x_1)[-(\varepsilon C + P)] \\ y_1(1 - y_1)(\varepsilon C + P) & (1 - 2y_1)[\eta D - C + x_1(\varepsilon C + P)] \end{bmatrix} \end{aligned} \tag{9}$$

There are one center point and four saddle points. Solving for the eigenvalue of $\left| \begin{bmatrix} \lambda & 0 \\ 0 & \lambda \end{bmatrix} - J_{(x_1^*, y_1^*)} \right| = 0$, we get $\lambda^2 + [-(\eta D - C)]\left[1 - \frac{-(\eta D - C)}{\varepsilon C + P}\right](P - C_1)(1 - \frac{P - C_1}{\varepsilon C + P})$.

Obviously, the eigenvalues λ_1, λ_2 corresponding to equilibrium point $E_1(x_1^*, y_1^*)$ are a pair of pure imaginary roots. Therefore, $E_1(x_1^*, y_1^*)$ is not an ESS. The evolutionary trajectory forms a closed-loop around the center point $E_1(x_1^*, y_1^*)$, as shown in Fig. 2.

Table 5. Stability analysis of possible equilibrium points in evolutionary game systems

Possible equilibrium points	$Det(J)$	$Tr(J)$	Stability
$O_1(0, 0)$	–	N	Saddle point
$P_1(0, 1)$	–	N	Saddle point
$M_1(1, 0)$	–	N	Saddle point
$N_1(1, 1)$	–	N	Saddle point
$E_1(x_1^*, y_1^*)$	+	0	Center point

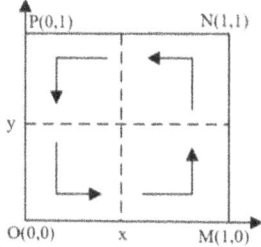

Fig. 2. Evolutionary phase diagram between park management departments and chemical enterprises

Proposition 1: The dynamics of strategic choices between chemical enterprises and park management department are characterized by a cyclical behavioral pattern.

Even if the strategy combination (supervise, implement) temporarily emerges, both parties struggle to maintain equilibrium. The underlying reason is that both the chemical enterprises and park management department often perceive greater short-term benefits in choosing "not supervise" and "not implement", leading to the (not supervise, not implement) combination, which is not a win-win scenario. This dynamic causes both sides to continually adjust their strategies in response to one another, resulting in a cyclical process of strategy evolution.

5 Numerical Analysis

5.1 Parameter Setting

To better elucidate the influence of each parameter on the strategy choice of the subjects, this study employs numerical analysis methods to verify the theoretical result using Python 3.11.7. Based on the hypotheses and existing literatures on setting parameters [20]. Three sets of parameter initial values are set as shown in Table 6. D_A and D have the same value in all three sets of experiments and are all set to 6 and 5, respectively.

Table 6. The initial values of the parameters

Set	D_B	η_A	η_B	L_A	L_B	C_A	C_B	$C_{A'}$	$C_{B'}$	P	C_1	C	ε	η
1	6	0.6	0.7	1.2	1	5	5	3	3	3	1.5	5	0.6	0.4
2	5	0.5	0.6	1.2	1	5	5	3	3	1	0.5	3	0.5	0.5
3	6	0.5	0.4	1.3	2	5.5	5.5	4	4	2.5	0.5	6	0.3	0.6

5.2 Model Validation

Based on the three sets of initial values in Table 6, the convergence of subject strategy in the game is verified. The step size is set to 0.001, with a total of 1000 iterations. The results are displayed in Figs. 3 and 4.

In the evolutionary game model of neighboring chemical enterprises, it can be seen from Fig. 3(a)-(c) that ESS increasingly converges to (implement, implement) as S_{NMEP} increases and S_{OMEP} decreases. Conversely, ESS converges to (not implement, not implement) as S_{NMEP} decreases and S_{OMEP} increases. This verifies the relationship between S_{NMEP} and S_{OMEP} with ESS.

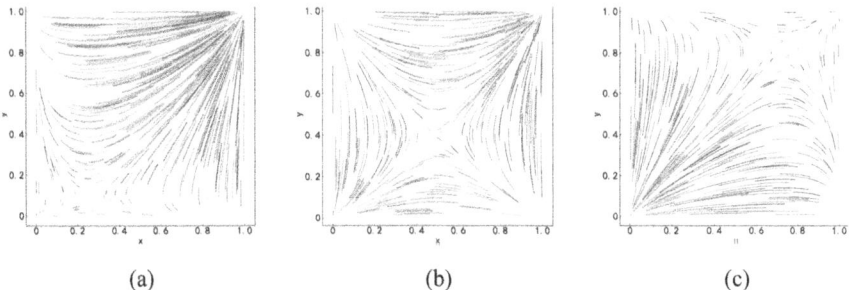

| (a) | (b) | (c) |

Fig. 3. Relationships between S_{NMEP} and S_{OMEP} with ESS between chemical enterprises

In the evolutionary game model between park management departments and chemical enterprises, the strategies of both parties continuously cycle, as shown in Fig. 4(a)-(c). This aligns with the theoretical analysis results presented in Sect. 4.

5.3 Simulation Analysis

Numerical Analysis Between Neighboring Chemical Enterprises. The second set of initial data is applied to more intuitively analyze the effect of initial strategy probability x, direct potential benefits D_i, potential benefits discount factor η_i, relative benefits L_i, and input costs (C_i or C_i') on evolutionary results. There is a similar situation between enterprise A and B. Only the strategy evolution process of enterprise A is analyzed here. We assume that initial probability is $x = 0.4$, $y = 0.1$.

(1) The effect of initial probability on strategy evolution.

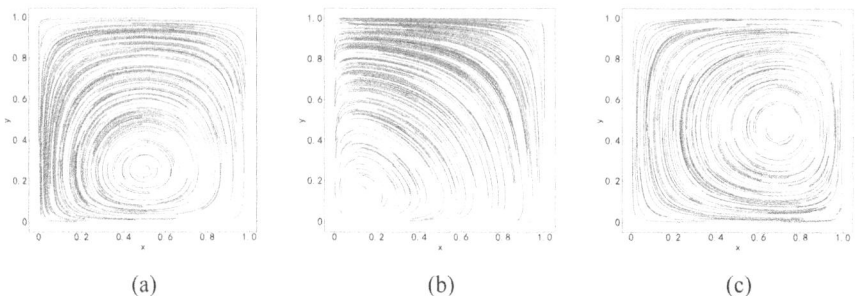

      (a)   (b)   (c)

Fig. 4. Relationships between S_{NMEP} and S_{OMEP} with ESS between park management departments and chemical enterprises

The numerical simulation results of initial probability x on strategy evolution are shown in Fig. 5(a). The horizontal coordinate t represents the dynamic evolution time, and the vertical coordinate x represents the probability of enterprises choosing the "implement" strategy. As x increases from 0.1 to 0.9, the slope of the curve decreases, indicating that the enterprise A has a higher probability of choosing "implement" strategy at the same time nodes. The time required to reach the evolutionarily stable strategy (ESS) decreases. In other words, there is a positive linear correlation between the initial probability of choosing "implement" and the subsequent willingness.

(2) The effect of direct potential benefits on strategy evolution.

With other parameters held constant, the impact of D on enterprise A's choice is shown in Fig. 5(b), where D takes values 3,4, 5, 6 and 7. When D is 6 or 7, enterprise A is more inclined to choose "implement". In contrast, for D is 3, 4 or 5, enterprise A tends to choose "not implement". Additionally, as D decreases, the probability x converges to 0 at a faster rate, indicating that chemical enterprises are inclined to choose the "implement" strategy only when the digital transformation of emergency management offers significant direct benefits.

(3) The effect of potential benefits discount factor on strategy evolution.

The effect of potential benefits discount factor η on the strategy of enterprise A is shown in Fig. 5(c), where η takes values 0.3, 0.4, 0.5, 0.6 and 0.7. When η is 0.6 or 0.7, the future potential benefits are relatively high, and the probability will gradually converge to 1. However, when η is 0.3, 0.4 or 0.5, the curve will converge to 0 at an increasingly faster rate as η decreases. This indicates that enterprises are more likely to choose the "implement" strategy only when the future prospects of digital transformation in emergency management are clearer and the potential benefits are substantial.

(4) The effect of relative benefits on strategy evolution.

The impact of changes in the simulation parameter L on the strategy of enterprise A is illustrated in Fig. 5(d). Apart from when L is 0.9, all other values of L lead to the enterprise being inclined to choose "implement" strategy, with the probability x gradually converging to 1. As the value of L increases, the probability of enterprise A implementing the strategy rises more rapidly. This indicates that enterprise A's willingness to implement digital emergency management strengthens with higher L. Thus, it is evident that direct profit L consistently influences strategy choice, encouraging the adoption of digital transformation in emergency management.

(5) The effect of input costs on strategy evolution.

Since the effect trend of C and C' on strategy evolution is similar, only the effect on the evolutionary results of C is discussed. The influence is illustrated in Fig. 5(e). When C is 4 or 5, the probability x gradually converges to 1. However, when C is 6, 7 or 8, the curve converges to 0 as C increases, with a faster rate of convergence. This indicates that when the input cost is very low, the probability of enterprise A choosing the "implement" strategy is higher. As the input costs rise, this probability gradually decreases, accelerating the decline. This suggests that high input costs significantly reduce enterprises' motivation to implement the digital transformation of emergency management, making them more inclined to maintain the status quo.

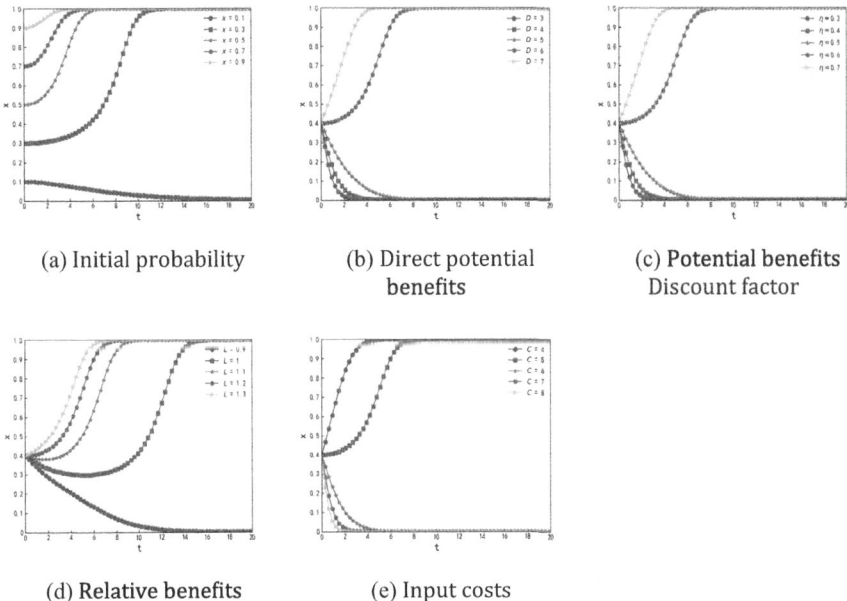

(a) Initial probability

(b) Direct potential
benefits

(c) Potential benefits
Discount factor

(d) Relative benefits

(e) Input costs

Fig. 5. The effect of parameters on strategy evolution between neighboring chemical enterprises

Numerical Analysis Between Park Management Departments and Chemical Enterprises. Through several calculations, it can be determined that the probability of park management departments and chemical enterprises initially choose "supervise" and "implement" is $Initial_0\left(x_1^*, y_1^*\right) = \left(-\frac{\eta D-C}{\varepsilon C+P}, \frac{P-C_1}{\varepsilon C+P}\right) = (0.5, 0.25)$. We adjust the initial probabilities up and down by 0.05, that is $Initial_1\left(x_1^*, y_1^*\right) = (0.55, 0.3)$ and $Initial_2\left(x_1^*, y_1^*\right) = (0.45, 0.2)$. The evolution results are shown in Fig. 6(a).

When the initial probability values of both subjects fluctuate, their strategy choices also fluctuate repeatedly. This indicates that the strategy choices of park management departments and chemical enterprises are always in a fluctuating state, which aligns with the theory results of stability analysis. This also underscores the recurrent and long-term nature of digital transformation of emergency management under the supervision of park management departments.

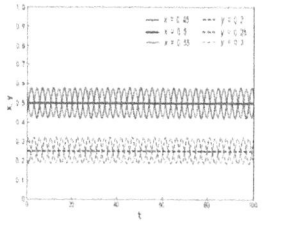
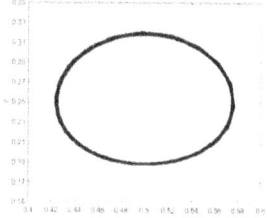

(a) Evolution process under different
initial values

(b) Evolution process of mixed
strategies

Fig. 6. The strategy evolution between park management departments and chemical enterprises

The evolution curve of system is presented in Fig. 6(b), when the initial strategy probabilities of park management departments and chemical enterprise are set as $Initial_1 (x_1^*, y_1^*) = (0.55, 0.3)$. The results indicate that the evolutionary process follows a closed-loop pattern, with periodic movement around the starting point. This demonstrates that the interaction between park management departments and chemical enterprises exhibits cyclical behavior in the game process, leading to dynamic equilibrium.

6 Conclusion

Industrial agglomeration has further exacerbated the risk of safety accidents in chemical enterprises. There is an urgent motivation to implement the digital transformation of emergency management. This study explores the mechanism of digital transformation in emergency management for chemical enterprises from the perspective of industrial agglomeration using evolutionary game theory. The results indicate that transformation decisions are influenced not only by direct potential benefits, potential benefits discount factor, relative benefits, and input costs, but also significantly by the decisions of neighboring enterprises and third-party park management departments. Neighboring chemical enterprises exhibit two stable evolutionary states: (implement, implement) or (not implement, not implement). However, when park management departments intervene with regulatory measures, there are no stable strategies in the entire evolutionary system, leading to periodic evolution of strategy choices over time and forming "regulatory dilemma". This implies that merely relying on regulatory measures may not achieve sustained and stable transformation outcomes.

Acknowledgments. This work was supported by the National Natural Science Foundation of China (72271041; 72434001), Key R & D (242400411089) and Promotion Projects in Henan Province (242102320060).

References

1. Qi, Q., et al.: New IT driven rapid manufacturing for emergency response. J. Manuf. Syst. **60**, 928–935 (2021)
2. Bhawsar, P., Chattopadhyay, U.: Evaluation of industry cluster competitiveness: a quantitative approach. Benchmarking. **25**(7), 2318–2343 (2018)
3. Aiello, G., et al.: Special section on modeling and simulation in disaster and emergency management. IEEE Trans. Eng. Manage. **67**(3), 516–518 (2020)
4. Kau, L.J., Chen, C.S.: A smart phone-based pocket fall accident detection, positioning, and rescue system. IEEE J. Biomed. Health Inform. **19**(1), 44–56 (2014)
5. Fedele, R., Merenda, M.: An IoT system for social distancing and emergency management in smart cities using multi-sensor data. Algorithms **13**(10), 254 (2020)
6. Ojha, A., Jindal, A., Chanak, P.: An intelligent indoor emergency evacuation system using IoT-enabled WSNs for smart buildings. IEEE Internet Things J. **11**(5), 8838–8847 (2023)
7. Park, S., et al.: Participatory framework for urban pluvial flood modeling in the digital twin era. Sust. Cities Soc. **108**, 105496 (2024)
8. Alghamdi, N.S., Alghamdi, S.M.: The role of digital technology in curbing COVID-19. Int. J. Environ. Res. Public Health **19**(14), 8287 (2022)
9. Zhou, C., et al.: COVID-19: challenges to GIS with big data. Geogr. Sustain. **1**(1), 77–87 (2020)
10. Sun, W., Bocchini, P., Davison, B.D.: Applications of artificial intelligence for disaster management. Nat. Hazards **103**(3), 2631–2689 (2020)
11. Kang, L., et al.: Risk warning technologies and emergency response mechanisms in Sichuan-Tibet Railway construction. Front. Eng. Manag. **8**(4), 582–594 (2021)
12. Harrison, S., Johnson, P.: Challenges in the adoption of crisis crowdsourcing and social media in Canadian emergency management. Gov. Inf. Q. **36**(3), 501–509 (2019)
13. Chen, M., et al.: Digital health interventions for COVID-19 in China: a retrospective analysis. Intell. Med. **1**(1), 29–36 (2021)
14. Dohale, V., et al.: Manufacturing strategy 4.0: a framework to usher towards industry 4.0 implementation for digital transformation. Ind. Manage. Data Syst. **123**(1), 10–40 (2022)
15. Dubey, R.: Unleashing the potential of digital technologies in emergency supply chain: the moderating effect of crisis leadership. Ind. Manage. Data Syst. **123**(1), 112–132 (2023)
16. Zhu, Y., et al.: Networked decision-making dynamics based on fair, extortionate and generous strategies in iterated public goods games. IEEE Trans. Netw. Sci. Eng. **9**(4), 2450–2462 (2022)
17. Daskalakis, C., Goldberg, P.W., Papadimitriou, C.H.: The complexity of computing a Nash equilibrium. Commun. ACM **52**(2), 89–97 (2009)
18. Friedman, D.: Evolutionary games in economics. Econometrica J. Econom. Soc. **59**, 637–666 (1991)
19. Sandholm, W.H.: Local stability under evolutionary game dynamics. Theor. Econ. **5**(1), 27–50 (2010)
20. Luo, J., Li, W., Zhao, Y.: Stability analysis of manufacturing enterprise service derivative based on evolutionary game theory. J. Syst. Eng. **31**(6), 761–771 (2016)

Research on the Credibility Evaluation Method of Online Medical Community Answer Content Based on Domain Knowledge Graph

Yunjiang Xi[1,3]([✉]), Jiaxiu Geng[1], YuShan Deng[1], Xiao Liao[2], and Juan Yu[4]

[1] School of Business Administration, South China University of Technology, Guangzhou 510641, China
yjxi@scut.edu.cn
[2] School of Internet Finance and Information Engineering, Guangdong University of Finance, Guangzhou 510521, China
[3] School of Management, Guangzhou City University of Technology, Guangzhou 510800, China
[4] Faculty of Economics and Management, Fuzhou University, Fuzhou 350108, China

Abstract. An intelligent evaluation method was proposed for the credibility of answers in online medical communities (OMCs). By applying the method to evaluate and classify the credibility of answers, this paper aimed to inspire users to adopt reliable health information, enhance the credibility of content, and support OMCs healthy development. The study constructed a content knowledge graph for answers in OMCs and a domain knowledge graph for diabetes. The concepts of entity regularity, relationship consistency coefficient, and relationship accuracy were introduced to calculate credibility scores for the triples in the community answers, which would be aggregated to evaluate the content credibility. Validation results from the xywy.com website show that our method effectively evaluated and classified the credibility of Q&A content, achieving intelligent identification and filtering of suspicious answers. The precision accuracy of credible answers is 92.5%, significantly improving efficiency and interpretability compared to manual scoring methods. This study optimized the current content review model in OMCs, enhanced content management efficiency and accuracy, and provided feasible tools and methods for monitoring the information quality in OMCs and delivering reliable medical knowledge services.

Keywords: Online medical community · Knowledge graph · Credibility evaluation · Deep learning

To advance the "Healthy China" strategy, the General Office of the State Council, in its Opinions on Promoting the Development of 'Internet + Healthcare', encourages services such as telemedicine and health consultations to enhance service efficiency and reduce costs. On March 22, 2024, a report published by the China Internet Network Information Center highlighted the rapid development of smart healthcare. The number of internet healthcare users increased by 51.39 million compared to 2022, and online medical communities have become a significant platform for accessing health information. However, concerns over medical safety remain critical, as misinformation may lead

X. Tang et al. (Eds.): KSS 2024, CCIS 2269, pp. 256–276, 2025.
https://doi.org/10.1007/978-981-96-0178-3_18

to serious consequences. To address the problems, this paper proposes a credibility test method for OMCs answers based on knowledge graph technology. Starting from the answer content itself, this study constructs a knowledge graph for medical community answers and a domain-specific knowledge graph, testing the consistency of community answer content with objective medical knowledge based on triples, and establishing consistency indicators as the basis for evaluating answer credibility.

1   Related Research

1.1   Research on Information Credibility

Information credibility originated in communication [1], encompassing various connotations such as authenticity, accuracy, professionalism, and reliability [2], which vary with research contexts and are influenced by the information itself, its source, and the dissemination medium [3, 4]. Before the advent of the internet, information credibility was mainly judged by its source [5]. With technological advancement, user-generated content and author characteristics have become more crucial criteria for evaluating information credibility [6].

Current research on factors influencing information credibility mainly focuses on social media and online communities [7]. In social media, users' personal profiles and cultural differences affect information credibility [8]. In online communities, factors such as users' online reputation, content objectivity, completeness, argument strength, evaluations of responses [9], and the relevance of Q&A influence answer credibility.

Credibility evaluation methods include user perception-based manual evaluations and content-based automated evaluations. Manual evaluations typically determine an indicator system from dimensions such as source, content, and dissemination medium, assigning reasonable weights to indicators using methods such as surveys and expert interviews [10] etc. Objective weight assignment methods include Bayesian inference [11] and D-S evidence theory [12] etc. Automated evaluations use data-driven models, treating credibility evaluation as binary or multi-class classification problems, using machine learning algorithms for model training and prediction [13].

Existing evaluation methods have limitations. Manual evaluations are heavily influenced by personal factors, while automated evaluations rely on features beyond the text, lacking deep semantic relationship extraction from the answer content itself. This paper proposes a knowledge graph-based method for testing information credibility, focusing on the information content.

1.2   Research on Knowledge Graph

A knowledge graph represents a knowledge base of concepts, entities, and their relationships in the objective world in a graph form [14], commonly used in Q&A or recommendation systems, risk identification, and credibility testing. In the medical and health field, knowledge graphs based on authoritative public data, such as breast tumor knowledge graphs [15] and diabetes knowledge graphs [16], assist in intelligent diagnosis and treatment. Literature and Electronic Medical Records (EMR)-based medical knowledge graphs are more domain-specific, but OMCs Q&A content updates quickly and is gaining increasing attention. Chen et al. proposed a domain knowledge support framework for online health communities, using KI-TM methods to extract and expand explicit knowledge from texts [17]. Zou Yuwei used disease introductions from medical websites as data sources, extracting entities, attributes, and relationships to semi-automatically construct a medical website knowledge graph [18].

Named entity recognition (NER) and relationship extraction (RE) are critical techniques for constructing knowledge graphs. With the development of deep learning, models like BiLSTM-CRF [19] are widely used for NER. RE techniques include Convolutional Neural Networks (CNN) [20], Recurrent Neural Networks (RNN) [21], and hybrid models incorporating attention mechanisms [22–24].

Current research on knowledge graphs in online medical Q&A communities focuses more on identifying entities and extracting knowledge, with less emphasis on evaluating the quality of extracted knowledge. This paper converts unstructured text from OMCs answers into structured triples, focusing on evaluating the credibility of the answer content based on the constructed knowledge graph.

2 Online Medical Community Answer Content Knowledge Graph Construction Framework

In the construction of answer content knowledge graph, this paper uses the current common Chinese knowledge graph construction method to build a technical framework of online community medical answer content knowledge graph based on pipeline mode and deep learning model, as shown in Fig. 1.

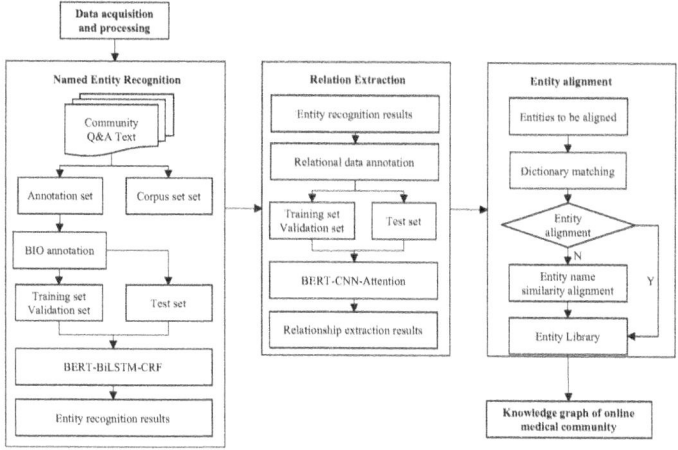

Fig. 1. A Framework for Constructing a Knowledge Graph of Answering Content in OMCs

NER is one of the basic sub-tasks of building a knowledge graph, that is, extracting named entities from text and classifying them into predefined categories. This paper uses deep learning models for NER. The model shown in Fig. 2, mainly includes two parts:

word embedding layer and feature extraction layer. In the word embedding layer, the character-level word vector is obtained by BERT-wmm. BERT-wmm uses the full-word masking mode for the characteristics of Chinese text, and the obtained word vector can contain the semantic information of the word at the same time. In the feature extraction layer, this paper uses the BiLSTM-CRF model. This method first extracts the features of the sentence sequence through the BiLSTM layer, and learns the deep representation of each word and the context relationship between the labels. The output features use the CRF model to establish the dependencies between the labels, obtain the labels and their corresponding probabilities, and output the entity labels corresponding to each word.

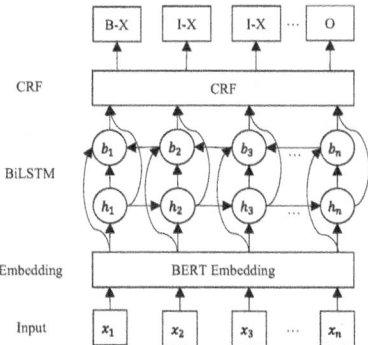

Fig. 2. NER Model (BERT-BiLSTM-CRF)

On the basis of NER, this study uses CNN-Attention model to classify the relationship between entity pairs to achieve the purpose of RE. As shown in Fig. 3, $x_1, x_2, \ldots, x_n$ are n input sentences. Each sentence contains entities e_1, e_2. After CNN convolution processing, the sentence vectors $l_1, l_2, \ldots, l_n$ are obtained, and then the weight α is assigned by the attention mechanism, so that the sentences with large relationship with the target get greater weight, and finally the relationship label Y is output.

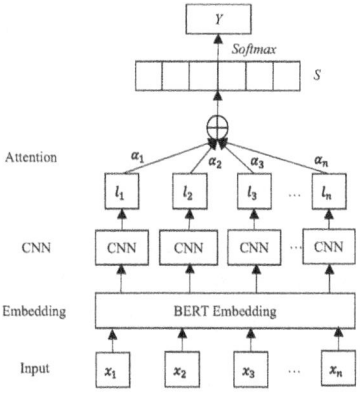

Fig. 3. Relation Extraction Model (BERT-CNN-Attention)

The text in the OMCs has the characteristics of irregular expression and data sparseness. An entity may have multiple expressions. Entity alignment can solve the inconsistency problems such as entity conflict and unclear point in heterogeneous data. In this study, two methods of dictionary matching and entity name similarity are used to align entities, as shown in the entity alignment section of Fig. 1.

3   OMCs Answer Content Credibility Test Method

Based on representing unstructured community answer texts as structured triples in the knowledge graph, a consistency check of entity pairs and relationships within the triples is conducted using an entity alignment method. Entity regularity and relationship consistency coefficients are introduced as metrics for measuring the consistency of triples. These metrics are combined to obtain a consistency score for the triples. Subsequently, considering the impact of relationship accuracy, a credibility score for the triples is derived and aggregated. Finally, the credibility of the answer content is assessed using the Sigmoid function.

3.1   Consistency Checking Algorithm Based on Triples

Each answer text $QA_i (i = 1, 2, .., n)$ in the medical community and the domain knowledge graph (FKG) are represented as sets of entity-relationship triples through knowledge extraction, as shown in Eqs. (1) and (2). Here, e_h^*, e_t^* denote the head and tail entities of a triple, respectively, with r_{lanel}^* representing the relationship between them. E^* represents the set of entities, and R^* represents the set of relationships.

$$QA_i = \{Triple_i^c = \left(e_h^c, r_{label}^c, e_t^c\right)_i | e_h^c, e_t^c \epsilon E^c, r_{label}^c \epsilon R^c, i = 0, 1, 2, \ldots, n\} \quad (1)$$

$$FKG = \{Triple_j^f = \left(e_h^f, r_{label}^f, e_t^f\right)_j | e_h^f, e_t^f \epsilon E^f, r_{label}^f \epsilon R^f, j = 0, 1, 2, \ldots, m\} \quad (2)$$

In OMCs, users engage in the processes of information seeking and adoption when asking questions or searching for health information. According to Information Adoption Theory [25] and Prominence-Interpretation Theory, this paper posits that the credibility of responses in OMCs refers to the authenticity and accuracy of the information itself, representing an objective metric. It indicates the extent to which the response content aligns with existing medical knowledge. Specifically, it measures the consistency and degree of agreement between the triples $Triple_i^c$ from community responses and the corresponding triples $Triple_j^f$ from the domain knowledge graph. Therefore, this study conducts consistency checks between triples from two different sources. The consistency check is divided into two steps: the consistency check of entity pairs is performed first and followed by, the relationship consistency check of the entity pairs if previous check is passed. The detailed process is illustrated in Fig. 4.

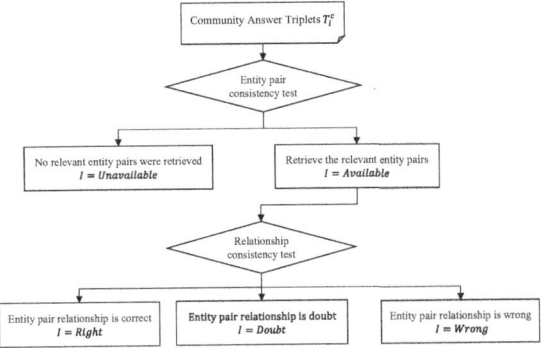

Fig. 4. Triple Consistency Checking Process

1. Entity pair consistency check

Due to the non-standard situation of different expressions of the same entity in the OMCs, this paper introduces the concept of entity specification to measure the difference between entities in the process of entity-to-entity consistency check.

For each entity e_i^c, it is aligned with the entity e_j^f of the same type in the domain knowledge graph, and the similarity between the two is calculated, and the entity $e_{max-min}^f$ with the highest similarity and the maximum similarity sim_{max} are returned, that is, $sim_{max} = MAX\left(Similar\left(e_i^c, e_j^f\right)\right)$. If the maximum similarity is greater than or equal to the threshold ε, the entity e_i^c is considered to be consistent with the entity $e_{max-min}^f$, and the entity specification $Specification(e_i^c)$ is the value of sim_{max}. Conversely, if sim_{max} is less than the threshold, it is considered that the entity consistent with the community answer entity e_i^c cannot be found in the domain knowledge graph, and the UNK symbol is returned, and the entity specification $Specification(e_i^c)$ is 0, that is, $Specification\left(e_i^c\right) =$
$$\begin{cases} sim_{max}, sim_{max} \geq \varepsilon \\ 0, sim_{max} < \varepsilon \end{cases}$$. The specific process is shown in Fig. 5.

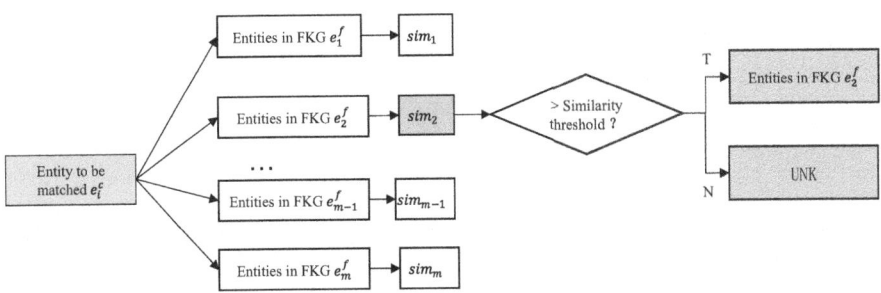

Fig. 5. Entity Matching Process

According to the process shown in Fig. 5, for a certain community answer triple $Triple_i^c = (e_h^c, r_{label}^c, e_t^c)_i$, the domain knowledge entity pairs e_h^c, e_t^c with the greatest similarity to the entity pairs $e_{h-match}^f, e_{t-match}^f$, and the normalization of the entity pairs can be obtained, as shown in Eq. (3), where α is the weight coefficient, and $\alpha = 0.5$.

$$Specification(e_h^c, e_t^c) = \alpha \times Specification(e_h^c)$$
$$+ (1 - \alpha) \times Specification(e_t^c) \quad (3)$$

2. Relationship consistency check

In the test of relationship consistency, this paper introduces the concept of relationship consistency coefficient to represent the degree of consistency between community answer entity pairs and domain knowledge graph entity pairs.

First, if there is an entity in the entity pair (e_h^c, e_t^c) that cannot be retrieved in the domain knowledge graph, its relationship consistency coefficient Consistency $Consistency(r_{label}^c, Unavailable)$ is 0.

When the entity pair (e_h^c, e_t^c) passes the entity pair consistency check, it shows that the consistency check of its relationship can be based on the existing domain knowledge graph, as shown in Fig. 6.

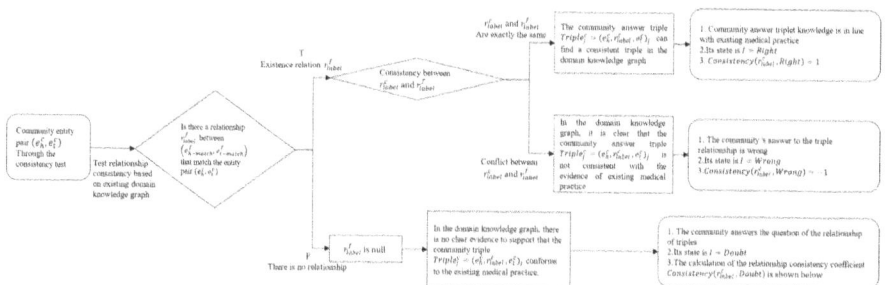

Fig. 6. Relationship Consistency Verification Steps

For the case of doubtful relationship, the relevant literature [25] pointed out that information consistency refers to the degree of consistency with other information in the community. According to the existing research [9], when users evaluate information, it is easy to compare it with other similar information. The more consistent with most information, the easier it is to be believed. Information consistency has a positive effect on the credibility of online health information [26]. Similarly, the more a triplet is mentioned in different answers of different doctors, the more evidence to support its establishment. Referring to the principle of TF-IDF, the specific calculation of its relationship consistency coefficient is shown in Eq. (4), where x represents the number of times the triplet is mentioned by different doctors in different answers. According to the L'Hôpital's limit theorem, with the increase of x, the relationship consistency coefficient tends to 1. When x is 0, it means that no other doctor supports the relationship of the triple, and the relationship consistency coefficient is -1, which conforms to the definition of the relationship consistency coefficient in this paper.

$$Specification(r_{label}^{c}, Doubt) = \begin{cases} xln(1 + \frac{1}{x}), x > 0 \\ -1, x = 0 \end{cases} \tag{4}$$

Based on the consistency weight coefficient of the entity pair normalization degree of the triple and its relationship, the consistency score of the triple is obtained, see Eq. (5), where $I = \{Unavailable, Right, Wrong, Doubt\}$. The consistency score represents the degree of consistency between a community answer triple and its corresponding domain knowledge graph triple.

$$Consistency\_score(Triple_i^c) = Consistency(r_{label}^c, I) \times Specification(e_h^c, e_t^c) \tag{5}$$

3.2 Answer Content Credibility Test Based on Triplet-Based Consistency Checking Algorithm

Through the consistency check, a consistency score for each triple is obtained, from which the credibility score based on the triple's consistency is derived. In this study, the credibility of the answer content in OMCs refers to the authenticity and accuracy of the content. Accuracy is also a component factor of the credibility of the triple content. Additionally, considering potential errors caused by the RE algorithm, this study introduces the concept of Relation Accuracy. Relation Accuracy represents the probability value that a triple in the community knowledge is valid, i.e., in the extracted triple $Triple_i^c = (e_h^c, r_{label}, e_t^c)_i$, it denotes the probability that the entity pair (e_h^c, e_t^c) has the relationship r_{label}, as expressed in Eq. (6). This is used as a weight adjustment factor for the credibility score of the triple, leading to the final credibility score of the triple, as shown in Eq. (7).

$$Relation\_accuracy(Triple_i^c) = \frac{e^{R_r}}{\sum_r^Q e^{R_r}} \tag{6}$$

$$\begin{aligned} Credibility\_score(Triple_i^c) \\ = Relation\_accuracy(Triple_i^c) \\ \times Consistecy\_score(Triple_i^c) \end{aligned} \tag{7}$$

where, $Relation\_accuracy(Triple_i^c) \in (0,1)$.

The credibility score of an answer can be determined by aggregating the credibility scores of all the triples within the response. To account for erroneous triples, a penalty term is introduced. If incorrect triples are present in the answer, the credibility score will decrease rapidly from its original value, tending towards negative infinity as the number of erroneous triples increases. This is represented in Eqs. (8) and (9), where n denotes the number of incorrect triples.

$$Credibility\_score(QA_i) = \sum_{i=1}^{M} Credibility\_score(Triple_i^c) + Penalty(I) \tag{8}$$

$$Penalty(I) = \begin{cases} -e^n, I = Wrong \\ 0, I \neq Wrong \end{cases} \tag{9}$$

Based on the obtained credibility scores, there is a significant degree of dispersion. Therefore, this paper applies the Sigmoid function to process the credibility scores, yielding the final credibility $Credibility(QA_i)$ for each answer. See Eq. (10).

$$Credibility(QA_i) = \frac{1}{1 + e^{-Credibility\_score(QA_i)}}, i = 1, 2, \ldots, n \tag{10}$$

Next, the rationality of the aforementioned credibility test algorithm is analyzed. The Sigmoid function maps credibility scores ranging from $(-\infty, +\infty)$ to the interval $(0, 1)$, meeting the numerical mapping requirements of this study. When the content credibility score is 0, it indicates that the content alone cannot determine whether the answer is credible, with a probability of 0.5. If the content credibility score is greater than 0, the probability that the answer is credible exceeds 0.5, and as the content credibility score increases, the credibility of the answer content approaches 1. Conversely, if the content credibility score is less than 0, the probability that the answer is credible is below 0.5, and as the score decreases further, the probability of being a credible answer asymptotically approaches 0.

4   Case Analysis of Credibility Test of Diabetes Answer Content in xywy.Com

4.1   Experimental Data

In this study, Python was utilized to collect 18793 Q&A text data related to diabetes from the "Q&A section" of xywy.com, spanning from June 1, 2016, to September 1, 2020. The collected data underwent preprocessing to remove entries with missing or duplicate doctor responses, resulting in 18521 valid data points. The question descriptions from the askers and the corresponding doctor responses were merged to serve as the data source for constructing the community Q&A knowledge graph.

4.2   Construction of Community Answer Content Knowledge Graph

Name Entity Recognition
The health needs of users within the community, derived from the experimental data using word cloud analysis. Based on the results of the word cloud analysis, and integrating the definitions of diabetes knowledge graph entity types from Table 6 and the literature [27], 12 types of entities have been identified as shown in Table 1.

NER is a supervised learning process that requires prior data annotation. In this study, the YEDDA annotation tool [28] was utilized to randomly sample and annotate Q&A texts using the BIO tagging scheme.

Due to the poor performance of neural network models in handling long sequences, the long texts were segmented at punctuation marks ("。 | ? | !") before model training.

Table 1. OMCs Diabetes Entity Category

ID	Category	Coding	Meaning
1	Disease	DIS	The name of disease
2	Symptom	SYP	Symptoms and signs of the disease
3	Body parts	BOD	All parts and organs of the human body
4	Test	TES	Inspection methods and indicators
5	Test Result	TER	The specific value of the index
6	Drug	DRU	The name of the drug
7	Amount of Drug	AMO	Dosage, several tablets at a time
8	Method of administration	MED	The time of medication,etc
9	Treatment	TRM	Non-drug treatments
10	Operation	OPR	The name of the surgery, etc
11	Drug effect	EFF	The role of drugs
12	Diet	DIE	Food and food characteristics

The processed sentences were kept under 120 characters in length. To address excessively short sentences resulting from irregular punctuation usage, sentences shorter than 10 characters were merged with the subsequent sentence. This standardization process yielded 10,541 training samples, 1,332 validation samples, and 1,233 test samples.

In evaluating the effectiveness of NER, precision and recall were combined to obtain the F1-score, which serves as the criterion for assessing the performance of the model.

In the word embedding layer of NER, this study compares two prevalent Chinese BERT models, BERT-base and BERT-wwm. Early stopping was employed to monitor the performance of the validation set during training, thereby mitigating the risk of model overfitting. The training results, as illustrated in Fig. 7, indicate that the model utilizing BERT-wwm demonstrates more stable performance.

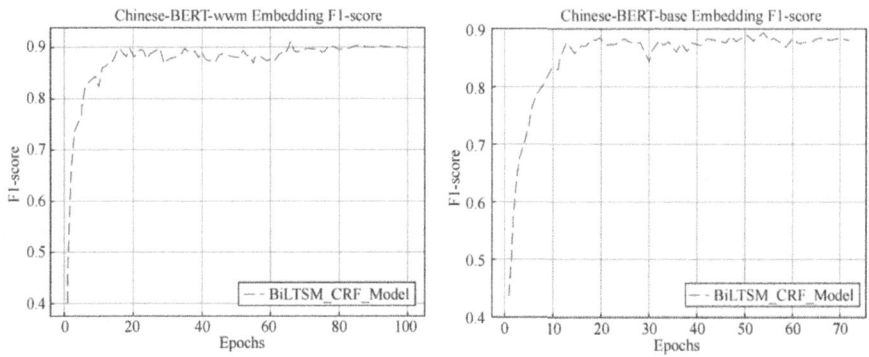

Fig. 7. Comparison of Effects Between BERT_wwm and BERT_base

In the feature extraction layer, four models CNN-LSTM [29], Bi-LSTM, BiGRU and BiGRU-CRF are selected to compare with BiLSTM-CRF model. The comparison results are illustrated in Fig. 8.

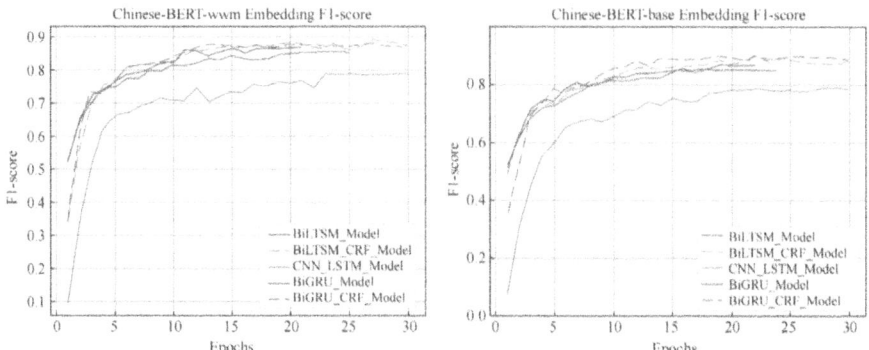

Fig. 8. Comparison of NER Models

NER experiments on OMCs texts were conducted using the BERT-BiLSTM-CRF model. The experimental results indicate that the model achieved an overall precision of 88.4%, a recall rate of 91.46%, and an F1-score of 89.83%. These results demonstrate that the model performs effectively, accurately identifying relevant entities within the texts of online medical Q&A communities.

Relation Extraction

Prior to RE, relationship categories were defined. Based on the results of NER, word cloud analysis, and the constraints on diabetes entity relationships as outlined in reference, this study ultimately defined 11 relationship categories as shown in Table 2.

This paper employs a pipeline approach for NER and RE, with the data annotation for RE conducted on the pre-processed data. The conventional data annotation method, involves labeling in the format of "'Entity 1', 'Entity 2', 'Relationship Label', 'Sentence'". However, this method is inadequate for handling cases where a sentence contains multiple entities and overlapping relationships. To address this issue, entity masking is applied to the data generated from conventional annotation. Specifically, Entity 1 and Entity 2 are concatenated with the symbol "$" at the beginning of the sentence, and the entities involved in the sentence are replaced with the symbol "#", effectively creating a cloze-style gap-filling format. The processed data is presented in Table 3.

Table 2. Relationship Types of Diabetes in OMCs

ID	Relationship Constraint Type	Coding
1	Test -> Disease	TES_DIS
2	Symptom -> Disease	SYP_DIS
3	Non-drug treatment -> Disease	TRM_DIS
4	Drug -> Disease	DRU_DIS
5	Undesirable diet -> Disease	N_DIE_DIS
6	Suitable diet -> Disease	Y_DIE_DIS
7	Body -> Disease	BOD_DIS
8	Amount of drug -> Drug	AMO_DRU
9	Method of administration -> Drug	MED_DRU
10	Drug effect -> Drug	EFF_DRU
11	Unknown	UNK

Table 3. Example of Annotation Data After Mask Processing

Sentence
Generally speaking, diabetes is polydipsia polyphagia polyuria, it is recommended to check fasting blood glucose (糖尿病\$多饮\$一般来说, ###是##多食多尿的,建议检查空腹血糖)
Generally speaking, diabetes is polydipsia polyphagia polyuria, it is recommended to check fasting blood glucose (糖尿病\$多食\$一般来说, ###是多饮##多尿的,建议检查空腹血糖)
Generally speaking, diabetes is polydipsia polyphagia polyuria, it is recommended to check fasting blood glucose (糖尿病\$多尿\$一般来说, ###是多饮多食##的,建议检查空腹血糖)
Generally speaking, diabetes is polydipsia polyphagia polyuria, it is recommended to check fasting blood glucose (糖尿病\$空腹血糖\$一般来说, ###是多饮多食多尿的,建议检查####)

After entity masking, the training samples for different entity pairs and relationships input into the model are varied, which is more conducive to the model's learning of sentence features and subsequent classification of relationship labels. As shown in Table 4, the model with entity masking processing demonstrates a 17.73% improvement in precision, an 18.25% improvement in recall, and an 18.21% improvement in F1-score compared to the model without entity masking.

In the RE phase, the BERT-wwm model, which demonstrated strong performance in the NER phase, was utilized for word embedding to generate character-level word vectors, serving as input for the subsequent convolutional neural network model. To enhance the accuracy of entity relationship extraction in OMCs, this study compared several state-of-the-art models, including CNN-Attention, BiLSTM-Attention, and BiGRU-Attention. The results, depicted in Fig. 9, indicate that the CNN-Attention model

Table 4. Comparison of Annotation Methods

Whether Mask processing is performed	Precision	Recall	F1-score
Yes	84.95%	85.83%	84.95%
No	67.22%	67.58%	66.74%

achieved superior performance on the dataset used in this study. Consequently, the experiments for RE were conducted using the BERT-CNN-Attention model.

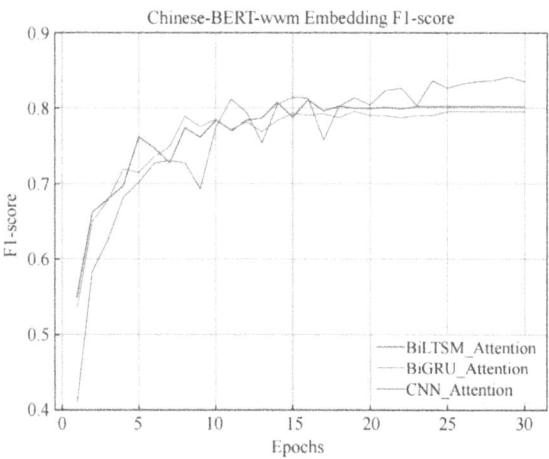

Fig. 9. Comparison of Three RE Models

An experiment was conducted using the BERT-CNN-Attention model for RE from texts in online medical Q&A communities. The results indicate that the model achieved an overall precision of 84.95%, a recall of 85.83%, and an F1-score of 84.95% through weighted averaging. These results demonstrate that the model effectively extracts relationships between entities in diabetes-related Q&A text within OMCs.

Entity Alignment

In the entity alignment phase, this study initially gathered ICD-10 disease classification dictionaries, as well as dictionaries for common drug names, bynames, and trade names, to construct a medical terminology glossary for dictionary-based matching. Table 5 presents examples of common name, byname, and trade name of common drugs. While dictionary matching can identify entity aliases, it cannot address issues such as abbreviations, short forms, and typographical errors. To resolve these issues, this study employs three methods—edit distance, overlap coefficient, and vector space model—to calculate the similarity between entity names. Through multiple experiments, the maximum similarity score from these three methods is used to determine the similarity between two entities, with a threshold of 0.7 applied for entity alignment.

Table 5. Examples of Common Name, Byname and Trade Name of Common Drugs (Part)

Common Name	Byname	Trade Name
Nifedipine tablets (硝苯地平片)	Nifedipine tablets (心痛定片)	-
Isosorbide dinitrate tablets (盐酸苯海索片)	Isosorbide dinitrate tablets (安坦)	-
Repaglinide tablets (瑞格列奈片)	-	Novonorm (诺和龙)
Glipizide tablets (格列吡嗪片)	-	Glipizide tablets(美吡达)
...	...	...

4.3   Construction of Domain Knowledge Graph

This study utilizes the first domestic text annotation dataset in the field of diabetes, jointly released by the Alibaba Tianchi platform and Ruijin Hospital in Shanghai, to construct a domain-specific knowledge graph. The dataset, drawn from authoritative Chinese journals in the diabetes field, covers various aspects including medication usage and treatment methods, spanning a period of seven years. It comprises 493 articles totaling approximately 4.94 million words. The dataset defines both the conceptual hierarchy and relational constraints for the diabetes domain knowledge graph, featuring 15 types of entities, as detailed in Table 6, and 10 types of relationships, as outlined in Table 7.

Table 6. Domain Knowledge Mapping Entity Type

ID	Category	Coding	Meaning
1	Disease	Disease	The name of the disease
2	Reason	Reason	Causes of disease
3	Symptom	Symptom	Symptoms of patients
4	Test	Test	Check item and indicator names
5	Test_Value	Test_Value	The specific value of the index, etc.
6	Drug	Drug	Drug
7	Frequency	Frequency	Frequency of taking medications
8	Amount	Amount	Dosage of the drug
9	Method	Method	Drug administration, such as oral, etc.
10	Treatment	Treatment	In addition to drug therapy, etc.
11	SideEff	SideEff	Adverse reaction of drug
12	Anatomy	Anatomy	Parts and organs of the human body
13	Level	Level	Including the severity of the disease etc.
14	Duration	Duration	Duration of symptoms
15	Operation	Operation	Operative name

Table 7. Relationship Constraints of Domain Knowledge Mapping

ID	Relationship Constrict Type	Coding
1	Test -> Disease	Test_Disease
2	Symptom -> Disease	Symptom_Disease
3	Treatment -> Disease	Treatment_Disease
4	Drug -> Disease	Drug_Disease
5	Anatomy -> Disease	Anatomy_Disease
6	Frequency -> Drug	Frequency_Drug
7	Duration -> Drug	Duration_Drug
8	Amount -> Drug	Amount_Drug
9	Method -> Drug	Method_Drug
10	SideEff-Drug	SideEff-Drug

To obtain the entity and relationship mapping set of the domain knowledge graph, regular expressions were employed to extract entities and relationships, resulting in the domain knowledge graph entity-relationship set $\{e_h, r, e_t\}$, where e_h, e_t represent entities, and r denotes the relationship type between the entities. Initially, a total of 189,561 entities and 92,059 relationships were identified. Variations in the textual representation of the same entity, such as "Type one diabetes (一型糖尿病)", "Type I diabetes (I型糖尿病)" and "Type 1 diabetes (1型糖尿病)" prompted the need for entity and relationship deduplication. Additionally, the entity alignment method previously mentioned was applied to align entities in the domain knowledge graph, resulting in a final set of 32,258 entities and 41,016 relationships. The processed entity-relationship triples were then stored in a Neo4j graph database, facilitating query operations through Cypher statements.

4.4   Community Knowledge Triplet Extraction

The entities within the remaining 67,512 unannotated community Q&A sentences from the corpus were predicted using the BERT-BiLSTM-CRF model trained in Section Name entity recognition. The results were extracted, categorized, and processed, with the total number of identified entities summarized in Table 8.

Table 8. Number of Entities Related to Diabetes Identified in OMCs

ID	Entity Label	Number
1	DIS	4753
2	SYP	4502
...	...	...
11	AMO	181
12	MED	129
	Total	21122

Similarly, after processing the remaining corpus using entity masking, it was input into the BERT-CNN-Attention model for RE. For each text, the model outputs the relationship labels r_i and corresponding probabilities p_i for each entity pair (e_h, e_t) in the text. Table 9 presents a partial output of the relation extraction model. A total of 80,749 relations were extracted, and the number of relations for part of type is shown in Table 10.

Table 9. Example of RE Model Output Results

Text number	Entity1	Entity 2	Relationship Label	Probability
1000	Eat more	Diabetes	SYP	0.885479
10005	Eye	Diabetes	BOD	0.999852
10060	Sight	Fundus disease	TED	0.961212
13973	Chewing	Acarbose	MED	0.996089

Table 10. Number of Diabetes Entity Relationships in OMCs

ID	Relationship Label	Number
1	UNK	20230
2	SYP_DIS	17061
...	...	...
11	EFF_DRU	785
	Total	80749

The distribution of entities and relationships within the domain knowledge graph was analyzed and found to be broadly consistent with community knowledge, though some discrepancies were noted. Consequently, this study focuses on six categories of relationships common to both: symptom and disease (SYP_DIS), test and disease (TES_DIS), drug and disease (DRU_DIS), anatomical site and disease (ANA_DIS), administration method and drug (MED_DRU), and treatment and disease (TRM_DIS). Consistency checks are performed on these relationships.

4.5 Answer Content Credibility Test

Consistency Checking

Once the structured triples have been obtained, the consistency of the entities within the community answer triples is first verified using entity alignment methods. The entity regularity is computed according to the matching process illustrated in Fig. 5. The resulting entity and entity pair regularities are presented in Table 11 and Table 12, respectively. To prevent matching errors caused by negation terms, such as "hyperglycemia" and "hypoglycemia", additional processing rules are implemented. For entity pairs that include negation terms, the entity pair regularity is set to 0.

Table 11. Partial Entity Standardization Results

ID	Community Entity	Entity Specification Degree	Domain Knowledge Graph Entity
1	Ponison (波尼松)	0.75	Ponison (波尼松)
2	Low blood sugar (血糖低)	0.928571	Hypoglycemia (低血糖)
3	More urine (尿的多)	0.81667	Polyuria (多尿)

Table 12. Example of Physical Inspection Results

Q_ID	Community Answers Entity		Domain Knowledge Entity		Norm Degree
1	Ponison	Diabetes	Ponison	Diabetes	0.875
1	Ponison	Acne	Ponison	UNK	0
1	Antidiabetic Drug	Diabetes	Antidiabetic Drug	Diabetes	1

For entity pairs that pass the entity consistency check, a further examination of relationship consistency is conducted to determine their relationship consistency coefficient. During the experimental process, three types of relationship consistency coefficients were identified: $I = Univailable$, $I = Right$, and $I = Doubt$. For entity pairs with uncertain relationships, where certain relationships are not covered by the domain knowledge graph, the relationship consistency coefficient is determined by applying information consistency theory and integrating responses from other doctors. The likelihood of an entity pair's existence increases with the number of mentions in answers from different doctors. Among the 9,751 entity pairs with uncertain relationships, 3,397 were found to be supported by other doctors, with the majority involving the symptom-disease (SYP_DIS) relationship.

Upon obtaining the entity regularity and relationship consistency coefficient, the consistency score for each diabetes-related triple in the xywy.com website is computed according to Eq. (5).

Credibility Score of Triplet

Based on Eqs. (5) and (6), and considering the impact of the RE algorithm, the credibility scores of the community knowledge triples were calculated. Statistical analysis revealed that the range of the credibility scores for the triples is [−0.9999871, 0.99999857].

According to Eqs. (8) and (10), the credibility scores of all triples in each answer are summarized, and the content credibility is obtained based on the Sigmoid function. Statistics show that the credibility score interval of the answer content is [−13.0121976, 19.7361269], and the content credibility is distributed between intervals (0, 1), which is in line with the previous analysis and definition. Table 13 gives the top 2 and the bottom 2 of the content credibility scores, and the 2 credibility scores are 0.

Table 13. Reliability Test Results of Diabetes Answers From Some Medical Websites

ID	Q_ID	Content Credibility Score	Content Credibility
1	4232	19.73613	0.999999997
2	4286	18.66895	0.999999992
3	5249	0	0.5
4	9725	0	0.5
5	10335	−8.98793	0.000125
6	1208	−13.0122	2.23E-06

4.6 Test and Analysis of Credibility Evaluation Results

The credibility of the answers obtained from the diabetes Q&A section of the xywy.com website was further analyzed. Based on the credibility scores of the answers, a threshold of 0.5 was applied to categorize the diabetes-related Q&A content into three categories: Credible, Unknown, and Suspicious. The distribution of these categories is as follows: 66% are Credible, 23% are Unknown, and 11% are Suspicious.

In this study, a selection of Q&A pairs was transformed into a questionnaire and evaluated by five authoritative experts in the field of diabetes. The experts categorized each response as either "Credible", "Unknown" or "Suspicious". Given the current limitations in technology for testing all types of errors and the challenge in measuring recall comprehensively, this paper primarily focuses on validating the accuracy of the proposed credibility test method. The results from the returned questionnaires indicate that the precision rate for identifying credible answers is 92.5%. This demonstrates that the credibility test method based on knowledge graphs effectively identifies credible content.

Analyze the characteristics of three types of answers using disease entity word cloud maps. Among them, the cloud map of disease entity words in credible answers is shown in Fig. 10. Analysis reveals that credible answers are highly correlated with diabetes and its complications, supported by substantial evidence suggesting their content is trustworthy. Unknown answers primarily pertain to diseases outside the realm of diabetes, with some content not assessable based on the diabetes domain knowledge graph. Suspicious answers predominantly involve complex foundational diseases, for which no evidence aligning with medical practice can be found in the domain knowledge graph, indicating numerous triples with dubious relationships.

Fig. 10. The Word Cloud Map of Disease Entities in Credible Answers

Analysis of answer texts with credibility scores exceeding 0.5 reveals that in these highly credible responses, patients provide precise and clear descriptions of their symptoms and test results. The content of the inquiries is highly relevant to diabetes, and the doctors' analyses are detailed and consistent with established medical practices for diabetes. In contrast, in responses with credibility scores below 0.5, patients typically inquire about other underlying conditions with low relevance to diabetes. The range of entities involved is broader, and the descriptions provided by patients are often unclear. Consequently, doctors are unable to make more accurate judgments, and supporting evidence from the domain knowledge graph or other responses within the community, which would align with established medical knowledge for diabetes, is lacking.

This study provides several insights and guidelines for patients using OMCs for consultation: (1) selecting the appropriate section for posting questions; (2) describing their condition as comprehensively as possible; and (3) clearly defining the purpose of the consultation. Additionally, for community administrators, the proposed method facilitates intelligent pre-screening of physician responses within the community. By generating a credibility score ranging from 0 to 1 for the answer content, the method further categorizes responses into Credible, Suspicious, and Unknown categories. These classifications are then displayed in the relevant sections of the webpage, assisting users in evaluating and making decisions about whether to accept the suggested health advice.

5 Conclusion

This paper proposed an intelligent credibility test method for OMCs answers based on domain knowledge graphs. To evaluate the credibility of OMCs answers, the consistency of triples extracted from Q&A with those in the domain knowledge graph was assessed and scored to determine the credibility score of the answer, which is then mapped to a [0, 1] interval. Using a threshold of 0.5, answers are categorized as Credible ((0.5, 1]), Suspicious ([0, 0.5)) and Unknown (0.5). Validation results based on data from the xywy.com demonstrate that the method realizes intelligent evaluation and classification of Q&A credibility. Compared to manual scoring methods, it improves both efficiency and interpretability, offering a novel approach for assessing the credibility of answers in OMCs. Additionally, the proposed method improves community management capabilities, supports the healthy development of medical communities and enhanced OMCs content quality.

This study has certain limitations. During the analysis of the credibility of answer content from the xywy.com website, the large volume of data hindered a more in-depth analysis of the results. Future research could consider dividing the data into different themes or metrics for a more comprehensive analysis. Additionally, this study is focused solely on the field of diabetes; future work could extend the methodology to cover a broader range of medical domains.

Acknowledgments. Supported in part by the National Natural Science Foundation of China under Grant 72171090, "Research On Quantitative Model of Information Credibility Evaluation for Virtual Health Communities and Its Applications", and in part by "2023 Annual Discipline Co-construction Project of Guangdong Provincial Philosophy and Social Science Planning: Research on Evaluation Model and Application of Information Quality in Question-and-Answer Community Based on Knowledge Graph (Project No.: GD23XGL044)".

References

1. Stacks, D.W., Salwen, M.B.: An integrated approach to communication theory and research (2009)
2. Hovland, C.I., Janis, I.L., Kelley, H.H.: Communication and persuasion (1953)
3. Pornpitakpan, C.: The persuasiveness of source credibility: a critical review of five decades' evidence. J. Appl. Soc. Psychol. **34**(2), 243–281 (2024)
4. Metzger, M.J., Flanagin, A.J., Eyal, K., et al.: Credibility for the 21st century: integrating perspectives on source, message, and media credibility in the contemporary media environment. Ann. Int. Commun. Assoc. **27**(1), 293–335 (2003)
5. Flanagin, A.J., Metzger, M.J.: Digital media and youth: unparalleled opportunity and unprecedented responsibility, pp. 5–27. MacArthur Foundation Digital Media and Learning Initiative, Cambridge (2008)
6. Kaplan, A.M., Haenlein, M.: Users of the world, unite! The challenges and opportunities of social media. Bus. Horiz. **53**(1), 59–68 (2010)
7. Hajli, M.N., Sims, J., Featherman, M., et al.: Credibility of information in online communities. J. Strateg. Mark. **23**(3), 238–253 (2015)

8. Morris, M.R., Counts, S., Roseway, A., et.al.: Tweeting is believing? Understanding microblog credibility perceptions. In: Proceedings of the ACM 2012 Conference on Computer Supported Cooperative Work, pp. 441–450 (2012)

9. Li, L., Yin, Q., Yan, Z.: The impacts of online-offline doctor service evaluation on patients' online consultation choice. Chin. J. Manage. **19**(04), 565–574 (2022)

10. Qian, M., Zhixuan, X., Wang, S.: Information service quality of online health platform based on user participation. J. China Soc. Sci. Tech. Inf. **38**(2), 132–142 (2019)

11. Li, B., Wang, Y., Zhou, K.: Measuring credibility of social media contents based on Bayesian theory. Data Anal. Knowl. Discov. **1**(06), 83–92 (2017)

12. Xukan, X., Lou, Y., Chengcheng, Y.: Research on the requirements of multi-source data fusion emergency decision-making based on Dempster-Shafer theory. Inf. Stud.: Theory Appl. **42**(08), 67–72 (2019)

13. Liu, Q., Liao, K., Tsoi, K.K., et al.: Acceptance prediction for answers on online health-care community. BMC Bioinform. **20**, 1–8 (2019)

14. Huang, H., Juan, Y., Liao, X., et al.: Review on knowledge graphs. Comput. Syst. Appl. **28**(06), 1–12 (2019)

15. Jin, S., Wang, S., Huang, Q., et al.: Study on the construction of knowledge graph based on breast cancer specialized disease database. J. Med. Inform. **44**(12), 65–70 (2023)

16. Liu, Y., Qi, M.: Research on the construction of medical knowledge graph based on diabetes prevention and treatment. J. Med. Inf. **33**(18), 11–14 (2020)

17. Chen, D., Zhang, R., Feng, J., et al.: Fulfilling information needs of patients in online health communities. Health Inf. Libr. J. **37**(1), 48–59 (2020)

18. Ye, Y., Lei, P., Fengzhen, H.: Entity extraction and graph construction based on Chinese medical text. J. China Pharmac. Univ. **54**(3), 363–371 (2023)

19. Zhang, F., Qin, Q., Jiang, Y., et al.: Named entity recognition for Chinese EMR with RoBERTa-WWM-BiLSTM-CRF. Data Anal. Knowl. Discov. **6**(2/3), 251–262 (2022)

20. Peng, Y., Rios, A., Kavuluru, R., et al.: Chemical-protein relation extraction with ensembles of SVM, CNN, and RNN models. arXiv Preprint arXiv:1802.01255 (2018)

21. Liu, C., Sun, W., Chao, W., Che, W.: Convolution neural network for relation extraction. In: Motoda, H., Wu, Z., Cao, L., Zaiane, O., Yao, M., Wang, W. (eds) ADMA 2013. LNCS, vol. 8347, pp. 231–242. Springer, Heidelberg (2013). https://doi.org/10.1007/978-3-642-53917-6_21

22. Vaswani, A., Shazeer, N., Parmar, N., et al.: Attention is all you need. In: Advances in Neural Information Processing Systems (2017)

23. Guo, X., Zhang, H., Yang, H., et al.: A single attention-based combination of CNN and RNN for relation classification. IEEE Access **7**, 12467–12475 (2019)

24. Zhang, X., Chen, F., Huang, R.: A combination of RNN and CNN for attention-based relation classification. Procedia Comput. Sci. **131**, 911–917 (2018)

25. Sussman, S.W., Siegal, W.S.: Informational influence in organizations: an integrated approach to knowledge adoption. Inf. Syst. Res. **14**(1), 47–65 (2003)

26. Zhu, Z., Bernhard, D., Gurevych, I.: A multi-dimensional model for assessing the quality of answers in social Q&A sites. In: ICIQ, pp. 264–265 (2009)

27. Yang, J.Q.: Construction of diabetes knowledge map based on Chinese natural language processing. Inner Mongolia University of Science and Technology (2020)

28. Yang J, Zhang Y, Li L, et al. YEDDA: A lightweight Collaborative Text Span Annotation tool. ArXiv Preprint ArXiv:1711.03759 (2017)

29. Wu, F., Liu, J., Wu, C., et al.: Neural Chinese named entity recognition via CNN-LSTM-CRF and joint training with word segmentation. In: The World Wide Web Conference, pp. 3342–3348 (2019)

A Comprehensive Framework for Sentiment Analysis and Cold-Start Recommendations in Vietnam Hospitality Sector

Xuan-Thang Tran[1,2]([✉]) [ID], Dang-Man Nguyen[1] [ID], Mau-Toan Nguyen[1] [ID],
and Van-Nam Huynh[1,3] [ID]

[1] Japan Advanced Institute of Science and Technology, Nomi, Japan
{txthang,mannd,nmtoan,huynh}@jaist.ac.jp
[2] Tay Nguyen University, Buon Ma Thuot, Vietnam
[3] School of Finance and Accounting, Industrial University of Ho Chi Minh City,
Ho Chi Minh City, Vietnam

Abstract. This paper presents a comprehensive study on the sentiment analysis of hotel reviews in Vietnam, utilizing various deep learning models. Among the conducted models, the Transformer model on the pre-trained 'all-mpnet-base-v2' embedding model achieved the highest accuracy of 85.73% and averaged F1-score of 0.8580, outperforming CNN, LSTM, BiLSTM, and BiLSTM-CNN on word2vec embeddings. Furthermore, we propose a cold-start recommendation system designed to assist new customers in finding suitable hotels based on their keywords. By employing a cosine similarity matrix over the embedding matrices obtained from sentence transformer model, the system calculates similarity scores between the input sentence and previous customer reviews. The quality of the recommendations is improved by integrating additional weights, including hotel ratings, total number of reviews, and sentiment derived from our sentiment analysis model. Experimental results demonstrate the effectiveness of our approach in providing accurate and relevant hotel recommendations.

Keywords: Sentiment analysis · deep learning models · cold-start recommendation · hospitality dataset

1 Introduction

Customer reviews and ratings play a crucial role in influencing consumer choices in the hospitality industry, particularly when selecting hotels and accommodations [26]. The rise of e-commerce and the sharing economy has fueled the growth of online booking platforms [1,3], such as Booking[1], TripAdvisor[2], and Traveloka[3]. Customer reviews on these platforms are crucial in shaping consumer decisions. Sentiment analysis, which extracts subjective information from

---

[1] https://www.booking.com/.
[2] https://www.tripadvisor.com/.
[3] https://www.traveloka.com/.

X. Tang et al. (Eds.): KSS 2024, CCIS 2269, pp. 277–292, 2025.
https://doi.org/10.1007/978-981-96-0178-3_19

text, is increasingly essential for understanding customer satisfaction and service quality.

Vietnam's economy is recovering after the impacts of COVID-19, with tourism witnessing a significant resurgence since late 2022. The country's rich tourism potential, highlighted by its extensive coastline and numerous UNESCO-recognized sites, drives this recovery [18]. The hotel industry plays a critical role in Vietnam's tourism, contributing over 5% to the national GDP [18]. According to Mordor Intelligence[4], the market size of the hospitality industry in Vietnam is projected to be USD 5.16 billion in 2024, growing to USD 9.91 billion by 2029, with a compound annual growth rate (CAGR) of 13.94% over the forecast period (2024–2029). Consequently, the Vietnamese government prioritizes improving service quality to attract more tourists, making it crucial to find effective methods to measure and enhance customer satisfaction with hotel services [18].

Despite various methods for measuring customer experience, no standardized formulas, algorithms, or processes have been established [18]. While many sentiment analysis techniques exist, their application to English-language reviews of Vietnamese hotels remains underexplored. Models like Convolutional Neural Networks (CNN), Long Short-Term Memory (LSTM), Bidirectional LSTM (BiLSTM), and BiLSTM-CNN hybrids have limitations in capturing semantic relationships and contextual dependencies. The advent of transformer-based models, such as BERT (Bidirectional Encoder Representations from Transformers), offers the potential for higher accuracy in sentiment analysis tasks. Another significant challenge in the hospitality industry is the cold-start problem in recommendation systems, where new customers without interaction history receive suboptimal suggestions. Leveraging sentiment analysis in recommendation systems can improve personalization and relevance.

This paper presents several key contributions. We develop an advanced sentiment analysis model using Transformers on top of the pre-trained sentence embedding '*all-mpnet-base-v2*', achieving 85.75% accuracy and outperforming traditional baselines like CNN, LSTM, BiLSTM, and BiLSTM-CNN with word2vec embeddings. We also propose a cold-start recommendation system using a cosine similarity matrix, calculating similarity scores between user input and existing reviews to recommend the top 10 hotels. To enhance recommendation quality, we integrate additional weights such as hotel ratings, review counts, and sentiment scores. These efforts aim to improve the effectiveness of sentiment analysis and recommendation systems in the hospitality industry, specifically addressing the challenges posed by English-language reviews of Vietnamese hotels.

2   Related Works

2.1   Advanced Sentiment Analysis with Deep Learning Techniques

Sentiment analysis (SA), a subfield of natural language processing (NLP), is the field of study that analyzes people's opinions, sentiments, evaluations, attitudes,

---

[4] https://www.mordorintelligence.com.

and emotions from written language [11]. It is one of the most active research areas in natural language processing and is also widely studied in data mining, Web mining, and text mining. SA has been extensively studied and applied across various domains. Early approaches to sentiment analysis often relied on lexicon-based methods, which utilized predefined lists of positive and negative words to determine the sentiment of a given text [5]. However, these methods were limited in their ability to handle complex linguistic phenomena such as sarcasm, context-dependent sentiments, and domain-specific language.

The advent of machine learning techniques marked a significant advancement in SA. Models like Support Vector Machines (SVM), maximum entropy, and Naive Bayes classifiers were used to classify text based on sentiment by learning from annotated datasets [13]. These models showed improved accuracy over lexicon-based approaches but still faced challenges in understanding context and semantic nuances.

With the rise of deep learning, more sophisticated models such as CNN and LSTM [4] networks have been developed to capture intricate patterns in text data. CNNs, originally designed for image recognition, have been adapted for text classification tasks by treating text as a one-dimensional sequence of words [7]. LSTMs, a type of recurrent neural network (RNN), are capable of capturing long-range dependencies in text, making them well-suited for sentiment analysis [10].

Several studies have demonstrated the efficacy of deep learning models in sentiment analysis [19,24]. For instance, Zhang et al. [23] employed a hybrid CNN-LSTM model to analyze sentiment in social media texts, achieving notable improvements in accuracy. Similarly, Sun et al. [17] utilized fine-tuned BERT for sentiment classification on eight widely-studied datasets, reporting superior performance on SA tasks.

In the context of hotel reviews, deep learning models have been applied to extract sentiments and opinions from user-generated content. Xu et al. [22] used an LSTM-based model to analyze sentiment in hotel reviews, highlighting the model's ability to capture sequential dependencies in text. The integration of review titles and contents using a deep learning-based fusion method has been shown to enhance document-level sentiment analysis performance by 2.68% to 12.36% on a TripAdvisor dataset compared to baseline methods [20]. Furthermore, transformer-based models like BERT have been employed to enhance sentiment analysis accuracy in hotel reviews for hotels in China, owing to their contextual understanding capabilities [21].

2.2 Recommendation Systems and Cold-Start Problem in Hospitality

Recommendation systems have become integral to various online platforms, providing personalized suggestions based on user preferences and behaviors. Collaborative filtering and content-based filtering are the two predominant approaches used in recommendation systems. Collaborative filtering relies on the collective preferences of users to make recommendations, whereas content-based filtering focuses on the characteristics of items to suggest similar ones [16].

The cold-start problem, which arises when there is insufficient data about new users or items, remains a significant challenge in recommendation systems. Hybrid approaches that combine collaborative and content-based methods have been proposed to address this issue [2]. Additionally, recent studies have explored the use of deep learning techniques to mitigate the cold-start problem. For example, Zhang et al. [25] proposed a neural network-based approach, namely Deep-Rec, that integrates users' interests and item features to improve recommendations for new users.

In the hospitality industry, recommendation systems are particularly valuable for suggesting hotels based on user preferences. Several studies have explored the use of textual reviews to enhance hotel recommendations. Zhang et al. [14] developed a recommendation system that leverages sentiment analysis of hotel reviews to improve the relevance of suggestions. Similarly, Jalan et al. [6] proposed a hybrid recommendation system that combines sentiment analysis and semi-supervised clustering to address the cold-start problem in hotel recommendations.

In Vietnam, research on hotel recommendation systems, especially for the cold-start problem, is still in its nascent stages. However, there is a growing recognition of its potential to enhance customer experience management and support the country's tourism development goals. This study aims to contribute to this emerging field by developing advanced sentiment analysis models and recommendation systems tailored to the Vietnamese hospitality industry.

3   Research Problem

This study investigates sentiment analysis at the document level for each review, including both the review title and review content. We conduct experiments and compare the results between several deep learning models and embedding models, using embeddings from both review titles and review contents combined, as well as each one separately.

Consider a set of n customer reviews $\mathcal{R} = \{r_1, r_2, \ldots, r_n\}$. For each review $r_i \in \mathcal{R}$, it includes a title and content pair $r_i = [T_i, C_i]$. Let $\mathcal{W}_T^i$ represent the contextualized word embedding of the review title T_i and $\mathcal{W}_C^i$ represent the contextualized word embedding of the review content C_i for review r_i, both transformed by a vector representation model.

A deep learning model will then be developed for document-level SA to predict customer opinions. The objective is formalized by finding a mapping function $F : \mathcal{R} \rightarrow \{\text{positive}, \text{neutral}, \text{negative}\}$ such that:

$$F(r_i) = \begin{cases} \text{positive}, & \text{if the review sentiment is positive} \\ \text{neutral}, & \text{if the review sentiment is neutral} \\ \text{negative}, & \text{if the review sentiment is negative} \end{cases}, \qquad (1)$$

where $F(r_i)$ corresponds to $F([\mathcal{W}_T^i, \mathcal{W}_C^i])$.

Next, we develop a recommendation system for cold-start users on top of the sentiment analysis model. This system takes a keyword sentence (input sentence) $\mathcal{K}$ given by new user, then maps it into the embedding model (same as in SA model) to obtain the embedding matrix $\mathcal{W}_\mathcal{K}$. Next, the similarity score between the input sentence embedding $\mathcal{W}_\mathcal{K}$ and the embeddings of previous customer reviews $\mathcal{W}_\mathcal{R}$ could be calculated:

$$Similarity(\mathcal{W}_\mathcal{K}, \mathcal{W}_\nabla) = \frac{\mathcal{W}_\mathcal{K} \cdot \mathcal{W}_\nabla}{\|\mathcal{W}_\mathcal{K}\| \|\mathcal{W}_\nabla\|}, \tag{2}$$

where $\mathcal{W}_\mathcal{K} \cdot \mathcal{W}_\nabla$ is the dot product between vectors $\mathcal{W}_\mathcal{K}$ and $\mathcal{W}_\nabla$, $\mathcal{W}_\nabla \in \mathcal{W}_\mathcal{R}$ indicates the embedding metric for each review, obtained by concatenating the title and content and mapping to the embedding model. $\|\mathcal{W}_\mathcal{K}\|$ and $\|\mathcal{W}_\nabla\|$ are the magnitudes, or lengths of vectors $\mathcal{W}_\mathcal{K}$ and $\mathcal{W}_\nabla$, respectively. The dot product can be expressed as a sum of the product of all vector dimensions and magnitudes as:

$$\begin{aligned}
\|\mathcal{W}_\mathcal{K}\| \|\mathcal{W}_\nabla\| &= \sqrt{\mathcal{W}_\mathcal{K} \cdot \mathcal{W}_\mathcal{K}} \sqrt{\mathcal{W}_\nabla \cdot \mathcal{W}_\nabla} \\
&= \sqrt{\sum_{i=1}^{d} \mathcal{W}_{\mathcal{K}_i} \mathcal{W}_{\mathcal{K}_i}} \sqrt{\sum_{i=1}^{d} \mathcal{W}_{\nabla_i} \mathcal{W}_{\nabla_i}} \\
&= \sqrt{\sum_{i=1}^{d} \mathcal{W}_{\mathcal{K}_i}^2 \sum_{i=1}^{d} \mathcal{W}_{\nabla_i}^2}
\end{aligned} \tag{3}$$

Thus, the cosine similarity is given by:

$$\begin{aligned}
Similarity(\mathcal{W}_\mathcal{K}, \mathcal{W}_\nabla) = \cos(\theta) &= \frac{\mathcal{W}_\mathcal{K} \cdot \mathcal{W}_\nabla}{\|\mathcal{W}_\mathcal{K}\| \|\mathcal{W}_\nabla\|} \\
&= \frac{\sum_{i=1}^{d} \mathcal{W}_{\mathcal{K}_i} \mathcal{W}_{\nabla_i}}{\sqrt{\sum_{i=1}^{d} \mathcal{W}_{\mathcal{K}_i}^2 \sum_{i=1}^{d} \mathcal{W}_{\nabla_i}^2}},
\end{aligned} \tag{4}$$

where d is the dimension of the embedding matrix. This approach ensures that the recommendation system can effectively provide relevant and personalized hotel suggestions even for users without a history of prior interaction.

In addition, we investigate the impact of other characteristics on the final score, including three key characteristics: "hotel ratings", "total number of reviews", and "review sentiment", which have been chosen for their critical role in influencing customer decisions in the hospitality industry.

- **Hotel Ratings:** reflects overall customer satisfaction and service quality, a primary metric users consider when selecting accommodations.
- **Total Number of Reviews:** A higher number of reviews indicates greater reliability and popularity, ensuring a more comprehensive assessment of the hotel.

– **Review Sentiment:** SA captures the qualitative aspects of customer feedback, allowing our system to account for the emotional tone and specific
experiences described in reviews. This ensures that hotels with fewer but
highly positive reviews can rank fairly.

The strength of our system lies in addressing the cold-start problem by combining these features with sentiment-driven analysis. Unlike traditional systems
that focus primarily on review volume or ratings [16], our approach balances
quantitative metrics with qualitative insights, providing more accurate and personalized recommendations. This integration enhances the relevance and quality
of the recommendations, ensuring that even newer or less-reviewed hotels receive
fair consideration.

4   Proposed Framework

The proposed framework consists of two main components: sentiment analysis
using deep learning and a cold-start recommendation system for new users. The
Fig. 1 below illustrates the proposed framework.

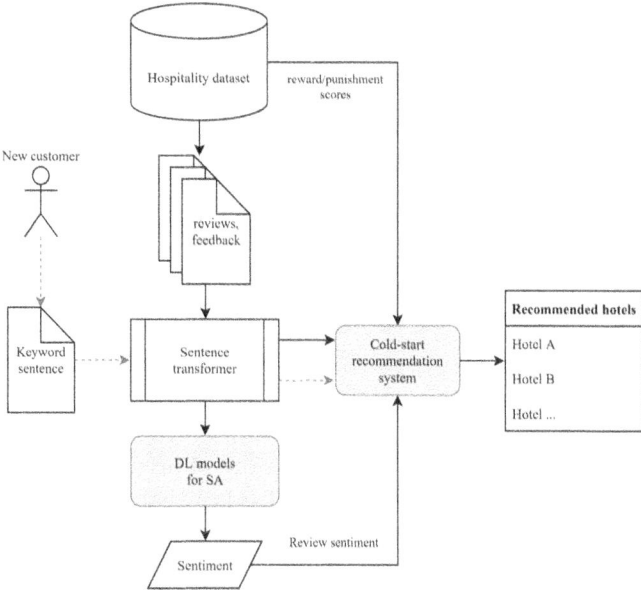

Fig. 1. Framework for SA and cold-start recommendation system for Hospitality

4.1   Sentiment Analysis for Hotel Reviews

The sentiment analysis component aims to classify the opinions expressed in customer reviews of hotel services. The raw dataset, collected from hotel booking websites, includes customer feedback in the form of review titles and review contents. The process involves the following steps:

- **Data Extraction:** Extract customer reviews, including titles and contents, from the raw dataset.
- **Embedding Generation:** Utilize well-known word embedding method, including Google word2vec, BERT-based sentence embedding to generate the numerical matrices of input sentence.
- **Model Application:** Apply several deep learning models, including CNN, LSTM, BiLSTM, BiLSTM-CNN, and Transformer, to classify the sentiment of reviews.

These models will be trained and evaluated to determine the most effective approach for sentiment classification in this context.

4.2   Hotel Recommendation System for New Users

The second component of the framework focuses on building a recommendation system for new users. This system leverages the sentiment analysis results and additional attributes from the original data to provide personalized hotel recommendations. The process involves the following steps:

- **Input Keyword Sentence:** The new user provides a keyword sentence describing their preferences.
- **Embedding Mapping:** Map the input keyword sentence into embedding matrices using the Sentence Transformer.
- **Similarity Calculation:** Calculate the cosine similarity score between the embedding of the new user's keyword and the embeddings of previous customer reviews.
- **Attribute Integration:** Incorporate important attributes from the original data, such as hotel value, total number of reviews received, and sentiment scores derived from the sentiment analysis model. These attributes are used to calculate additional reward/punishment weights for the similarity scores.
- **Recommendation Generation:** Sort the hotels based on the adjusted similarity scores and list the top 10 recommended hotels for the user.

This structured approach ensures that both sentiment analysis and the recommendation system are systematically developed and integrated, providing a comprehensive solution for analyzing customer feedback and generating personalized hotel recommendations.

5   Experiments

5.1   Experimental Settings

Dataset. In this study, we utilized a real-world dataset from TripAdvisor [9,20], comprising 527,770 English reviews from over 290,000 travelers, covering 8,297 hotels in Vietnam. The raw dataset included 24 attributes, but for this study, we retained only 7 key attributes: *'hotel ID'*, *'hotel name'*, *'hotel value'* (average rating), *'number of reviews'* (total received reviews), *'review title'* (a short summary), *'review content'* (detailed evaluation of hotel services), and *'review rating'* (ranging from 1 to 5 stars). For document-level sentiment analysis tasks on this hospitality dataset, reviews with *1-star* to *2-star* ratings are considered *negative*, *3-star* ratings are considered *neutral*, and *4-star* to *5-star* ratings are considered *positive*.

After removing duplicate, redundant records, and null values, the dataset consists of 475,375 reviews. Figure 2 shows the distribution of reviews by city from the TripAdvisor dataset. Hanoi and Ho Chi Minh City have the highest number of reviews, comprising 23.1% (108,866 reviews) and 20.3% (96,686 reviews) of the total reviews, respectively. These cities are major tourist and business hubs in Vietnam, attracting a large number of visitors. Da Nang, known for its beaches and as a gateway to UNESCO World Heritage sites, accounts for 8.6% (41,316 reviews), while Quang Nam, home to the ancient town of Hoi An, contributes 10.6% (50,199 reviews). Other notable cities include Khanh Hoa (5.3%), Lam Dong (3.7%), and Binh Thuan (3.5%). The category 'Others' includes various smaller cities, making up 7.6% of the total dataset. This distribution highlights the popularity of major tourist destinations in Vietnam.

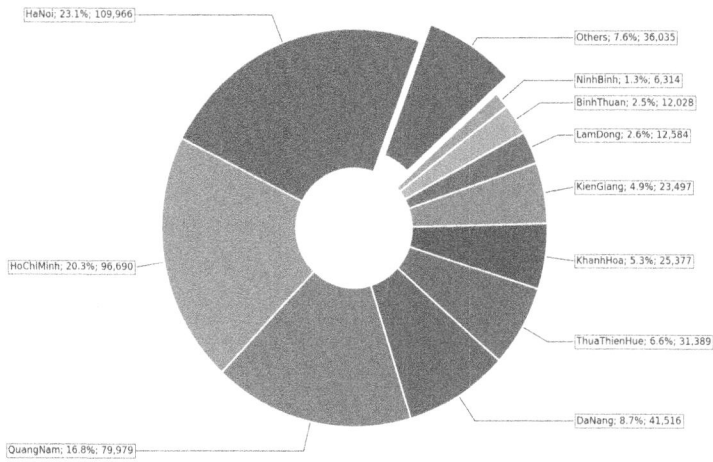

Fig. 2. Distribution of reviews by city from the TripAdvisor dataset

As described in [20], the data set is unbalanced, with reviews of 4 and 5 stars that comprise 87. 0% of the total dataset. The remaining reviews include 3-star ratings, which account for almost 7. 3%, and 2-star and 1-star ratings, which together cover only 5.7%. To mitigate the effects of data imbalance and accurately estimate the model's performance, we randomly selected 20,000 reviews for each label, resulting in a subset of 60,000 reviews. Then, this subset was randomly divided into a training set (80%) and a testing set (20%).

Model Specifications. For the sentiment analysis (SA) task, we conducted experiments using two different embedding methods: Google Word2vec and BERT-based Sentence Transformers. The deep learning models employed in these experiments include:

- **CNN:** Configured with window sizes of 3, 4, and 5, each with 100 features.
- **LSTM:** Configured with 2 hidden layers, each consisting of 128 nodes.
- **BiLSTM:** Using the same configuration as LSTM for forward and backward directional computation.
- **BiLSTM - CNN:** Combining the configurations of BiLSTM and CNN.
- **Transformer:** Utilizing the benchmark BERT-based model for sentence classification as featured on the MTEB leaderboard [12]. The Transformer consists of 12 layers, each with 12 self-attention heads and a hidden size of 768. The feedforward network within each layer has an intermediate size of 3072. Input sequences were tokenized with a maximum sequence length of 512.

Additionally, we conducted experiments on a parallel model described in [20], which utilizes an average fusion method to combine review titles and review contents, demonstrating the contribution of both components to the accuracy of the SA task. To evaluate the performance of the sentiment analysis (SA) models, each experiment was run five times with five epochs per run, and the results were averaged. The batch size was set to 128. The sparse categorical cross-entropy loss function was used, along with the Adam optimizer [8], which had a learning rate of 0.001 and a dropout rate of 0.5.

For the recommendation system designed for cold-start users, the same embedding method that yielded the best results in the SA task was employed. The system processes an input keyword sentence provided by a new user and maps it into an embedding matrix using the Sentence Transformer. The cosine similarity score, ranging between 0 and 1, was calculated between the input keyword embedding and the embeddings of previous customer reviews. Additionally, reward/punishment weights based on several key attributes, such as hotel value, total number of reviews, and sentiment scores from the SA model, were calculated to adjust the similarity scores. The top-10 recommended hotels were identified based on these adjusted scores. These results will be discussed in detail, analyzing how the attributes and adjusted scores influence the final recommendations, ensuring that the recommendation system provides relevant and high-quality suggestions for new users.

Table 1. Experimental results on SA task for Tripadvisor dataset

DL Model	Embedding model	Performance	
		Accuracy	F1-score
CNN	word2vec	0.8238	0.8245
LSTM		0.7429	0.7371
BiLSTM		0.8087	0.8078
Hybrid BiLSTM - CNN		0.8258	0.8252
Transformer	Pre-trained 'bert-base-uncased'	0.8149	0.8146
	Pre-trained 'all-mpnet-base-v2'	**0.8573**	**0.8580**

5.2 Empirical Results

Sentiment Analysis on Hospitality Dataset

The experimental results for the SA task in the TripAdvisor data set are summarized in Table 1. The table presents the performance of various deep learning models using different embedding methods, evaluated in terms of accuracy and F1-score. The experimental results demonstrate that different deep learning models and embedding techniques yield varying levels of performance on the sentiment analysis task. The CNN model with *word2vec* embeddings achieved an accuracy of 82.38% and an F1-score of 0.8245. This strong performance is attributed to CNN's ability to capture local patterns in text through convolutional filters. On the other hand, the LSTM model, recorded an accuracy of 74.29% and an F1-score of 0.7371, making it the lowest performing model among those evaluated. This indicates that while LSTMs are effective at capturing sequential dependencies, they may struggle with the complexity and variability of the dataset compared to CNN model.

The BiLSTM model, configured similarly to the LSTM, achieved higher performance with an accuracy of 80.87% and an F1-score of 0.8078. The bidirectional architecture of BiLSTM allows it to capture context from both directions, enhancing its effectiveness in sentiment analysis tasks. The BiLSTM - CNN hybrid model demonstrated the highest performance among the word2vec models, with an accuracy of 82.58% and an F1-score of 0.8252. This model combines the strengths of BiLSTM in capturing sequential dependencies and CNN in capturing local patterns, leading to superior performance.

Transformer-based models, utilizing pre-trained *'bert-base-uncased'* [12] and *'all-mpnet-base-v2'* [15] embeddings, showed substantial improvements over traditional models. The *'bert-base-uncased'* model achieved an accuracy of 81.49% and an F1-score of 0.8146, outperforming most traditional models. The *'all-mpnet-base-v2'* model, however, outperformed all other models, achieving the highest accuracy of 85.73% and an F1-score of 0.8580. These results highlight the superior ability of transformer models to capture complex contextual information, due to their self-attention mechanisms and large-scale pre-training.

The comparison indicates that while traditional models like CNN and BiL-STM - CNN perform well, transformer models, particularly those based on the sentence transformer such as *'all-mpnet-base-v2'*, offer significant improvements in both accuracy and F1-score. This underscores the importance of leveraging advanced embedding techniques and transformer architectures for sentiment analysis tasks in large, diverse datasets.

Cold-Start Recommendation System for New User

In this subsection, we describe the cold-start recommendation system designed for new users. We utilize the pre-trained *'all-mpnet-base-v2'* model as the embedding method, identified as the best-performing model in our sentiment analysis task. The system maps the input keyword sentence provided by a new user into an embedding matrix. The similarity score between the input sentence and previous customer reviews is then calculated using Eq. 4 in Sect. 3.

Hotel Averaged Rating Effect. The recommendation system leverages cosine similarity between input keywords and previous reviews, but the average rating of the hotel is also an important parameter. Hotel value, or average rating value, is calculated by dividing the total number of stars from all reviews by the total number of reviews for the hotel. We incorporate this rating effect into the similarity score using the following equation:

$$weight = (2 \times Q/5) \times hotel\_rating - Q, \tag{5}$$

where Q is a parameter that controls the extent of the rating effect on the final score. Table 2 illustrates the impact of different Q values on the cosine similarity score for various hotel ratings.

Table 2. Percentage effects of Q on cosine similarity value for each hotel rating

Hotel rating	Q			
	10	20	50	100
1	−6	−12	−30	−60
2	−2	−4	−10	−20
3	2	4	10	20
4	6	12	30	60
5	10	20	50	100

Hotel Total Reviews Effect. A hotel with a high average rating of 5, but only received 6 reviews may be less reliable than one with a slightly lower average rating (4.6, for example) but with 2000 reviews. Thus, we introduce a multiplier to adjust the rating effect based on the number of reviews, calculated as follows:

$$multiplier = e^{-t \times 0.68/total\_reviews}, \tag{6}$$

where t is a threshold value. The *multiplier* would be $e^{-0.68} \approx 0.5$ when threshold is equal to the total reviews of the hotel. This *multiplier* ranges between $(0, 1)$ and reduces the impact of ratings for hotels with fewer reviews.

Review Sentiment Effect. To further refine the similarity score, we adjust the hotel rating in Eq. 5 based on the sentiment of the review. This adjustment ensures that reviews classified as positive have a greater impact on the similarity score, while negative reviews diminish the influence of the hotel rating:

$$refined\_hotel\_rating = \begin{cases} (hotel\_rating + 7)/2, & \text{if review is } positive \\ hotel\_rating, & \text{if review is } neutral \\ hotel\_rating/2, & \text{if review is } negative \end{cases} \quad (7)$$

The *hotel_rating* in Eq. 5 will be replaced by the *refined_hotel_rating* to account for the sentiment effect as followed:

$$refined\_weight = (2 \times Q/5) \times refined\_hotel\_rating - Q \quad (8)$$

By incorporating the *refined_hotel_rating*, the recommendation system ensures a more nuanced adjustment of the similarity scores, reflecting both the sentiment of individual reviews and the overall quality of the hotel.

Thus, the similarity score between keyword input $\mathcal{K}$ and previous reviews will be adjusted as:

$$Adjusted\_Simiarity(\mathcal{W}_\mathcal{K}, \mathcal{W}_\nabla) = Similarity(\mathcal{W}_\mathcal{K}, \mathcal{W}_\nabla)$$
$$+ \frac{Similarity(\mathcal{W}_\mathcal{K}, \mathcal{W}_\nabla) \times refined\_weight \times multiplier}{100} \quad (9)$$

This combination of these adjustments allows the recommendation system to provide more accurate and reliable hotel suggestions by considering not only the similarity between the input keywords and the reviews, but also the overall quality and reliability of the hotels.

To evaluate the recommendation system, we consider the following sample keyword input: ***"Great location with friendly staff in Ho Chi Minh"***. The parameter Q is set to 10, and the review threshold t is set to 200. Next, the system calculates the initial similarity scores, which indicate how well the reviews of each hotel match the keyword input. After adjusting for factors such as hotel value, number of reviews, and review sentiment, adjusted similarity scores provide a refined ranking of the top-10 recommended hotels. The results of the cold-start recommendation system for a new user given the keyword sentence are presented in Table 3.

The hotel with ID 3245017 ranks first with an initial similarity score of 0.8726, which increases to 0.9707 after adjustment. This hotel has a high number of reviews (2075) and a positive sentiment, which contribute to its top ranking. The second-ranked hotel (ID 5508253) also shows a significant increase in its score from 0.8592 to 0.9648, reflecting its higher hotel value (4.5) and a substantial

Table 3. Top-10 Recommended Hotels for the new user given the keyword input. The superscripts $^{(1)}$, $^{(2)}$, and $^{(3)}$ indicate the ranking of the recommended hotels based on the initial similarity score.

Hotel ID	Review	Sentiment	Hotel Value	# Reviews	Similarity	Adjusted Similarity	Rank
3245017	Good location helpful staff. Nice modern stylish hotel [...] , friendly helpful staff and great food.my 1st trip to Ho Chi Minh and was impressed. [...]	Pos	4	2075	$0.8726^{(1)}$	0.9707	1
5508253	Good spot in Ho Chi Minh. A clean and comfortable well priced hotel with very friendly and helpful staff especially Jacky the services manager, [...]	Pos	4.5	2401	0.8592	0.9648	2
3320049	great location. This hotel was in a great location in Ho Chi Minh [...]. The staff were friendly and helpful, the breakfast buffet was substantial.	Pos	4.5	672	$0.8719^{(2)}$	0.9645	3
833276	Friendly staff, excellent location. This is situated at the quieter end of Bui Vien street [....] (although you're in Ho Chi Minh City so why would you?) [...].	Pos	4.5	2431	0.8561	0.9613	4
2555547	Great location. Great location and services.[...] it is located in the heart of Ho Chi Minh old quotas. Staffs make it with help needed promptly [...]	Pos	5	1136	0.8369	0.9409	5
3794814	Convenient Location and Great Staff. Enjoyed my stay at this hotel - convenient to walk to many sights in Ho Chi Minh City. Staff were very pleasant and helpful. [...]	Pos	4	323	$0.8701^{(3)}$	0.9386	6
454970	Great Location very friendly.. One of the best value for money hotels in Ho Chi Minh City , great locations close to everything [...]	Pos	4	784	0.8509	0.9367	7
302836	Great location, but not quiet. [...] Customer service was amazing and the suite size was great. [...]. The location was amazing, center of Ho Chi Minh [...]	Pos	4	3066	0.8396	0.9360	8
2282888	The friendliest staff. Don't loose more time. Friendly staff, good location and [...] who wants to experience Ho Chi Minh City in a comfortable way... :):):)	Pos	4.5	583	0.8475	0.9347	9
10151324	Good hotel in a good location. A modern hotel in a convenient, quiet side street in a good location. The staff were very professional.[..] when in Ho Chi Minh City	Pos	4.5	701	0.8411	0.9311	10

number of reviews (2401). Hotels with IDs 3320049, 833276, and 2555547 follow closely, each having positive sentiments and high hotel values. These hotels also benefit from a substantial number of reviews, which enhance their credibility and ranking. For example, hotel 2555547, with a perfect hotel value of 5, sees its similarity score increase from 0.8369 to 0.9409.

Interestingly, hotels with fewer reviews, such as hotel 3794814 (323 reviews) and hotel 2282888 (583 reviews), still make it to the top-10 list due to their strong initial similarity scores and positive sentiments. This highlights the importance of incorporating sentiment and relevance of the content in the reviews alongside traditional metrics such as the number of reviews and hotel ratings.

In summary, the adjustments made to the similarity scores based on hotel value, the number of reviews, and sentiment provide a more holistic assessment of each hotel's suitability for the user's query. The refined scores ensure that the recommended hotels not only match the keywords, but also offer high-quality and well-reviewed experiences, thus improving user satisfaction.

6   Conclusion and Future Works

This study investigated sentiment analysis and cold-start recommendation systems using a real-world hospitality dataset. The sentiment analysis was conducted using various deep learning models and embedding methods. The transformer on pre-trained model '*all-mpnet-base-v2*', achieved the highest accuracy of 85.82% and an F1-score of 0.8666, outperforming traditional models like CNN, LSTM, BiLSTM and BiLSTM-CNN on word2vec embedding.

For the cold-start recommendation system, the sentence embedding based on '*all-mpnet-base-v2*' pre-trained model was employed to map input keywords into embedding matrices. The cosine similarity scores between the input and previous reviews were calculated and adjusted based on the value of the hotel, the number of reviews and the sentiment scores extracted from SA model. The system effectively provided relevant and high-quality hotel recommendations, as evidenced by the top-10 hotels for the sample input keyword "*Great location with friendly staff in Ho Chi Minh*".

The study faced limitations such as data imbalance, variations in review quality, and challenges with the cold-start problem for new hotels. Future work will focus on enhanced text preprocessing, hybrid models, user-specific personalization and robust evaluation for this recommendation system with diverse datasets. These efforts aim to improve model performance and expand the applicability of the system in the hospitality domain.

References

1. Alrawadieh, Z., Law, R.: Determinants of hotel guests' satisfaction from the perspective of online hotel reviewers. Int. J. Cult. Tour. Hospit. Res. **13**(1), 84–97 (2019)

2. Burke, R.: Hybrid recommender systems: survey and experiments. User Model. User-Adap. Inter. **12**, 331–370 (2002)
3. Chalupa, S., Petricek, M.: Understanding customer's online booking intentions using hotel big data analysis. J. Vacat. Mark. **30**(1), 110–122 (2024)
4. Hochreiter, S., Schmidhuber, J.: Long short-term memory. Neural Comput. **9**(8), 1735–1780 (1997)
5. Hu, M., Liu, B.: Mining and summarizing customer reviews. In: Proceedings of the Tenth ACM SIGKDD International Conference on Knowledge Discovery and Data Mining, pp. 168–177 (2004)
6. Jalan, K., Gawande, K.: Context-aware hotel recommendation system based on hybrid approach to mitigate cold-start-problem. In: 2017 International Conference on Energy, Communication, Data Analytics and Soft Computing (ICECDS), pp. 2364–2370. IEEE (2017)
7. Kim, Y.: Convolutional neural networks for sentence classification (2014). https://arxiv.org/abs/1408.5882
8. Kingma, D.P., Ba, J.: Adam: a method for stochastic optimization. arXiv preprint arXiv:1412.6980 (2014)
9. Le, Q.H., Mau, T.N., Tansuchat, R., Huynh, V.N.: A multi-criteria collaborative filtering approach using deep learning and Dempster-Shafer theory for hotel recommendations. IEEE Access **10**, 37281–37293 (2022)
10. Li, D., Qian, J.: Text sentiment analysis based on long short-term memory. In: 2016 First IEEE International Conference on Computer Communication and the Internet (ICCCI), pp. 471–475. IEEE (2016)
11. Liu, B.: Sentiment Analysis and Opinion Mining. Springer, Heidelberg (2022). https://doi.org/10.1007/978-3-031-02145-9
12. Muennighoff, N., Tazi, N., Magne, L., Reimers, N.: MTEB: massive text embedding benchmark. arXiv preprint arXiv:2210.07316 (2022). https://doi.org/10.48550/ARXIV.2210.07316
13. Pang, B., Lee, L., Vaithyanathan, S.: Thumbs up? Sentiment classification using machine learning techniques. arXiv preprint cs/0205070 (2002)
14. Ray, B., Garain, A., Sarkar, R.: An ensemble-based hotel recommender system using sentiment analysis and aspect categorization of hotel reviews. Appl. Soft Comput. **98**, 106935 (2021)
15. Reimers, N., Gurevych, I.: Sentence-BERT: sentence embeddings using Siamese BERT-networks. In: Proceedings of the 2019 Conference on Empirical Methods in Natural Language Processing. Association for Computational Linguistics (2019). https://arxiv.org/abs/1908.10084
16. Ricci, F., Rokach, L., Shapira, B.: Introduction to recommender systems handbook. In: Ricci, F., Rokach, L., Shapira, B., Kantor, P. (eds.) Recommender Systems Handbook. Springer, Boston (2010). https://doi.org/10.1007/978-0-387-85820-3_1
17. Sun, C., Qiu, X., Xu, Y., Huang, X.: How to fine-tune BERT for text classification? In: Sun, M., Huang, X., Ji, H., Liu, Z., Liu, Y. (eds.) CCL 2019. LNCS (LNAI), vol. 11856, pp. 194–206. Springer, Cham (2019). https://doi.org/10.1007/978-3-030-32381-3_16
18. Thu, H.N.T., Ngoc, V.H., Minh, T.T., Binh, G.N.: Determining criteria for evaluating the quality of Vietnamese hotel through guest's online reviews. Int. J. Bus. Inf. Syst. **44**(2), 249–267 (2023)
19. Tran, T., Ba, H., Huynh, V.-N.: Measuring hotel review sentiment: an aspect-based sentiment analysis approach. In: Seki, H., Nguyen, C.H., Huynh, V.-N., Inuiguchi, M. (eds.) IUKM 2019. LNCS (LNAI), vol. 11471, pp. 393–405. Springer, Cham (2019). https://doi.org/10.1007/978-3-030-14815-7_33

20. Tran, X.T., Dang, D.T., Nguyen, N.T.: Improving hotel customer sentiment prediction by fusing review titles and contents. In: Nguyen, N.T., et al. (eds.) ACIIDS 2023. LNCS, vol. 13996, pp. 323–335. Springer, Cham (2023). https://doi.org/10.1007/978-981-99-5837-5_27

21. Wen, Y., Liang, Y., Zhu, X.: Sentiment analysis of hotel online reviews using the BERT model and ERNIE model-data from china. PLoS ONE **18**(3), e0275382 (2023)

22. Xu, G., Meng, Y., Qiu, X., Yu, Z., Wu, X.: Sentiment analysis of comment texts based on biLSTM. IEEE Access **7**, 51522–51532 (2019)

23. Zhang, J., Li, Y., Tian, J., Li, T.: LSTM-CNN hybrid model for text classification. In: 2018 IEEE 3rd Advanced Information Technology, Electronic and Automation Control Conference (IAEAC), pp. 1675–1680. IEEE (2018)

24. Zhang, L., Wang, S., Liu, B.: Deep learning for sentiment analysis: a survey. Wiley Interdisc. Rev.: Data Min. Knowl. Discov. **8**(4), e1253 (2018)

25. Zhang, W., Du, Y., Yoshida, T., Yang, Y.: DeepRec: a deep neural network approach to recommendation with item embedding and weighted loss function. Inf. Sci. **470**, 121–140 (2019)

26. Zhao, X., Wang, L., Guo, X., Law, R.: The influence of online reviews to online hotel booking intentions. Int. J. Contemp. Hosp. Manag. **27**(6), 1343–1364 (2015)

Mining Complementary Relationships of Items for Diversified Recommendation

Wei Qian[1], Xianneng Li[1,2]([✉]), Shuang Zheng[1], Deqiang Hu[1,2], and Yang Yu[1,2]

[1] School of Economics and Management, Dalian University of Technology, Dalian 116024, China
`xianneng@dlut.edu.cn`
[2] Institute for Advanced Intelligence, Dalian University of Technology, Dalian 116024, China

Abstract. Nowadays, recommender systems are playing an important role in online platforms. Beyond accuracy, diversity has been recognized as an important metric to evaluate recommendation performance, where re-ranking is one of the main tools to pursue diversity in the recommendation. However, existing re-ranking strategies generally introduce diversity at the expense of sacrificing accuracy performance, since the trade-off between accuracy and diversity is not trivial to achieve. In addition, complementary relationship exists extensively between items. Complementary items themselves belong to different categories but have a very large correlation. Therefore, we introduce a novel diversified re-ranking model named CRDR by using complementary relationships of items, which increase the positions of items in the recommendation list that are complementary to the items previously purchased by the user, so as to improve the diversity of recommendations while ensuring accuracy. Specifically, we first mine complementary relationships in co-purchase items with graph attention networks. Then, we add our complementary scores to the MMR model and use a greedy strategy to solve the problem. Extensive experiments on two different datasets demonstrate that our CRDR model achieve a balance between accuracy and diversity over traditional state-of-the-art methods.

Keywords: Complementary relationships · Diversified recommendation · Graph neural network · Re-ranking

1 Introduction

Nowadays, recommender system has been widely used in electronic commerce [18]. Most of the research on recommender systems mainly focuses on personalized recommendation [19], where accuracy, precision-recall, and click-through rate (CTR) are the main metrics [10]. Personalized recommender systems usually assume that consumer demand is stable and aim to improve accuracy, which will lead to a high degree of homogenization of recommended items. This may not bring good satisfaction to the consumer and lead to the filter bubble problem

© The Author(s), under exclusive license to Springer Nature Singapore Pte Ltd. 2025
X. Tang et al. (Eds.): KSS 2024, CCIS 2269, pp. 293–304, 2025.
https://doi.org/10.1007/978-981-96-0178-3_20

[14]. The purchasing decision-making process of consumers is often extremely complex. Although the past preferences of consumers are very important, environmental factors and the inherent characteristics of consumers can also affect consumer behavior. These factors together lead to the uncertainty of consumer demand, which needs the recommender systems to take scientific and reasonable consideration.

Besides accuracy, researchers and businesses have realized that diversity is also an important metric to evaluate the performance of recommendation results. A diversified recommender system can recommend diversified and novel items, which can solve the above problems well. It should be emphasized that diversified recommendation is only a means to meet the needs of users, and the accuracy of recommendation should not be ignored when we try to optimize diversified recommendations. Unfortunately, there is a dilemma between accuracy and diversity [31]. The relationship between accuracy and diversity is mostly contradictory, and the increase in diversity is often accompanied by a decrease in accuracy. Therefore, much research on diversified recommendations will sacrifice some accuracy while improving the diversity of recommendation results. The contradiction between diversity and accuracy is still a very challenging issue.

Complementary and substitute item theory is very famous in economics. As for complementary item theory, products are considered complementary if lowering the price of one product leads to an increase in the sales of the other [2,23,24]. Complementary items widely exist on e-commerce websites, such as tennis balls and tennis rackets. According to [25], consumers may be influenced by the items they have previously purchased, owned, and used when purchasing new items, so the user's current consumption may be to better satisfy a need along with items purchased before. For example, suppose consumers have bought tennis rackets before. In that case, the current consumers may need to buy tennis balls to meet the demand of consumers to play tennis together with tennis rackets. Therefore, the recommendation of complementary items can hit the needs of users with a greater probability and ensure the accuracy of the recommendation. In addition, complementary items are themselves of different categories, which means that the recommendation of complementary items can also increase diversity. This study combines complementary item theory with the diversified recommendation and uses complementary relationships between items to improve the position of items that have complementary relationships with users' historical purchases in the Top-N item list, so as to improve the diversification degree of the item list and ensure that the recommendation is accurate.

Hence, we proposed a diversified recommender model based on the item's complementary relationship which contains three stages, the ranking stage, the complementary relationship modeling stage, and the re-ranking stage. In our model, we first mine complementary relationships through item data that users have co-purchased historically. During the recommendation processes, we use the traditional recommender model as the ranking model to get accurate item lists. Then we introduce a novel re-ranking model after the ranking model named **C**omplementary **R**elationships of Items for **D**iversified **R**ecommendation (CRDR) which improve the position of items that are complementary to items user purchased historically. In this way, we can improve the diversity of the recommendation list while ensuring accuracy. Through a set of experiments using *Amazon datasets* and *Meituan Maicai datasets*, we have confirmed that our model can not only improve the diversity of recommendations but also improve the probability of users buying items our model recommended.

To summarize, the contributions of this paper are as follows:

- We mine complementary relationships between items and model user-item complementary score on this basis.
- We introduce diversified re-ranking model CRDR by using user-item complementary relationship and achieve a balance between accuracy and diversity of recommendations.
- Considering users' inconsistent understanding of diversity, we model differences in the diversity of recommendations for users.

2 Complementary Relationship Modeling

In order to take advantage of the items' complementary relationship to improve the diversity of recommendations, we first need to mine complementary relationship between items. In economics, complementary items are defined as certain consumption dependence between two kinds of items, that is, the consumption of one kind of items must match the consumption of another kind of items. However, the complementary relationships are not easy to quantify. Currently, research on the recommendation of complementary items [15,20,27] proposes that consumers' co-purchase behavior of items often reflects the complementary relationship between items, that is, mutually complementary items are often jointly purchased by consumers. According to these ideas, we try to model complementary relationship between items by mining the data of consumers' co-purchase behavior with neural networks and graph embedding techniques.

First of all, we use the co-purchase relationship of items to build a item graph where the nodes of the graph are items and the edges between the nodes represent the two nodes that have been jointly purchased by a user before. The characteristics of the nodes are used as input to a universal embedding module $FFN(\cdot)$ and mapped to the initial item vector. Then, on the basis of item graph, we use the graph embedding to generate the high-dimensional item embedding from the initial vector. Specially, we follow the idea of [13] which used GAT [26] model to fine-tune the parameters of the above embedding module $FFN(\cdot)$. The model is as follows:

$$\theta_i = FFN\left(\gamma_i\right) = \sigma\left(\gamma_i W^{(1)} + b^{(1)}\right) \tag{1}$$

$$\theta_i \leftrightarrow \theta_{N_i} = GAT\left(\theta_i, \theta_{j1}, \ldots, \theta_{jm}\right) \tag{2}$$

where γ_i is the raw feature vector for the item's catalog. θ_i is the original embedding generated by embedding module $FFN(\cdot)$. θ_{N_i} is the output of the GAT model. $\theta_{j1}, \theta_{j2}, \ldots, \theta_{jm}$ are the embeddings of item θ_i's neighbor nodes. $W^{(1)}$ is weight matrices with the corresponding biased term $b^{(1)}$. $\sigma(\cdot) = ReLU(\cdot)$ is a non-linear activation function. We want θ_i and θ_{N_i} to be as close as possible.

Second, we use a multi-layer neural network to build a classification model which predicts whether there are complementary relationships between items. The input is the high-dimensional embedding θ_i of the two items obtained by fine-tuned $FFN(\cdot)$, and the model is as follows:

$$CP(\theta_{i1}, \theta_{i2}) = \sigma\left(\sigma\left(\left(\theta_{i1} \bigoplus \theta_{i2}\right) W^{(1)} + b^{(1)}\right) W^{(2)} + b^{(2)}\right) W^{(3)} + b^{(3)} \quad (3)$$

where $\sigma(\cdot) = ReLU(\cdot)$ is a non-linear activation function. $W^{(1)}, W^{(2)}, W^{(3)}$ are weight matrices with the corresponding biased terms $b^{(1)}, b^{(2)}, b^{(3)}$. As for the positive sample, we directly use the behavior data of users' co-purchase of items mentioned above [13,20]. The corresponding negative sample should be that there is no complementary relationship between the two items (including substitution relationship and no relationship). Therefore, we use random negative sampling to generate negative samples. After obtaining positive and negative samples in the above way, the model can be trained, and we use the Cross Entropy loss function and Adam Optimizer.

On the basis of this model, we need to further get the user-item complementary relationship. Since the historical purchase data of the user can reflect the items owned by the user to a large extent, we can identify the complementary relationship between the current item and the items purchased by the user before, and obtain the complementary score between the current item and the user by taking the average method. The formula is as follows:

$$Com_{ui} = \frac{\sum_{k \in L_u} CP(\theta_i, \theta_k)}{|L_u|} \quad (4)$$

where θ_i is the embedding of the current item. θ_k is the embedding of items purchased by the user before. L_u is the item set purchased by the user. Com_{ui} is the output which is the complementary score between item θ_i and the user u.

In this way, we successfully mine the complementary relationship between users and items. And the following diversified recommendation model is carried out on this basis.

3   Diversified Re-ranking Modeling

We propose our CRDR model which achieve a balance between accuracy and diversity for the recommendation. CRDR use the user-item complementary relationship to improve the diversified re-ranking algorithm and also model the heterogeneity of users' diversification preferences. The main structure of the overall model is shown in Fig. 1 below.

As the most important metric in the recommender system, the accuracy of recommendation is always the first thing to be guaranteed. Therefore, we

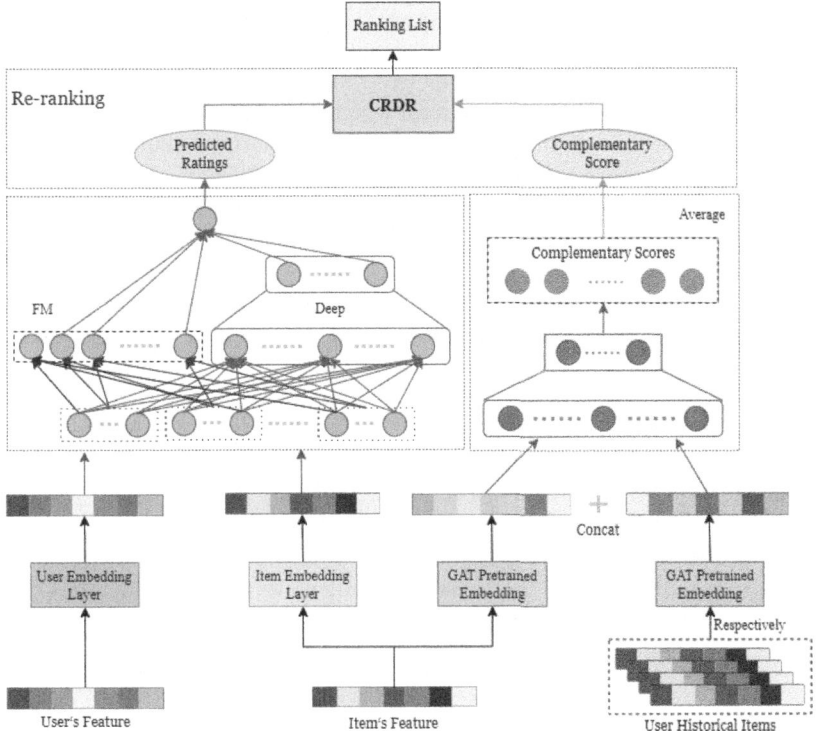

Fig. 1. The CRDR model.

need accuracy as the cornerstone while improving diversity. Generally, the recommender system consists of three stages: recall, ranking, and re-ranking. Our CRDR model is a re-ranking model on the whole. Thus, we first need to use the traditional ranking model for accuracy modeling to generate the initial recommendation list. We chose the DeepFM model [11] as the ranking algorithm, which combine a simple linear model with a complex deep learning model so that the model has both "memory ability" and "generalization ability" to improve recommendations. After building the ranking model, we can get the ranking item list for each user and the corresponding ranking score. And then the next step is our re-ranking model CRDR. But, the re-ranking stage can only take up very little time in practice which means it is unacceptable to take all the items for re-ranking. So we choose the top 50 items from the ranking result as the input of the subsequent re-ranking model for each user, which not only ensure the accuracy of the model but also save a lot of time cost.

We introduce our CRDR model based on the recommendation item list generated by the ranking model, which will re-rank the position of items based on the complementary relationship of items purchased in users' historical behavior. As for diversified re-ranking algorithms, MMR is one of the classic models. MMR is very good at improving the diversity of an item list, but it usually causes a massive decline in accuracy. So we choose MMR model as the base model and integrate the items' complementary relationship to balance the accuracy and diversity. On the basis of the diversified re-ranking algorithm MMR, we add a complementary score of the user and item which is generated from Sect. 2. Thus, the final result consists of three parts to achieve a better balance between the accuracy and diversity of recommendation results, modeled as follows:

$$
\begin{aligned}
CRDR = & Argmax_{D_i \in R \setminus S} [\lambda * (Sim_1(D_i, U) + Com(D_i, U)) \\
& - (1 - \lambda) * max_{D_j \in S} Sim_2(D_i, D_j)]
\end{aligned}
\tag{5}
$$

where D_i, D_j are items from items set R which is the original item list generated by the ranking model. S is the final diversified item list generated by the re-ranking model CRDR. λ is the equilibrium parameter. The above formula is an NP-hard problem, so we refer to the MMR model to solve the model in a greedy way. This method cannot guarantee the optimal result, but it can find a relatively optimal result with a faster solving speed.

In addition, considering the heterogeneity of diversification needs of each person, we establish personalized diversification algorithms, that is, using different equilibrium parameters λ for each user. We calculate the diversification degree of the set of goods purchased by consumers to make a personalized adjustment of lambda value for each user. The model is as follows:

$$
User_{div} = \frac{2 * \sum\limits_{D_i, D_j \in L_u, D_i \neq D_j} (1 - similarity(D_i, D_j))}{|L_u| * (|L_u| - 1)}
\tag{6}
$$

$$
similarity(D_i, D_j) = \frac{\theta_i \bullet \theta_j}{\|\theta_i\| \|\theta_j\|}
\tag{7}
$$

$$
\lambda = down + (1 - User_{div}) * (up - down)
\tag{8}
$$

where D_i, D_j are items from the item set which has been purchased by users historically with the corresponding embedding θ_i, θ_j. $similarity(\cdot)$ is the cosine distance. L_u is the item set purchased by the user. $up, down$ are the upper and lower limits of equilibrium parameter λ adjustment respectively.

Through the above modeling methods, the diversified needs of users can be well modeled and differentiated recommendations can be realized. In addition, the diversified recommendation algorithm based on the complementary relationship of items also includes the module $Com(D_i, U)$ for calculating the complementary score between users and items, which is also closely related to users. The complementary score of different users is also different, which means that the complementary score also affects the diversity of recommendation results.

4 Evaluation

We use two datasets to verify our CRDR model. The first one is Amazon grocery dataset which is crawled from Amazon.com by Julian McAuley [44]. We find that the co-purchase data in Amazon dataset already have strong complementary relationships. It means that this dataset is very suitable for mining complementary relationships of items. The second dataset is Meituan Maicai which is the largest O2O company in China. Maicai(where users can buy vegetables and products) is a new business for Meituan where users can buy multiple items at the same time where we can get users' co-purchase data. Different from the Amazon dataset, there are a lot of noisy data in this dataset. So it is much harder to mine the complementary relationship from the co-purchase data. But at the same time, this dataset can verify that whether our model is suitable for more general scenarios. Due to the confidential agreement with the company, we are not allowed to make this dataset publicly available.

As for evaluating our model, we use two categories of metrics. The first one is the Accuracy metric including Hit, NDCG, Precision, and Recall. The other one is the diversity metric including Diversity, Category. We also choose several baselines to compare. They are as follows: **1) DeepFM** [11]: a traditional ranking model which has been widely used in many recommender systems. **2) MMR** [3]: a re-ranking model that adopts a greedy strategy to generate the Top-N result list. First, the items with the highest relevance are selected firstly. Then, each time, the items with a high matching degree to the query and low maximum similarity to the already selected items are selected. **3) DPP** [4]: a high-performance probabilistic model that converts complex probability calculations into simple deterministic calculations, calculates the probability of each subset through the determinant of the kernel matrix, and then finds the subset with the highest relevance and diversity in the product set by estimating the maximum a posteriori probability.

Table 1. Top5 Results of Amazon

model	hit@5	NDCG@5	Diversity@5	Precision@5	Recall@5	Category@5
DeepFM	0.34837	0.23713	0.24587	0.09516	0.17318	0.02684
MMR	0.28623	0.20198	0.4739	0.07366	0.1314	**0.06774**
DPP	0.27882	0.19975	**0.48285**	0.07135	0.1264	0.0533
CRDR	**0.48045**	**0.33679**	0.26241	**0.1333**	**0.26612**	0.04964

We first conducted the experiment on the Amazon dataset, and the results are in Table 1 and Table 2. Then, we carried out the experiment on Meituan Maicai dataset, and the results are in Table 3 and Table 4. We can find that CRDR model has achieved similar results in the two datasets through the experimental results, that is, compared with the accuracy model DeepFM, the CRDR model has achieved certain improvements in the index of diversity and the accuracy index, while compared with the diversified recommendation model MMR and DPP, MLC model is significantly inferior to them in terms of diversity indicators, but both models sacrifice the accuracy of results and deviate from the original

Table 2. Top10 Results of Amazon

model	hit@10	NDCG@10	Diversity@10	Precision@10	Recall@10	Category@10
DeepFM	0.50617	0.2879	0.24298	0.08297	0.29225	0.041916
MMR	0.45955	0.25756	0.37194	0.07116	0.24888	**0.08548**
DPP	0.44921	0.25431	**0.37675**	0.0684	0.23927	0.07859
CRDR	**0.65374**	**0.39275**	0.2277	**0.11053**	**0.42265**	0.0712

Table 3. Top5 Results of Meituan

model	hit@5	NDCG@5	Diversity@5	Precision@5	Recall@5	Category@5
DeepFM	0.30085	0.20827	0.33628	0.07683	0.05101	0.13533
MMR	0.18364	0.1483	0.77872	0.03976	0.0271	0.22831
DPP	0.15946	0.13637	**0.84076**	0.03351	0.0223	**0.2438**
CRDR	**0.31038**	**0.2152**	0.33894	**0.07961**	**0.053**	0.13017

intention of diversity recommendation. In addition, as mentioned above, the co-purchasing behavior data set of Amazon data set is of high quality, while the co-purchasing behavior data set of Meituan Food shopping data set is relatively poor, mixed with a lot of noise data, so the effect is not as significant as that of Amazon data set. This result also verifies the generalization performance of this model, which is suitable for industrial scenarios.

5 Related Work

5.1 Diversified Recommendation

The fundamental starting point of the recommender system is to meet the personalized needs of users, which has been the most important work for a long time. Rendle [22] proposed a factorization machine (FM), which can automatically learn cross-feature combinations by using implicit features, and achieved good results. Guo et al. [11] proposed the DeepFM model on the basis of wide&deep model [5], used the FM model in the wide part, and realized the weight sharing between FM and Deep part.

Table 4. Top10 Results of Meituan

model	hit@10	NDCG@10	Diversity@10	Precision@10	Recall@10	Category10
DeepFM	0.37216	0.22852	0.39534	0.05394	0.07169	0.22831
MMR	0.22906	0.16098	0.70315	0.0263	0.03549	0.3936
DPP	0.21783	0.15296	**0.74906**	0.02471	0.03228	**0.42975**
CRDR	**0.38388**	**0.23571**	0.40402	**0.05602**	**0.07443**	0.2407

Recently, diversity has been considered an indispensable component of recommender systems and has gradually attracted the attention of both researchers and businesses. There are two main strategies for improving diversity. The first one is to re-rank recommendation lists after obtaining predicted ratings by existing rating prediction approaches. MMR [3] is a classic diversified re-ranking algorithm. Pathak et al. [21] used a post-filtering strategy to propose a new method to improve the diversity of recommendations based on a clustering algorithm. Le Wu et al. [28] proposed a REC model to enable the recommender system to capture users' unpopular interests based on collaborative filtering. Adomavicius et al. [1] introduced many techniques which can generate more different recommendations among all users while maintaining relatively reliable recommendation accuracy and improving the overall diversity of recommendations in their study. Hulu [4] used the determinant point process to improve the diversity of the recommender system and has achieved good results in the experiment. Another method is to directly optimize the recommender algorithm and introduce the diversity measure to build a new recommender system. Anupriya et al. [9] modified the matrix decomposition model by adding regularization factors that minimized variance. Mahmut et al. [16] applied penalty functions to the objective function. PLUS [8] uses power functions to suppress user-like influences caused by popular items. COUSIN [7] builds a profile of similarity between users and commodities, and uses a regression model to quantify the association strength between users and commodities. Yu Ting et al. [30] proposed an adaptive trust-aware recommendation model based on the user-item binary network from the perspective of users' trust to balance the accuracy and diversity of recommender systems. Most of these researches are focusing on improving diversity and some even have sacrificed accuracy. They also tend to ignore that each user has a different understanding of diversity.

5.2   Complementary Product Recommendation

Complementary items can have a great influence on consumers' consumption behavior. [29] suggest that complementary apparel items should be coordinated together on the website to produce favorable consumer shopping outcomes. Research of complementary item bundles selling strategy shows that consumers evaluate bundles with complementary items more favorably than they evaluate bundles with non-complementary items [17]. [6] finds that consumers may make complementary choices in consumption episodes.

Recommender systems are widely used to recommend relevant items given item features and user-item behaviors. Different from most work focusing on modeling user-item relationships or similarity-based item-item relationships, some research has dived deep into discovering complementary relationships among items. The most straightforward way for Complementary Product Recommendation(CPR) is based on frequent pattern mining and association rules [12]. Some recent works in this direction seek to classify whether two items are complementary or substitutable. Two representative examples are Sceptre [20] and PMSC [27]. These models are limited to traditional algorithms and basically do not use a deep learning model based on neural networks. Junheng Hao et al. [13] innovatively proposed the design of a complementary product recommendation model P-Companion based on AutoEncoder of deep learning model. However, they mainly operate on item level and lack diversity consideration in modeling. These models do not use the information of users, but only predict the complementary relationship between items through the information of items, and do not combine it with the existing recommender system which is also what these models need to achieve.

6   Conclusion

In this study, we introduce a new diversified recommendation model CRDR which uses the item's complementary relationship to improve recommendation diversity reasonably while ensuring the accuracy of the recommendation. To utilize an item's complementary relationships, we build a neural network to classify whether two items have complementary relationships. Simultaneously, we use the traditional algorithm DeepFM as a ranking model and propose our CRDR model during the re-ranking process and get the final items to recommend. We conduct experiments on two different datasets, and the results demonstrate that our CRDR model achieve a balance between accuracy and diversity.

Acknowledgement. This research was supported by the National Natural Science Foundation of China (NSFC) under Grant 72071029, 72231010. This research was also partially supported by Meituan.

Disclosure of Interests. The authors have no competing interests to declare that are relevant to the content of this article.

References

1. Adomavicius, G., Kwon, Y.: Improving aggregate recommendation diversity using ranking-based techniques. IEEE Trans. Knowl. Data Eng. **24**(5), 896–911 (2011)
2. Bucklin, R.E., Russell, G.J., Srinivasan, V.: A relationship between market share elasticities and brand switching probabilities. J. Mark. Res. **35**(1), 99–113 (1998)
3. Carbonell, J., Goldstein, J.: The use of MMR, diversity-based reranking for reordering documents and producing summaries. In: Proceedings of the 21st Annual International ACM SIGIR Conference on Research and Development in Information Retrieval, pp. 335–336 (1998)
4. Chen, L., Zhang, G., Zhou, E.: Fast greedy map inference for determinantal point process to improve recommendation diversity. In: Advances in Neural Information Processing Systems, vol. 31 (2018)
5. Cheng, H.T., et al.: Wide & deep learning for recommender systems. In: Proceedings of the 1st Workshop on Deep Learning for Recommender Systems, pp. 7–10 (2016)
6. Dhar, R., Simonson, I.: Making complementary choices in consumption episodes: highlighting versus balancing. J. Mark. Res. **36**(1), 29–44 (1999)

7. Gan, M.: COUSIN: a network-based regression model for personalized recommendations. Decis. Support Syst. **82**, 58–68 (2016)
8. Gan, M., Jiang, R.: Improving accuracy and diversity of personalized recommendation through power law adjustments of user similarities. Decis. Support Syst. **55**(3), 811–821 (2013)
9. Gogna, A., Majumdar, A.: DiABLO: optimization based design for improving diversity in recommender system. Inf. Sci. **378**, 59–74 (2017)
10. Gunawardana, A., Shani, G.: A survey of accuracy evaluation metrics of recommendation tasks. J. Mach. Learn. Res. **10**(12) (2009)
11. Guo, H., Tang, R., Ye, Y., Li, Z., He, X.: DeepFM: a factorization-machine based neural network for CTR prediction. arXiv preprint arXiv:1703.04247 (2017)
12. Han, J., Cheng, H., Xin, D., Yan, X.: Frequent pattern mining: current status and future directions. Data Min. Knowl. Disc. **15**(1), 55–86 (2007)
13. Hao, J., et al.: P-companion: a principled framework for diversified complementary product recommendation. In: Proceedings of the 29th ACM International Conference on Information & Knowledge Management, pp. 2517–2524 (2020)
14. Herlocker, J.L., Konstan, J.A., Terveen, L.G., Riedl, J.T.: Evaluating collaborative filtering recommender systems. ACM Trans. Inf. Syst. (TOIS) **22**(1), 5–53 (2004)
15. Kang, W.C., Wan, M., McAuley, J.: Recommendation through mixtures of heterogeneous item relationships. In: Proceedings of the 27th ACM International Conference on Information and Knowledge Management, pp. 1143–1152 (2018)
16. Karakaya, M.Ö., Aytekin, T.: Effective methods for increasing aggregate diversity in recommender systems. Knowl. Inf. Syst. **56**, 355–372 (2018)
17. Karataş, M., Gürhan-Canli, Z.: When consumers prefer bundles with noncomplementary items to bundles with complementary items: the role of mindset abstraction. J. Consum. Psychol. **30**(1), 24–39 (2020)
18. Linden, G., Smith, B., York, J.: Amazon.com recommendations: item-to-item collaborative filtering. IEEE Internet Comput. **7**(1), 76–80 (2003)
19. Lu, J., Wu, D., Mao, M., Wang, W., Zhang, G.: Recommender system application developments: a survey. Decis. Support Syst. **74**, 12–32 (2015)
20. McAuley, J., Pandey, R., Leskovec, J.: Inferring networks of substitutable and complementary products. In: Proceedings of the 21th ACM SIGKDD International Conference on Knowledge Discovery and Data Mining, pp. 785–794 (2015)
21. Pathak, A., Patra, B.K.: A knowledge reuse framework for improving novelty and diversity in recommendations. In: Proceedings of the Second ACM IKDD Conference on Data Sciences, pp. 11–19 (2015)
22. Rendle, S.: Factorization machines. In: 2010 IEEE International Conference on Data Mining, pp. 995–1000. IEEE (2010)
23. Russell, G.J., Bolton, R.N.: Implications of market structure for elasticity structure. J. Mark. Res. **25**(3), 229–241 (1988)
24. Russell, G.J., Petersen, A.: Analysis of cross category dependence in market basket selection. J. Retail. **76**(3), 367–392 (2000)
25. Shocker, A.D., Bayus, B.L., Kim, N.: Product complements and substitutes in the real world: the relevance of "other products". J. Mark. **68**(1), 28–40 (2004)
26. Veličković, P., Cucurull, G., Casanova, A., Romero, A., Lio, P., Bengio, Y.: Graph attention networks. arXiv preprint arXiv:1710.10903 (2017)
27. Wang, Z., Jiang, Z., Ren, Z., Tang, J., Yin, D.: A path-constrained framework for discriminating substitutable and complementary products in e-commerce. In: Proceedings of the Eleventh ACM International Conference on Web Search and Data Mining, pp. 619–627 (2018)

28. Wu, L., Liu, Q., Chen, E., Yuan, N.J., Guo, G., Xie, X.: Relevance meets coverage: a unified framework to generate diversified recommendations. ACM Trans. Intell. Syst. Technol. (TIST) **7**(3), 1–30 (2016)

29. Yoo, J., Kim, M.: Online product presentation: the effect of product coordination and a model's face. J. Res. Interact. Market. (2012)

30. Yu, T., Guo, J., Li, W., Wang, H.J., Fan, L.: Recommendation with diversity: an adaptive trust-aware model. Decis. Support Syst. **123**, 113073 (2019)

31. Zhou, T., Kuscsik, Z., Liu, J.G., Medo, M., Wakeling, J.R., Zhang, Y.C.: Solving the apparent diversity-accuracy dilemma of recommender systems. Proc. Natl. Acad. Sci. **107**(10), 4511–4515 (2010)

A Novel Two-Stage Approach for Customer Satisfaction Analysis

Chunlan Liang[1(✉)] and Fuying Jing[2]

[1] School of Mathematics and Statistics, Chongqing Technology and Business University, Chongqing 400067, China
liangchunlan@ctbu.edu.cn
[2] National Research Base of Intelligent Manufacturing Service, Chongqing Technology and Business University, Chongqing 400067, China

Abstract. Online customer reviews generate an electronic word-of-mouth (eWOM) effect in the form of text reviews and ratings, which is an effective way for operators to understand customer experience and satisfaction. Collecting customer data through interviews or questionnaires to explore key factors affecting customer satisfaction is an empirical practice based on constrained time and sample size. However, there are a large number of online customer reviews, and customers can provide detailed feedback on product defects and improvement needs through text without being constrained by questionnaire questions. Therefore, how to effectively mine the dimensional information of customer satisfaction using online comment data and establish a satisfaction analysis framework is a noteworthy issue. This paper develops a two-stage method using text mining and multi-criteria decision making (MCDM) technology. Consider combining unsupervised topic modeling techniques with supervised recurrent neural networks, combining comment text and auxiliary data into the topic model for topic information extraction to obtain satisfaction evaluation attributes. Then, the multi-attribute satisfaction analysis model was extended, which fully considers the qualitative forms of customer judgments and preferences.

Keywords: Text Mining · Deep Learning · Customer Satisfaction Analysis · MCDM

1 Introduction

Customer satisfaction is a key indicator that measures the extent to which a product's realized value meets or exceeds customer expectations [1]. Angelella et al. [2] pointed out that customer satisfaction assessment plays a crucial role in organizational structure, revealing and elucidating customer preferences to reveal various aspects of corporate strategy. Moreover, some researchers believe that customer satisfaction may enhance a company's competitiveness, identify potential market opportunities, guide new measures to improve product or service quality, and have a positive impact on brand assets [3–5].

In recent years, the research on customer satisfaction in the literature has aimed to explore measurement, influencing factors, and improvement measures. Packard and

Berger [6] conducted a study based on 200 real customer service phone samples provided by online retailers to explore how employee language affects consumer behavior and attitudes. Some researchers have also collected data through questionnaires and telephone surveys to study the customer satisfaction spillover effects between product manufacturers and service providers [7]. Researchers have provided various solutions for evaluating customer satisfaction, especially with the widespread application of structural equation modeling. In addition, the American Customer Satisfaction Index (ACSI) is often combined with a layered Bayesian model to derive stable path coefficients. These coefficients effectively capture the "communality" within the industry and the "heterogeneity" outside the industry [8]. Conklin et al. [9] used tools such as cooperative game theory and risk analysis, combined with Carnot theory, to study the relationship between product quality and customer satisfaction. Meanwhile, research has also shown that customer satisfaction can lead to customer loyalty [7].

From a multi-criteria perspective, the assessment of customer satisfaction has garnered attention. MCDM stands out as one of the most potent methods for addressing the intricacies of conflict management [10]. Some common MCDM methods, such as VIKOR [11], TOPSIS [12], etc., have been used to measure product satisfaction. Although these methods analyze customer satisfaction in multiple dimensions, they do not consider the qualitative form of customer preferences.

The text mining method of online comments for customer satisfaction analysis has been a relatively new research method in recent years. Ahani et al. [13] determined the preferences of travelers based on their comments and ratings on TripAdvisor and segmented them based on their satisfaction. In the literature [14], the technical attributes of online text comments and customer participation in the comment community are used to predict overall customer satisfaction. Researchers collected tourist review data from TripAdvisor and analyzed customer satisfaction at historical sites by combining latent Dirichlet allocation (LDA) text mining methods and adaptive neurofuzzy inference systems [15]. Jia et al. [16] collected 8691 comments from Ctrip and extracted the influencing factors of customer satisfaction in the homestay industry through qualitative and quantitative analysis. Mejia et al. [17] used text from online comments on Yelp.com and employed non-negative matrix factorization to identify and extract the service dimensions of restaurants. Researchers have extracted consumer opinions on health topics from restaurant reviews and then studied the impact of food regulation on changes in consumer opinions [18].

Although the above methods all use text mining techniques and machine learning methods to obtain potential information from online comment data and then conduct subsequent empirical analysis, most methods separate the text data and ratings of online comments for research. Existing research directly applies the LDA model to text data, which is an unsupervised topic modeling technique that ignores other auxiliary data attached to the text (such as labels, ratings, or categories), which may be useful supervisory signals.

In recent research, the technology for modeling topics in online comment data has shifted towards topic models based on deep learning. TopicRNN has been proposed, which integrates the advantages of RNN and latent topic models [19]. Miao et al. [20] proposed alternative neural methods for topic modeling, which allow for training through backpropagation within the framework of neural variational inference. The Topic Attention Model (TAM) was introduced by Wang and Yang [21], who developed a backpropagation inference method using document specific topic ratios and global topic vectors learned from neural topic models. Yang et al. [22] proposed a novel supervised deep topic modeling method called sDTM to extract potential topic information from online comments. However, there is relatively little research on using these deep learning text mining methods for customer satisfaction analysis.

A two-stage method was developed using text mining and MCDM technology in this paper. In addition, this paper combines unsupervised topic modeling techniques with supervised recurrent neural networks and incorporates both comment text and auxiliary data into the topic model for topic information extraction in order to obtain satisfaction evaluation attributes. Then extend a multi-attribute satisfaction analysis model to conduct satisfaction analysis on each evaluation attribute.

The structure of this paper is the following: The Preliminaries section introduces the basic process of a neural topic model and variational autoencoder. Next, a two-stage method was developed using text mining and MCDM technology. And then, we conducted specific data analysis and discussed the results. Finally, we end this paper in the Conclusion section.

2 Preliminaries

2.1 Neural Topic Model

A topic model based on neural networks mainly utilizes deep learning to reconstruct the text production process of the topic model and adds sparse constraints on topic vocabulary in the modeling process to generate more expressive topic words. The neural topic model was proposed by Miao [20] and is a topic model based on the standard variational autoencoder (VAE). This model aims to extract potential features (i.e., topics) from the word vector space of documents and generate documents based on them.

2.2 Variational Autoencoder

The goal of VAE is to model the true posterior distribution $p(z|x)$ of the latent variable z, and through Bayes law, the posterior probability can be represented by the likelihood $p(x|z)$, prior distribution $p(z)$, and the marginal distributions $p(x)$ as follows:

$$p(z|x) = \frac{p(x|z)p(z)}{p(x)} \tag{1}$$

The denominator can be calculated using the following formula in principle:

$$p(x) = \int p(x|z)p(z)dz \tag{2}$$

However, due to the fact that the integration in Eq. (2) requires traversing all values of z, it is often difficult to calculate in high-dimensional space. Therefore, the true posterior distribution $p(z|x)$ is generally not directly solved, but the variational posterior distribution $q_\phi(z|x)$ (ϕ is the variational parameter) is solved, and the difference between $q_\phi(z|x)$ and $p(z|x)$ is continuously reduced to achieve the goal of approximating $p(z|x)$. $q_\phi(x)$ usually chooses distribution families that are easy to calculate, such as Gaussian distributions.

Essentially, this method transforms inference problems into optimization problems, solving $p(z|x)$ by minimizing the Kullback Leibler divergence between the variational distribution $q_\phi(z|x)$ and the true distribution $p(z|x)$. The KL divergence between distribution $q_\phi(z|x)$ and $p(z|x)$ is defined as Eq. (3):

$$
\begin{aligned}
D_{KL}[q_\phi(z|x)||p(z|x)] &= \sum_z q_\phi(z|x) \log \frac{q_\phi(z|x)}{p(z|x)} \\
&= E_{q_\phi(z|x)} \left[\log \frac{q_\phi(z|x)}{p(z|x)} \right] \\
&= E_{q_\phi(z|x)} \left[log(q_\phi(z|x)) - log(p(z|x)) \right]
\end{aligned}
\tag{3}
$$

The posterior distribution in Eq. (3) is replaced by the Bayesian rule (Eq. (1)) to obtain the following equation:

$$
\begin{aligned}
D_{KL}[q_\phi(z|x)||p(z|x)] &= E_{q_\phi(z|x)} \left[log(q_\phi(z|x)) - \log \frac{p(x|z)p(z)}{p(x)} \right] \\
&= E_{q_\phi(z|x)} \left[\log(q_\phi(z|x)) - \log(p(x|z)) - \log(p(z)) + \log(p(x)) \right] \\
&= \log(p(x)) - \underbrace{E_{q_\phi(z|x)} \left[\log(p(x,z)) - \log(q_\phi(z|x)) \right]}_{ELBO(\phi)}
\end{aligned}
\tag{4}
$$

The part in the right parenthesis of Eq. (4) is the lower bound of variation of $\log p(x)$, , denoted as ELBO (Evidence Lower Bound). After the given data x, the distribution $p(x)$ of x can be regarded as a constant; therefore, minimizing the optimization objective of KL divergence on the left side of Eq. (4) is equivalent to maximizing ELBO. Using the technique of interpolation, ELBO in Eq. (4) can be rewritten as Eq. (5):

$$
\begin{aligned}
ELBO(\phi) &= E\left[\log(p(x,z)) - \log(p(z)) + \log(p(z)) - \log(q_\phi(z|x)) \right] \\
&= E\left[\log(p(x|z)) + \log(p(z)) - \log(p(z)) + \log(p(z)) - \log(q_\phi(z|x)) \right] \\
&= E\left[\log(p(x|z)) + \log(p(z)) - \log(q_\phi(z|x)) \right] \\
&= E\left[\log(p(x|z)) \right] - D_{KL}\left[q_\phi(z|x)||p(z) \right]
\end{aligned}
\tag{5}
$$

Therefore, maximizing ELBO is equivalent to maximizing the right side of Eq. (5), where the first term is the likelihood function, which will force the decoder to restore the generated sample $\hat{x}$ as much as possible to the input sample x, usually measured by cross entropy; The second term is a regularization term regarding the distribution of z, which will force the variational posterior distribution $q_\phi(z|x)$ to approach the prior distribution $p(z)$.

The Fig. 1 shows the architecture of VAE, where $q_\phi(z|x)$ serves as the encoder, mapping data x to the mean $\mu(x)$ and variance $\sigma(x)$ of the distribution of hidden variables z, sampling $z \sim N(\mu(x), \Sigma(x))$ from this distribution, and $p_\theta(x|z)$ serves as the decoder to generate sample $\hat{x}$ through the hidden variables. The parameters ϕ and θ in the distribution correspond to the weight parameters in the encoder and decoder networks, respectively.

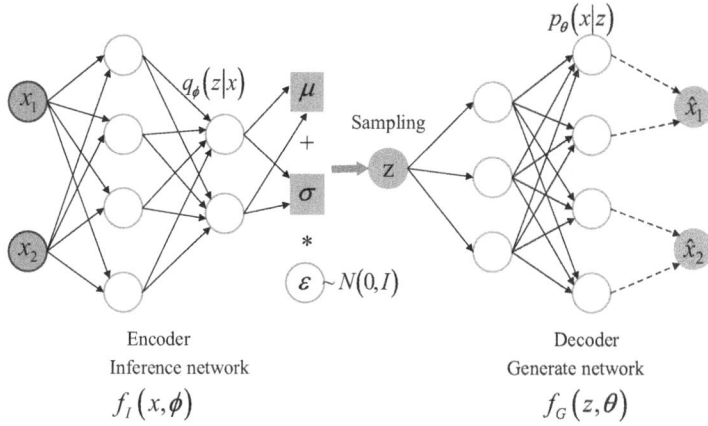

Fig. 1. VAE structure

3   Methodology

A two-stage method was developed using text mining and MCDM technology. The topic modeling technique of text mining helps to better understand the potential semantic information of customer online comments and extract dimensions of satisfaction. Considering that machine learning methods can effectively predict satisfaction, there are still certain shortcomings in guiding enterprise decision-making. Therefore, in the context of multi-criteria decision-making, this paper extends a multi-attribute satisfaction analysis model that fully considers the qualitative forms of customer judgments and preferences.

3.1   Method for Obtaining Customer Satisfaction Dimension Information

In the topic model of deep learning, the neural topic model (NTM) is a representative and unsupervised topic modeling technique. However, most studies separate textual data from auxiliary data, which may be useful supervisory signals but are overlooked by unsupervised topic models [22]. This paper uses the following feature word extraction methods to obtain the dimensional attributes of customer satisfaction.

This paper assumes that document **d** in the dataset is associated with auxiliary labels l, which can be categorical or numerical. Integrate NTM with deep sequence models to utilize auxiliary labels. Choose to use RNN, whose input is a series of word markers in the document. Convert the input word markers into a word embedding sequence $d = \{e_1, e_2, ..., e_N\}$, generated by a trainable word embedding matrix $M \in \mathbb{R}^{|V_{RNN}| \times W}$, where $|V_{RNN}|$ is the size of the vocabulary and W is the dimension of the word embedding.

If document **d** and its auxiliary labels are given, the goal is to maximize the likelihood of the joint margin:

$$\hat{\ell} = \left[\log p_\Theta(d|\hat{t}) + \log p_\Psi(l|\hat{t})\right] - KL(q_\Phi(t|d)||p(t)) \qquad (6)$$

Because the auxiliary label l is independent of **d** and only depends on the RNN input and topic mixture **t**, the above joint marginal distribution can be decomposed into:

$$\hat{\ell} = \left[\log p_\Theta(d|\hat{t}) + \log p_\Psi(l|\hat{t})\right] - KL(q_\Phi(t|d)||p(t)) \qquad (7)$$

The above integration is difficult to directly optimize and can be approximated using variational reasoning. Derive logarithmic likelihood and its ELBO (Evidence Lower Bound):

$$\log p_{\Theta,\Phi,\Psi}(l,d) = \log \int_t \frac{p_\Theta(t)}{q_\Phi(t|d)} q_\Phi(t|d) p_\Psi(l|t) p_\Theta(d|t) dt$$

$$= \log E_{q_\Phi(t|d)}\left[\frac{p_\Theta(t)}{q_\Phi(t|d)} p_\Psi(l|t) p_\Theta(d|t)\right]$$

$$= E_{q_\Phi(t|d)}\left[\log p_\Theta(d|t) - \log \frac{q_\Phi(t|d)}{p_\Theta(t)} + \log p_\Psi(l|t)\right]$$

$$\geq E_{q_\Phi(t|d)}\left[\log p_\Theta(d|t) + \log p_\Psi(l|t)\right] - KL(q_\Phi(t|d)||p(t)) \qquad (8)$$

where Ψ represents all parameters from the RNN attention model, Θ represents all parameters in the generative network, and Φ represents all parameters in the inference network. Therefore, the expectation for a single sample can be estimated using the following equation:

$$\hat{\ell} = \left[\log p_\Theta(d|\hat{t}) + \log p_\Psi(l|\hat{t})\right] - KL(q_\Phi(t|d)||p(t)) \qquad (9)$$

The goal is to maximize ℓ relative to Θ, Φ, Ψ. Maximize the target $\hat{\ell}$ through random gradient ascent and jointly update all parameters.

This method can enhance the discovery of potential semantics in text and can also be used for topic extraction in text analysis.

3.2 Extension of a Multi-attribute Satisfaction Analysis Model Based on Preference Disaggregation

Measuring and analyzing customer satisfaction is achieved through the application of the Multi-criteria Satisfaction Analysis (MUSA) method, which is described in the literature [23]. Typically, this method focuses on measuring user satisfaction for a single company, product, or service. However, this paper primarily centers on extending the MUSA approach to examine user satisfaction assessments across multiple products or services.

After using the feature word extraction method in Sect. 3.1 to obtain the satisfaction evaluation attributes, the obtained satisfaction evaluation attributes are represented as attribute set $X = (X_1, X_2, ..., X_n)$, specific attribute i is represented as monotonic variable X_i in the attribute set, and the overall customer evaluation is defined as variable Y. Given the customer's Y and X_i, evaluate the overall satisfaction function Y^* and partial satisfaction function X_i^*, and follow the ordered regression analysis equation [24]:

$$\begin{cases} Y^* = \sum_{i=1}^{n} b_i X_i^* \\ \sum_{i=1}^{n} b_i = 1 \end{cases} \tag{10}$$

the weighing factor of the criterion i is represented by b_i, and the number of model criteria is represented by n, where the marginal value functions Y^* and X_i^* are normalized at the range [0,100]. The model's restrictions are as follows in more detail:

$$\begin{cases} y^{*1} = x_i^{*1} = 0 \\ y^{*\alpha} = x_i^{*\alpha_i} = 1 \end{cases} for \ i = 1, 2, ..., n \tag{11}$$

where the numbers for the global and marginal value levels, respectively, are α and α_i. The ordinal regression equation, incorporating a dual-error variable within the framework discussed earlier, is presented as follows:

$$\tilde{Y}^* = \sum_{i=1}^{n} b_i X_i^* - \sigma^+ + \sigma^- \tag{12}$$

where σ^+ and σ^- indicate the overestimation and underestimation errors, respectively, and $\tilde{Y}^*$ is the estimation of the global value function Y^*.

The following transformations are used based on the qualitative regression analysis and after the monotonicity requirements for Y^* and X_i^* are lifted:

$$\begin{cases} z_m = y^{*m+1} - y^{*m} & for \ m = 1, 2, ..., \alpha - 1 \\ w_{ik} = b_i x_i^{*k+1} - b_i x_i^{*k} & for \ k = 1, 2, ..., \alpha_i - 1, \ i = 1, 2, ..., n \end{cases} \tag{13}$$

where y^{*m} is the y^m level's value and x_i^{*k} is the x_i^k level's value.

Therefore, the constructed multi-attribute customer satisfaction analysis model is:

$$\begin{aligned} &\min F = \sum_{j=1}^{M} \sum_{t=1}^{T} \sigma_{tj}^+ + \sigma_{tj}^- \\ &s.t. \begin{cases} \sum_{i=1}^{n} \sum_{k=1}^{q_{tji}-1} w_{ik} - \sum_{m=1}^{q_{tj}-1} z_m - \sigma_{tj}^+ + \sigma_{tj}^- = 0, j = 1, 2, ..., M, t = 1, 2, ..., T \\ \sum_{m=1}^{\alpha-1} z_m = 100 \\ \sum_{i=1}^{n} \sum_{k=1}^{\alpha_i-1} w_{ik} = 100 \\ z_m, w_{ik}, \sigma_{tj}^+, \sigma_{tj}^- \geq 0, \forall m, i, j, k, t \end{cases} \end{aligned} \tag{14}$$

where q_{tj} and q_{tji} are the global and partial satisfaction judgments of the jth customer regarding the tth product. The numbers M and T represent the quantity of customers and products, respectively.

4 Data Analysis and Results

4.1 Data Collection

In order to achieve the research objectives and evaluate the proposed methods, a web crawler program was developed to collect data from comprehensive review websites. The focus of this study is on online customer reviews of new energy vehicles at Autohome (autohome.com). On the Autohome website, customers can comment on service quality, product performance, and other levels of new energy vehicles. Therefore, text mining methods targeting each user or group of users can discover the main dimensions of satisfaction. This article collected reviews of 5 new energy vehicles based on the same price range, namely the BYD-Yuan PLUS, BYD-Song PLUS New Energy, Haval-Haval Xiaolong, Deep Blue Car-Deep Blue S7, and Chang'an Auchan-Auchan Z6 Smart iDD. The collection of these five brands of new energy vehicles is $X = \{x_1, x_2, x_3, x_4, x_5\}$. Using web scraping technology, we obtained 58266 customer reviews from January 2023 to February 2024, including review time, text reviews, numerical ratings, overall ratings, etc. The overall rating displayed overall customer satisfaction.

After collecting data through crawlers, the raw data obtained on the page is not standardized and cannot be directly analyzed. In order to avoid errors caused by non-standard data, a series of data preprocessing tasks such as data cleaning, the removal of stop words, and high frequency word statistics are needed. For example, redundant comments, comments with HTML tags and emoticons, and spaces.

4.2 Data Analysis

After the text preprocessing is completed, the satisfaction dimension information of the collected online comment data is obtained by using the feature word extraction method of text mining in Sect. 3.1, and the topic is summarized. After extracting the "topic-word" distribution, topic recognition is performed, and the content of each topic is determined based on the vocabulary under each topic, while ignoring topics that express repetition and those with ambiguous content. The final result of identifying comment topics is the first 7 topics, which are used as dimensional attributes for satisfaction calculation. The final result of identifying the comment topic is shown in Table 1.

After obtaining the dimensional attributes of satisfaction, a multi-attribute satisfaction analysis model is used to calculate customer satisfaction based on the rating data of customer online comments. To aggregate the preferences expressed by users, the new attribute (a_8: word-of-mouth) introduced is used as the global satisfaction variable, as word-of-mouth is generally a combination of other attributes, that is, the overall level of customer perception of the product. Solve the variables in model (14). In the optimal solution discovered, the sum of errors is 0. The great degree of fitting and stability of the supplied end findings is demonstrated by the reliability evaluation analysis. Table 2 displays the main results of the model (criteria weights and average satisfaction indices).

The weights of satisfaction criteria show the relative importance of the assessed satisfaction dimensions. According to Table 2, it can be seen that the weights based on user perception, a_1, a_2, a_3, and a_6, are relatively high compared to other evaluation

Table 1. Topic-word distribution

Symbol	Topic	Topic content
a_1	Durability and electricity consumption	Normal temperature durability, low temperature durability, high temperature durability, high-speed durability, etc
a_2	Charging	Charging compatibility, low-temperature charging, charging immunity, etc
a_3	Safety	Collision safety, hydroelectric safety, battery system waterproof, field safety, etc
a_4	Dynamic	Power performance, speed, vehicle speed, etc
a_5	Operation and control	Control, steering wheel, braking, turning performance, driving performance, etc
a_6	Comfort	Ride comfort, seats, sound insulation, and interior space, etc
a_7	Convenience and quality	Appearance, configuration, cost-effectiveness, interior design, etc

Table 2. Main results

Criterion	Weight (%)	Average satisfaction index (%)				
		x_1	x_2	x_3	x_4	x_5
a_1	26.5	94.34	94.34	93.40	95.28	93.40
a_2	15	22.67	22.67	22.67	42.00	22.67
a_3	22	48.86	40.34	40.34	40.34	40.34
a_4	9.5	92.11	92.11	69.74	92.11	92.11
a_5	9	90.23	91.76	90.33	92.05	91.27
a_6	12	68.75	89.58	86.46	90.625	68.75
a_7	6	74.65	72.35	74.48	81.93	73.07
Global satisfaction a_8	-	68.75	69.375	66.625	73.25	66.625

attributes. A spider diagram has been developed and is shown in Fig. 2 with regard to the impact of model criteria on word-of-mouth.

The weighting factors for each of the model criteria are shown in Fig. 2. The criteria a_1 turned out to have the greatest impact on word-of-mouth (global satisfaction). Additionally, it is shown that a_2, a_3, and a_6 are additional crucial influencing factors. The global and partial value functions for customer satisfaction, Y^* and X_i^*, are referred to as additive and marginal value or utility functions. Particularly, the collective value function Y^* illustrates the value system of the customers' preferences and shows how satisfaction criteria are affected. The marginal value functions of global satisfaction (word-of-mouth) and partial satisfaction (the model criteria), based on the data from Table 2, are reported in Fig. 3.

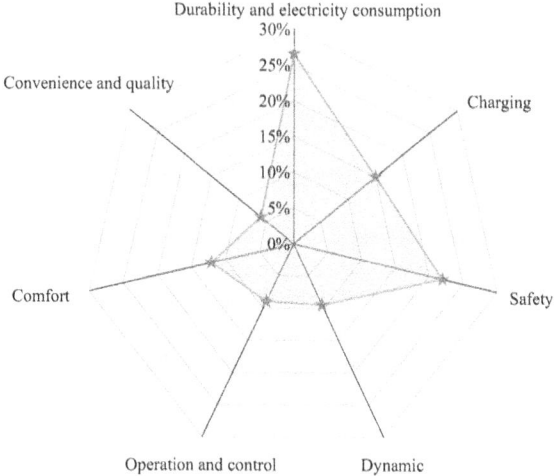

Fig. 2. Importance of the criteria

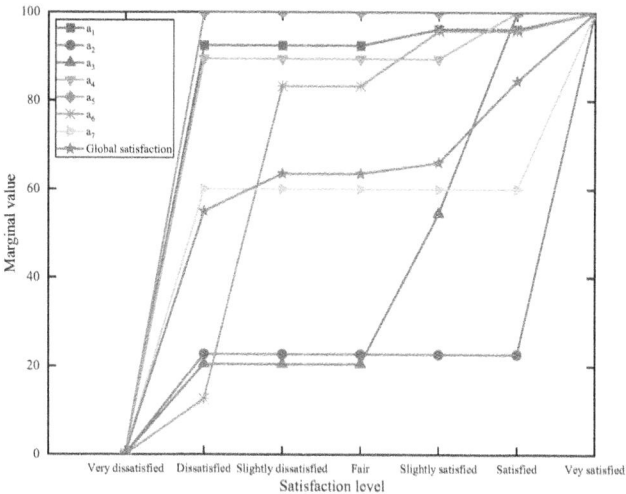

Fig. 3. Marginal value function

The results of data analysis are presented, and the following points are proposed:

(1) The fourth vehicle stands out with a relatively high level of customer satisfaction, boasting an average global satisfaction index of 73.25%. The remaining four new

energy vehicles exhibit similar global satisfaction levels, with scores of 68.75%, 69.375%, 66.625%, and 66.625%, respectively.

(2) The criterion of a_1 emerges as the most influential factor in terms of customer satisfaction, carrying a substantial weight of 26.5%. This criterion consistently yields a notably high satisfaction index across all the new energy vehicles. In comparison, the criteria a_2 and a_3 carry weights of 15% and 22%, respectively. However, it is worth noting that these criteria, while important, yield lower satisfaction indices. This observation is further corroborated by the relevant action chart, highlighting the significance of a_1 as a key driver of satisfaction among customers.

The analysis of user satisfaction with new energy vehicles is reflected in the level of satisfaction with the evaluation attributes. For each new energy vehicle, it is possible to distinguish between their competitive advantages, competitive disadvantages, and where improvements are needed. According to the above data analysis, the relative action diagram of five new energy vehicles can be drawn. This diagram combines the average satisfaction indexes and their respective weights, thereby facilitating the identification of priority areas for improvement.

5   Conclusion

Customer satisfaction analysis is a great help to enterprises in developing strategies and marketing plans and can help enterprises make effective decisions. The customer reviews posted online represent eWOM, and online reviews have become an important source of information for customers to obtain product information and influence their purchasing decisions in the service industry. This paper is based on online review data to obtain the dimensional attributes of customer satisfaction and considers the qualitative forms of customer judgment and preferences. Therefore, a two-stage method was developed using text mining and MCDM technology in this paper. This method can effectively mine the potential dimension information of customer satisfaction and enables the examination of competitive advantages, value functions, attribute weights, and the calculation of average satisfaction indices for each product.

References

1. Bi, J.W., Liu, Y., Fan, Z.P., Cambria, E.: Modelling customer satisfaction from online reviews using ensemble neural network and effect-based Kano model. Int. J. Prod. Res. **57**(22), 7068–7088 (2019)
2. Angilella, S., Corrente, S., Greco, S., Słowiński, R.: MUSA-INT: multicriteria customer satisfaction analysis with interacting criteria. ACS Omega **42**(1), 189–200 (2014)
3. Subramanian, N., Gunasekaran, A., Yu, J., Cheng, J., Ning, K.: Customer satisfaction and competitiveness in the Chinese E-retailing: structural equation modeling (SEM) approach to identify the role of quality factors. Expert Syst. Appl. **41**(1), 69–80 (2014)
4. Miles, P., Miles, G., Cannon, A.: Linking servicescape to customer satisfaction: exploring the role of competitive strategy. Int. J. Oper. Prod. Manag. **32**(7), 772–795 (2012)
5. Torres, A., Tribó, J.A.: Customer satisfaction and brand equity. J. Bus. Res. **64**(10), 1089–1096 (2011)

6. Packard, G., Berger, J.: How concrete language shapes customer satisfaction. J. Consum. Res. **47**(5), 787–806 (2021)
7. Chai, K.H., Ding, Y., Xing, Y.: Quality and customer satisfaction spillovers in the mobile phone industry. Serv. Sci. **1**(2), 93–106 (2009)
8. Terui, N., Hasegawa, S., Chun, T., Ogawa, K.: Hierarchical Bayes modeling of the customer satisfaction index. Serv. Sci. **3**(2), 127–140 (2011)
9. Conklin, M., Powaga, K., Lipovetsky, S.: Customer satisfaction analysis: identification of key drivers. Eur. J. Oper. Res. **154**(3), 819–827 (2004)
10. Deng, Y., Chan, F.T.S.: A new fuzzy dempster MCDM method and its application in supplier selection. Expert Syst. Appl. **38**(8), 9854–9861 (2011)
11. Opricovic, S., Tzeng, G.H.: Extended VIKOR method in comparison with outranking methods. Eur. J. Oper. Res. **178**(2), 514–529 (2007)
12. Siddiquie, R.Y., Khan, Z.A., Siddiquee, A.N.: Prioritizing decision criteria of flexible manufacturing systems using fuzzy TOPSIS. J. Manuf. Technol. Mana. **28**(7), 913–927 (2017)
13. Ahani, A., Nilashi, M., Yadegaridehkordi, E., et al.: Revealing customers' satisfaction and preferences through online review analysis: the case of Canary Islands hotels. J. Retail. Consum. Serv. **51**, 331–343 (2019)
14. Zhao, Y., Xu, X., Wang, M.: Predicting overall customer satisfaction: big data evidence from hotel online textual reviews. Int. J. Hosp. Manag. **76**, 111–121 (2019)
15. Nilashi, M., Fallahpour, A., Wong, K.Y., Ghabban, F.: Customer satisfaction analysis and preference prediction in historic sites through electronic word of mouth. Neural Comput. Appl. **34**(16), 13867–13881 (2022)
16. Jia, M., Kim, H.S., Tao, S.: B&B customer experience and satisfaction: evidence from online customer reviews. Serv. Sci. **16**(1), 42–54 (2023)
17. Mejia, J., Mankad, S., Gopal, A.: Service quality using text mining: measurement and consequences. Manuf. Serv. Oper. Manag. **23**(6), 1354–1372 (2021)
18. Puranam, D., Narayan, V., Kadiyali, V.: The effect of calorie posting regulation on consumer opinion: a flexible latent Dirichlet allocation model with informative priors. Mark. Sci. **36**(5), 726–746 (2017)
19. Dieng, A.B., Wang, C., Gao, J., Paisley, J.: TopicRNN: a recurrent neural network with long-range semantic dependency. arXiv:1611.01702 (2016)
20. Miao, Y., Grefenstette, E., Blunsom, P.: Discovering discrete latent topics with neural variational inference. In: Proceedings of the 34th International Conference on Machine Learning, pp. 2410–2419. PMLR (2017)
21. Wang, X., Yang, Y.: Neural topic model with attention for supervised learning. In: Proceedings of the 23rd International Conference on Artificial Intelligence and Statistics, pp. 1147–1156. PMLR (2020)
22. Yang, Y., Zhang, K., Fan, Y.: SDTM: a supervised Bayesian deep topic model for text analytics. Inf. Syst. Res. **34**(1), 137–156 (2023)
23. Grigoroudis, E., Siskos, Y.: Preference disaggregation for measuring and analysing customer satisfaction: the MUSA method. Eur. J. Oper. Res. **143**(1), 148–170 (2002)
24. Greco, S., Mousseau, V., Słowiński, R.: Ordinal regression revisited: multiple criteria ranking using a set of additive value functions. Eur. J. Oper. Res. **191**(2), 416–436 (2008)

Predicting the Decision-Making Performance Based on Self-attention and Long-Short Term Memory Network

Erbiao Yuan[1], Guangfei Yang[1,2(✉)], Yuhe Zhou[1], and Lian Liu[2]

[1] Institute of Systems Engineering, Dalian University of Technology, Dalian 116024, China
gfyang@dlut.edu.cn, miyaacca_zyh@163.com
[2] Operation Software and Simulation Research Institute, Dalian Naval Academy,
Dalian 116021, China

Abstract. Decision-making requires a combination of intuition, perception and deep thinking skills, which is crucial in personal work and life. Improving decision-making performance is quite beneficial for enhancing people's competitiveness. This paper predicts the decision-making performance and further analyzes the impact of indoor environment on it. Unlike previous studies, which were mostly empirical, our paper discusses the decision-making performance from a data-driven perspective. We collect data using Chinese chess game and indoor environmental sensors as experimental tools, and propose a hybrid model SA–LSTM, which integrates self-attention (SA) mechanism and long-short term memory (LSTM) network. The results demonstrate that our model can accurately predict the decision-making performance. Compared with other five models, SA–LSTM has lower error and better goodness of fit. Specifically, the root mean square error and mean absolute error of our model are reduced by 11% and 14% compared with the second best performing model LSTM. In addition, SA–LSTM model shows greater robustness. Further interpretability analysis suggests the impact of environmental features on the prediction of decision-making performance, and the SA mechanism enables a deeper exploration of their relationship, thereby emphasizing the significance of environmental variables and enhancing prediction accuracy.

Keywords: Decision-making · Deep learning · indoor environment · Self-attention · Long-short term memory

1 Introduction

Decision-making is of crucial significance in life and work. People with high decision-making ability are often favored in the labor market, as high-skill jobs are becoming increasingly important, especially with explosion of generative AI technology [1, 2]. Decision-making not only affects the short-term performance and long-term success of individuals, but also plays an import role in the development of companies [3, 4]. A poor decision may cause the company to fall into trouble, and even lead to the company's subsequent decline and bankruptcy [5, 6]. Therefore, for decision-makers, mastering

X. Tang et al. (Eds.): KSS 2024, CCIS 2269, pp. 317–329, 2025.
https://doi.org/10.1007/978-981-96-0178-3_22

the changes in their own decision-making performance and identifying key factors can improve their performance in decision-making. But the performance of decision-making is not solely contingent upon the decision-maker themselves, but rather influenced by a multitude of other factors. The complexity of relationships hinders decision-makers from fully comprehending the efficacy of their decisions. Fortunately, in recent years, advances in data science have made it possible to accurately predict the decision-making performance.

Previous researches have shown that individual characteristics, such as age and decision-making style, have an impact on decision-making performance [7–9]. Nonetheless, changing these characteristics of decision-makers is not easy and often requires long-term effort. In addition to individual characteristics, in the past decade, some scholars have demonstrated that the air quality has a negative effect on decision-making performance [10–12]. Since most of people's activities are indoors, indoor air quality has a more direct impact on the performance. And [13] found that the change in temperature led to statistically significant changes in cognitive performance. These evidence indicated that environmental conditions could be important predictors of high-skill task performance. During the decision-making process, decision-makers are often exposed to a dynamic and evolving environment. Considering that the conditions of decision-makers themselves will not change much in a short period of time, it is feasible to predict the subsequent performance of decision-making by exploring the relationship between environmental conditions and decision-making performance.

Despite the importance and usefulness of decision-making performance prediction, there is still a lack of relevant researches. The most critical challenge is the objective measurement of decision-making performance. Typically, researchers will set up a task to simulate the decision-making process, such as the Iowa Gambling Task [14, 15]. In this task, participants are required to choose cards from a deck with different payoffs for different cards, and the net payoffs from choosing cards multiple times are used to understand how decision-making performance changes. Similarly, the Balloon Analogue Risk Task was also used to study individual differences in risky decision-making [16]. Although these tasks are widely used in decision-related research, they are also very costly in human and material resources. Besides, the amount of data available through them is quite limited, making it difficult to extract information from these data and then establish machine learning models. In our experiment, we used a new tool to simulate decision-making scenarios, which is Chinese chess.

Since its first application in cognitive psychology in 1973 [17], chess has been widely applied in the study of human behaviour. Although cognitive science remains the main area that scholars have studied with chess [18–20], in the last few years some scholars have extended it to the study of work efficiency and decision-making problems [10, 21]. Finding the optimal move in chess is a complex task that requires intuition, perception, and high problem-solving abilities, which have similar skill requirements for real-word human activities such as cognition, work and decision-making mentioned above [22]. The appeal of chess lies not only in its provision of a realistic environment, but also in the advancement of AI technology on chess supporting in-depth research in cognition and decision-making. Chinese chess is a highly popular chess game in China. Although there are differences between Chinese chess and chess in terms of structural design and rules, they are consistent in the skill requirements for participants. Therefore, it is feasible to use Chinese chess to study decision-making performance.

In this paper, we proposed a deep learning model SA–LSTM to predict the decision-making performance, which combined the self-attention (SA) mechanism and long-short term memory (LSTM) network. Performance data for each move of participants and environmental data monitored by sensors were collected. The general steps in our experiment are as follows. Firstly, by constructing a neural network as a similarity measure, the SA mechanism was applied to the input data. Next, cross validation technique and Bayesian optimization algorithm were used to determine the optimal hyper-parameters of the model. Finally, LSTM model was built with the optimal parameters and then predict decision-making performance. The experimental results demonstrated that the SA–LSTM model can predict the decision-making performance well and outperforms other benchmark models, with higher prediction accuracy and stronger robustness.

The main contributions of our work can be summarized as follows. Firstly we use Chinese chess game as an experimental tool to overcome the difficulty of measuring and collecting decision-making performance data, and this provides support for applying data-driven methods to study decision-making. Secondly by combining the SA mechanism with machine learning models, accurate prediction of decision-making performance is achieved. The impact of environmental variables on decision-making performance is explained from the data-driven perspective, and it is found that the SA mechanism can better explore the relationship between them.

The remainder of the paper is organized as follows. The data and methods used in this paper are introduced in the next section. Then, in Sect. 3, we analyze and discuss the results. Finally, we summarize and draw conclusions in Sect. 4.

2 Data and Methods

2.1 Decision-Making Performance Measurement

In recent years, the advancement of AI technology has led to the emergence of some excellent Chinese chess-related AI products. Pikafish, a Chinese chess engine, was applied to evaluate each move of participants because of its open source and excellent performance. At its core is a neural network architecture called efficiently updatable neural network (NNUE) [23] It was invented for shogi, and later ported to other board games such as chess and Chinese chess. Nowadays, it is widely used in academic researches [10, 19–21]. Based on the remaining pieces and their positions on the board, NNUE calculates the evaluation value, known as the pawn metric. The pawn metric reflects the relative advantage for each decision-maker. If the pawn metric is greater than 0, the decision-maker with the white pieces is in a better position and is more likely to win the game. This also represents that the decision-maker has better decision-making performance in this move.

In our experiment, we developed a Chinese chess web game that can automatically record the moves of participants, and then calculate the pawn metric for each move by calling Pikafish engine. The experiment was conducted in a conference room, and 16 undergraduate and postgraduate students participated. After processing, we finally obtained 3040 valid moves.

2.2 Indoor Environment Measurement

The environment has a non-negligible impact on human decision-making performance. Compared to other influencing factors, the indoor environment in which decision-makers are located is easier to change, providing an avenue for enhancing the decision-making performance. In the experiment, we utilized sensors to collect environmental data within the experimental site [24]. The sensors are capable of simultaneously monitoring temperature, relative humidity, CO_2, PM2.5, PM10, formaldehyde, and VOC levels. The exterior and interior schematics of the sensor are depicted in Fig. 1.

(a)   (b)

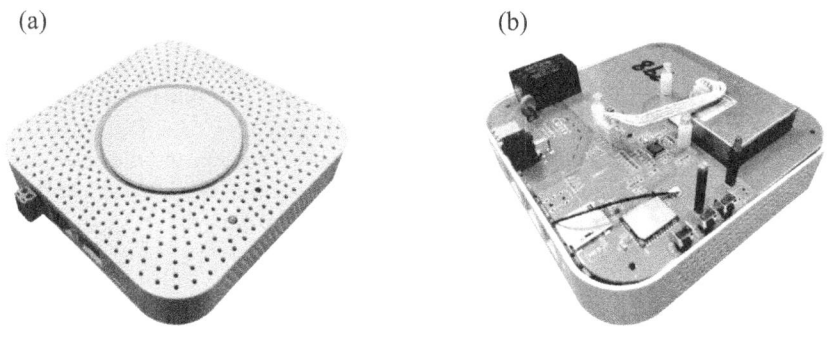

(a) External schematic diagram.   (b) Internal schematic diagram.

Fig. 1. Indoor environmental sensor.

2.3 SA–Based LSTM Model

LSTM model is widely used in environmental management, energy prediction and fault detection [25–27] because of its excellent performance in sequence problems. As shown in Fig. 2, by introducing the gating mechanism and a long-term memory unit c, the LSTM model effectively mitigates the issues of gradient vanishing and exploding that are inevitable in the recurrent neural network. Nonetheless, when more historical data is required or the input data dimension is too large, LSTM still fails to effectively capture key information, resulting in poor prediction results.

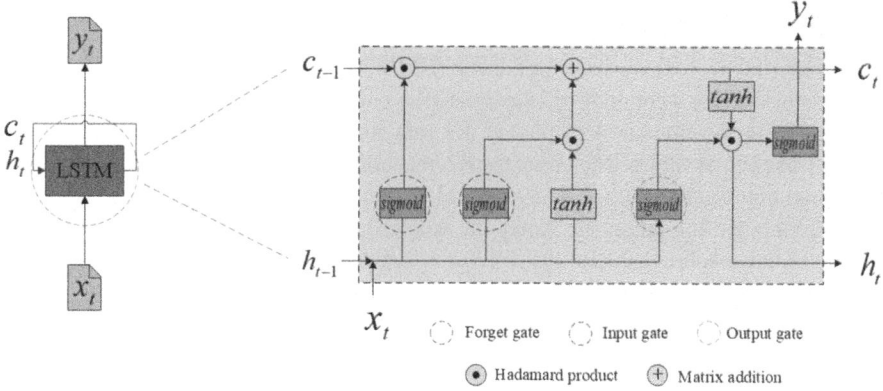

Fig. 2. Core schematic of LSTM model.

Since the attention mechanism was proposed by Bahdanau et al. [28], its combination with neural network has achieved remarkable success [29–31]. The concept of attention mechanism involves the allocation of weights. As shown in Eq. (1), the similarity between the data q and k is first measured, and then softmax normalization is performed to obtain the weights. And the higher the similarity, the greater the weight. The SA mechanism is inherently a specific instance of the attention mechanism, wherein q and k are identical.

$$Weight(q, k) = \frac{exp(sim(q, k))}{\sum exp(sim(q, k))} \tag{1}$$

In our experiment, we integrated the SA mechanism into the LSTM model. Since neural networks can be used to model any function, we attempted to fit the similarity function with a neural network. As depicted in Fig. 3, the SA module was applied before the LSTM model to mine key information in the input data.

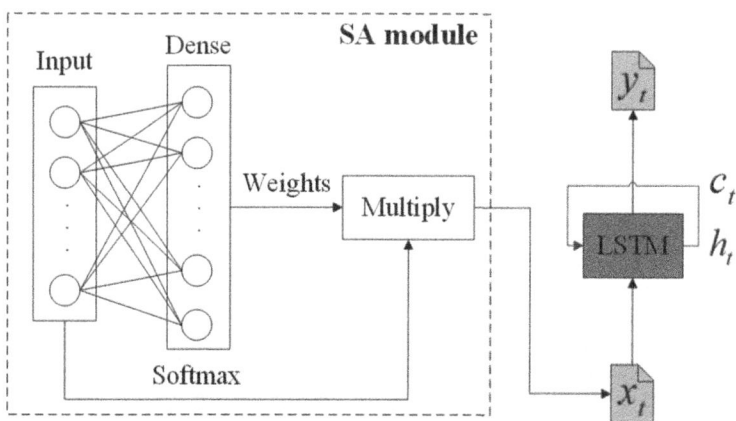

Fig. 3. Application of SA mechanism in LSTM.

2.4 Model Evaluation Metrics

In the experiment, root mean square error (RMSE), mean absolute error (MAE) and goodness of fit (R2) were applied to evaluate the prediction performance of all models. The calculation formula are Eq. (2)–(4). R2 reflects how well the output values of the model fits the actual values. The larger the R2, the more accurately the model can simulate the change of the actual value. MAE and RMSE can directly reflect the prediction error of the model, and the larger the value, the worse the prediction performance. MAE is the sum of the absolute value of the difference between the predicted values and actual values, so MAE can describe the difference more realistically and intuitively.

$$RMSE = \sqrt{\frac{1}{n} \sum_{i=1}^{n} (y_i - \hat{y}_i)^2} \tag{2}$$

$$MAE = \frac{1}{n} \sum_{i=1}^{n} |y_i - \hat{y}_i| \tag{3}$$

$$R^2 = 1 - \frac{\sum_{i=1}^{N} (\hat{y}_i - y_i)^2}{\sum_{i=1}^{N} (\bar{y}_i - y_i)^2} \tag{4}$$

where y_i represents the actual value, $\hat{y}_i$ represents the predicted value, and $\bar{y}_i$ represents the average value.

3   Results and Discussions

3.1   The Prediction of Decision-Making Performance

The hyperparameters of our model mainly include the number of hidden layers and the number of neurons in each layer. To ensure optimal predictive performance, we combine the 5 times 10-fold cross-validation technique and the Bayesian optimization algorithm [32] to determine the best parameters of the model. The search process and parameter values are shown in Fig. 4. The Bayesian search was iterated 120 times, and the optimal parameters obtained were 1 hidden layer and 94 neural units in this layer.

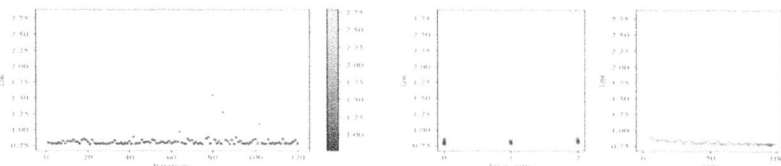

Fig. 4. Results of Bayesian search.

SA–LSTM model was constructed with the optimal parameters obtained from Bayesian optimization and predictions were made. Five other machine learning models were also applied for comparison, namely decision tree (DT) model, random forest (RF) model, multi-layer perceptron (MLP) regressor, SA–MLP model and LSTM model. Figure 5 showed the results of predicting the decision-making performance for 428 piece moves. Overall, our model has the best predictive performance, especially when there is a turnaround in decision-making performance, our model is able to capture these changes well.

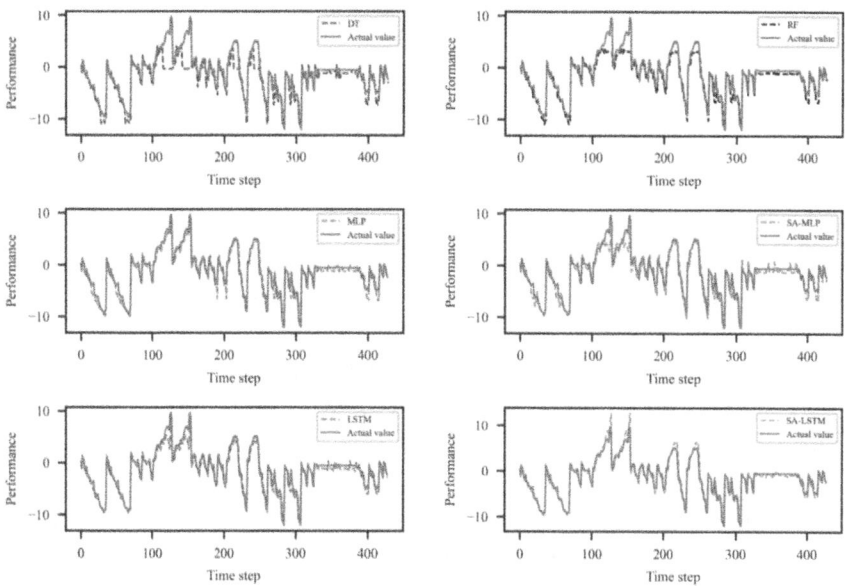

Fig. 5. Prediction results of the models.

3.2 The Comparisons and Analysis

To further verify the applicability of our model SA–LSTM in predicting decision-making performance, we conducted a detailed comparison of its prediction errors with the five models mentioned above. Each model was run 50 times to avoid errors due to the randomness of the parameters initialization. Figure 6 intuitively demonstrated that the SA–LSTM model significantly outperformed the other models in terms of all metrics, with average values of RMSE, MAE and R2 of 0.95, 0.67 and 94%. Overall, the DT and RF model have large errors and poor goodness of fit, so tree models may not be suitable for prediction of decision-making performance. In neural network models, the performance of recurrent neural network models surpassed that of fully connected neural network models. Although the SA mechanism can be well integrated with neural networks, the effectiveness was not consistent across different models. The introduction of SA mechanism in MLP model made the prediction performance worse, but the combination of SA and LSTM reduced RMSE by 11% and MAE by 14%.

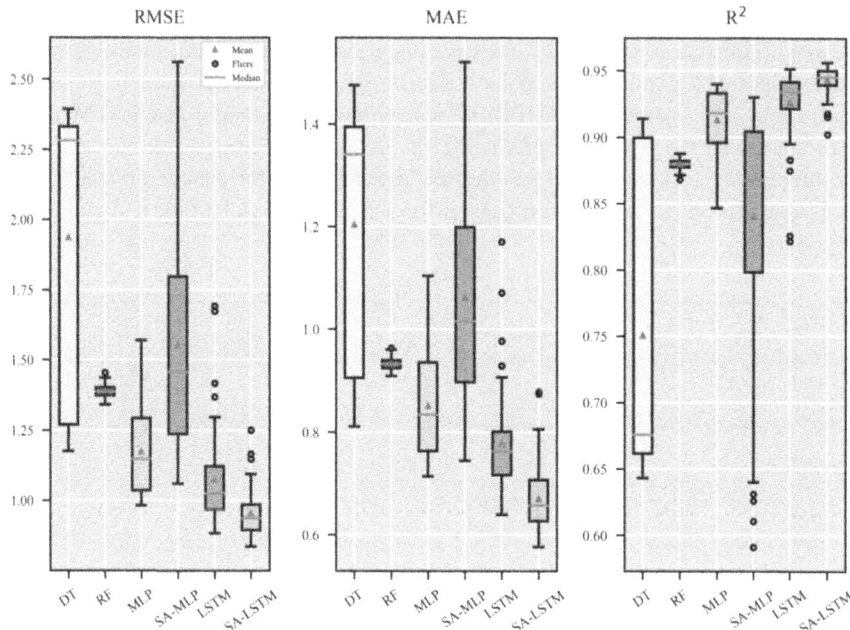

Fig. 6. Prediction accuracy of machine learning models.

Moreover, as can be seen in Fig. 6, the SA–LSTM model has a shorter chamber, indicating a more stable performance. Then the error bar was plotted to further exhibit enhanced robustness of our model. The length of the bars in Fig. 7 represents the standard deviation of prediction accuracy. Although the prediction accuracy of the RF model has the smallest standard deviation in 50 independent repeated experiments, the average accuracy is not ideal. And the SA–LSTM model achieves high average prediction accuracy while having a small standard deviation, with RMSE having a standard deviation of 0.09, MAE having a standard deviation of 0.07, and R2 having a standard deviation of 0.01. This indicates that compared with the LSTM model, the introduction of the SA mechanism not only improves the prediction accuracy of our model, but also enhances its robustness.

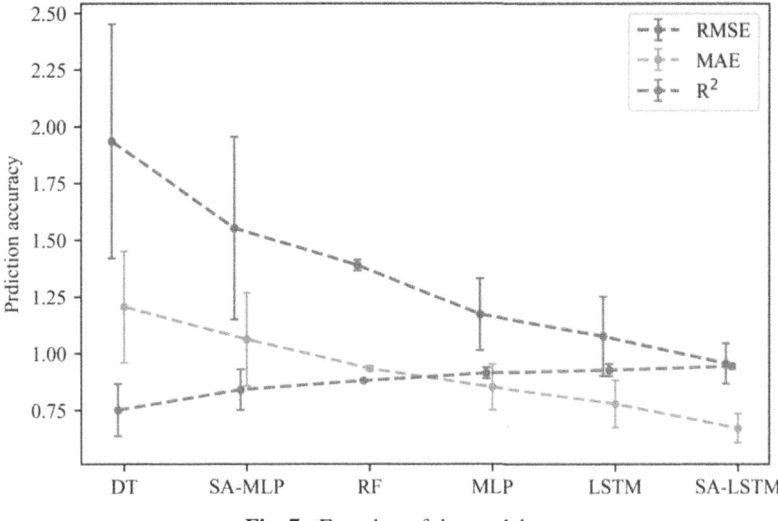

Fig. 7. Error bar of the models.

3.3 Relationships Between Decision-Making Performance and the Environment

In this section we discussed the role of each feature in the prediction of decision-making performance. Figure 8 depicted the SHapley Additive exPlanation (SHAP) values of the top 20 variables across all samples in the SA–LSTM model. It reflected the magnitude of influence of each variable in the prediction task at a macro level. The color of the dot represents the size of the variable value, the redder the value the larger. The horizontal coordinate is the SHAP value, and a positive value indicates that the variable has a positive impact. Figure 9 describes the SHAP values of variables in a single sample, which provides a more detailed demonstration of the influence of variables in decision-making performance prediction. And red indicates positive influence, blue indicates negative influence, and the length of the arrow reflects the strength of the influence.

There are several findings from the figures. As can be seen from the figures, the most important features are decision-making performance, CO2, and temperature, but the impact of these features varies at different moments. The sample distribution of decision-making performance at t − 1 moment is the most scattered, so its influence is the greatest. What follows is that CO2 at t − 1 and t − 2 moment has a significant impact on the prediction results. But the effect of CO2 at these two moments is opposite. This may be related to the fact that people breathe CO2. Usually people consume more energy when they think carefully and make decisions, and then CO2 emissions increase; on the other hand, when the CO2 concentration exceeds the threshold, it will have an inhibitory effect on human cognition and decision-making [33]. Similarly, the temperature at moment t − 2 and t − 3 also exhibit contrasting effects, which may be associated with decision-making performance and the threshold for human temperature perception [34].

Furthermore, we explored the changes in SHAP values after incorporating the SA mechanism into the LSTM model. The horizontal coordinate in Fig. 10 is the sum of the absolute SHAP values of all samples, and it actually reflects the overall impact of a

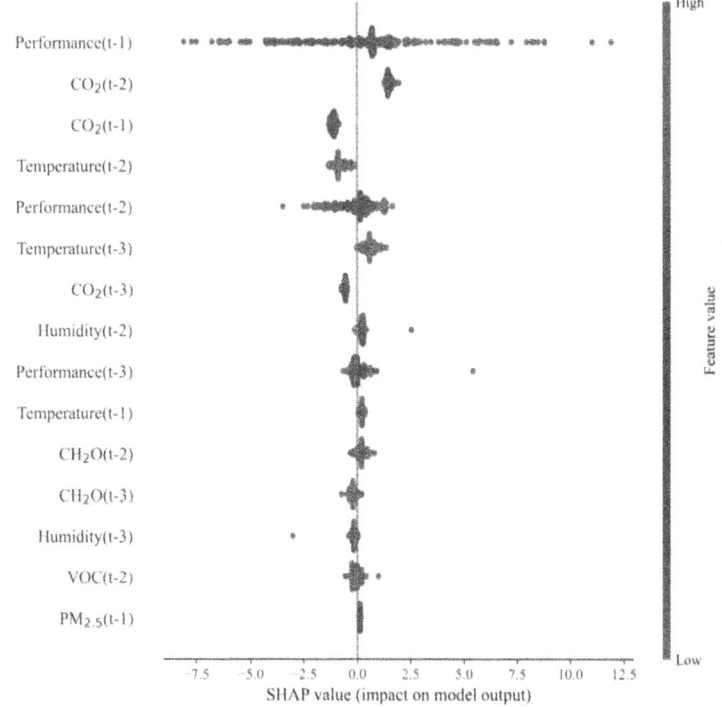

Fig. 8. The beeswarm plot of all samples.

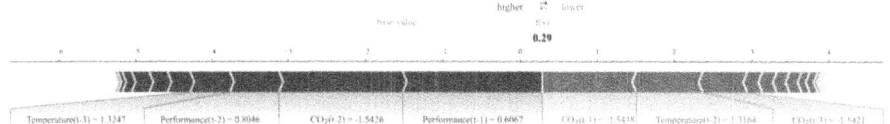

Fig. 9. The force plot of single sample.

variable on the prediction of decision-making performance. For the feature of decision-making performance, both SA–LSTM and LSTM models have a large SHAP value, and they are almost the same, indicating that the past decision-making performance in the two models has a great influence on the next decision-making performance. However, the SHAP values on environmental features are not the same. In the SA–LSTM model, the SHAP values of CO_2 and temperature are larger. We believe that the introduction of SA mechanism enables the LSTM model to capture the role of environmental information in the prediction of decision-making performance, thus giving higher weight to these environmental features. This is also the reason why the SA–LSTM model can achieve higher accuracy.

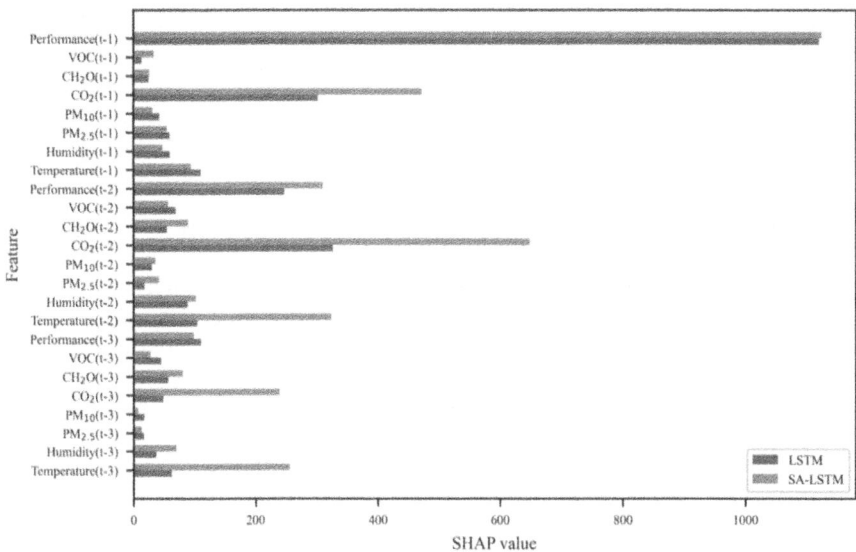

Fig. 10. Comparison of SHAP value between SA–LSTM and LSTM model.

4 Conclusions

With the rapid development of generative AI, decision-making ability is becoming more and more important, especially in the labor market. We attempt to accurately predict decision-making performance in order to provide support for further improving people's decision-making performance. In this paper, we studied decision-making performance from the data-driven perspective, and proposed a deep learning model SA–LSTM. The results showed that our proposed model could predict the decision-making performance well, and had significant advantages in prediction accuracy and robustness compared with the benchmark models. The interpretability analysis of the model revealed that indoor environmental variables, particularly indoor CO_2 concentration, indoor temperature, and indoor relative humidity, significantly influenced the prediction of decision-making performance. Additionally, by assigning greater significance to environmental variables, the SA mechanism module in our model enhanced the impact of the environmental features, thereby substantiating the superior performance of the SA–LSTM model. Our findings will offer support for improving decision-making performance by changing indoor environmental conditions for a short period of time. We also hope to provide a new idea and perspective for decision-making related research. However, our research still has some limitations. Firstly, our data is not sufficient. Secondly, individual characteristics that will not change in the short term have not been incorporated into the model. We will conduct further experiments to address these issues in the future.

Acknowledgments. This work was supported by the National Natural Science Foundation of China (42071273), Fundamental Research Funds for the Central Universities (DUT24YG147).

References

1. Acemoglu, D., Restrepo, P.: Automation and new tasks: how technology displaces and reinstates labor. J. Econ. Perspect. **33**(2), 3–30 (2019). https://doi.org/10.1257/jep.33.2.3
2. Deming, D.J.: The Growing Importance of Decision-Making on the Job, National Bureau of Economic Research Working Paper Series, No. 28733 (2021). https://doi.org/10.3386/w28733
3. Ireland, R.D., Miller, C.C.: Decision-making and firm success. Acad. Manag. Perspect. **18**(4), 8–12 (2004). https://doi.org/10.5465/ame.2004.15268665
4. Nash, L., Stevenson, H.: Success that lasts. Harv. Bus. Rev. **82**(2), 102–109 (2004)
5. Abramson, C., Currim, I.S., Sarin, R.: An experimental investigation of the impact of information on competitive decision making. Manag. Sci. **51**(2), 195–207 (2005). https://doi.org/10.1287/mnsc.1040.0318
6. Zhang, D.: Subsidy expiration and greenwashing decision: is there a role of bankruptcy risk? Energy Econ. **118**, 106530 (2023). https://doi.org/10.1016/j.eneco.2023.106530
7. Bruine De Bruin, W., Parker, A.M., Fischhoff, B.: Explaining adult age differences in decision-making competence. J. Behav. Decis. Mak. **25**(4), 352–360 (2012). https://doi.org/10.1002/bdm.712
8. Korniotis, G.M., Kumar, A.: Do older investors make better investment decisions? Rev. Econ. Stat. **93**(1), 244–265 (2011)
9. Bavoľár, J., Orosová, O.: Decision-making styles and their associations with decision-making competencies and mental health. Judgm. Decis. Mak. **10**(1), 115–122 (2015). https://doi.org/10.1017/S1930297500003223
10. Künn, S., Palacios, J., Pestel, N.: Indoor air quality and strategic decision making. Manag. Sci. (2023)
11. Huang, J., Xu, N., Yu, H.: Pollution and performance: do investors make worse trades on hazy days? Manag. Sci. **66**(10), 4455–4476 (2020). https://doi.org/10.1287/mnsc.2019.3402
12. Chew, S.H., Huang, W., Li, X.: Does haze cloud decision making? A natural laboratory experiment. J. Econ. Behav. Organ. **182**, 132–161 (2021)
13. Graff Zivin, J., Hsiang, S.M., Neidell, M.: Temperature and human capital in the short and long run. J. Assoc. Environ. Resource Econ. **5**(1), 77–105 (2018)
14. Li, D., Li, Y., Gao, S., He, Y.: Exact and approximate calculation and uncertain decision-making in children and adults: evidence from fuzzy trace theory. Pers. Individ. Differ. **225**, 112661 (2024). https://doi.org/10.1016/j.paid.2024.112661
15. Kovacs, I., Richman, M.J., Janka, Z., Maraz, A., Ando, B.: Decision making measured by the Iowa Gambling Task in alcohol use disorder and gambling disorder: a systematic review and meta-analysis. Drug Alcohol Depend. **181**, 152–161 (2017)
16. Lauriola, M., Panno, A., Levin, I.P., Lejuez, C.W.: Individual differences in risky decision making: a meta-analysis of sensation seeking and impulsivity with the balloon analogue risk task. J. Behav. Decis. Mak. **27**(1), 20–36 (2014)
17. Chase, W.G., Simon, H.A.: Perception in chess. Cogn. Psychol. **4**(1), 55–81 (1973)
18. Charness, N.: The impact of chess research on cognitive science. Psychol. Res. **54**, 4–9 (1992)
19. Moxley, J.H., Ericsson, K.A., Charness, N., Krampe, R.T.: The role of intuition and deliberative thinking in experts' superior tactical decision-making. Cognition **124**(1), 72–78 (2012)
20. Strittmatter, A., Sunde, U., Zegners, D.: Life cycle patterns of cognitive performance over the long run. Proc. Natl. Acad. Sci. **117**(44), 27255–27261 (2020)
21. Künn, S., Seel, C., Zegners, D.: Cognitive performance in remote work: evidence from professional chess. Econ. J. **132**(643), 1218–1232 (2022)

22. Gerdes, C., Gränsmark, P.: Strategic behavior across gender: a comparison of female and male expert chess players. Labour Econ. **17**(5), 766–775 (2010)

23. Nasu, Y.: Efficiently updatable neural-network-based evaluation functions for computer shogi. In: The 28th World Computer Shogi Championship Appeal Document, p. 185 (2018)

24. Yang, G., Yuan, E., Wu, W.: Predicting the long-term CO_2 concentration in classrooms based on the BO–EMD–LSTM model. Build. Environ. **224**, 109568 (2022). https://doi.org/10.1016/j.buildenv.2022.109568

25. Shi, H., Wei, A., Xu, X., Zhu, Y., Hu, H., Tang, S.: A CNN-LSTM based deep learning model with high accuracy and robustness for carbon price forecasting: a case of Shenzhen's carbon market in China. J. Environ. Manag. **352**, 120131 (2024). https://doi.org/10.1016/j.jenvman.2024.120131

26. Dao, F., Zeng, Y., Qian, J.: Fault diagnosis of hydro-turbine via the incorporation of Bayesian algorithm optimized CNN-LSTM neural network. Energy **290**, 130326 (2024). https://doi.org/10.1016/j.energy.2024.130326

27. Sun, H., Cui, Q., Wen, J., Kou, L., Ke, W.: Short-term wind power prediction method based on CEEMDAN-GWO-Bi-LSTM. Energy Rep. **11**, 1487–1502 (2024). https://doi.org/10.1016/j.egyr.2024.01.021

28. Bahdanau, D., Cho, K., Bengio, Y.: Neural machine translation by jointly learning to align and translate, arXiv preprint arXiv:1409.0473 (2014). https://doi.org/10.48550/arXiv.1409.0473

29. Li, Y., Zhu, Z., Kong, D., Han, H., Zhao, Y.: EA-LSTM: evolutionary attention-based LSTM for time series prediction. Knowl.-Based Syst. **181**, 104785 (2019)

30. Liu, J., Wang, G., Duan, L., Abdiyeva, K., Kot, A.C.: Skeleton-based human action recognition with global context-aware attention LSTM networks. IEEE Trans. Image Process. **27**(4), 1586–1599 (2017)

31. Dai, C., Liu, X., Lai, J.: Human action recognition using two-stream attention based LSTM networks. Appl. Soft Comput. **86**, 105820 (2020)

32. Shahriari, B., Swersky, K., Wang, Z., Adams, R.P., De Freitas, N.: Taking the human out of the loop: a review of Bayesian optimization. Proc. IEEE **104**(1), 148–175 (2015)

33. Fan, Y., Cao, X., Zhang, J., Lai, D., Pang, L.: Short-term exposure to indoor carbon dioxide and cognitive task performance: a systematic review and meta-analysis. Build. Environ. 110331 (2023)

34. Lan, L., Tang, J., Wargocki, P., Wyon, D.P., Lian, Z.: Cognitive performance was reduced by higher air temperature even when thermal comfort was maintained over the 24–28 C range. Indoor Air **32**(1), e12916 (2022)

A Domain Knowledge-Based Railway Equipment Fault Diagnosis Framework

Qilan Li, Shanwei Cao, and Lingling Zhang(✉)

School of Economics and Management, University of Chinese Academy of Sciences,
Beijing, China
zhangll@ucas.ac.cn

Abstract. Effective equipment fault diagnosis plays a crucial role in ensuring the safe and stable industrial process. However, intelligent fault diagnosis of equipment remains challenging due to sparse fault data. This study proposes a novel knowledge-based railway equipment fault diagnosis framework (KEFD), which forms a diagnosis knowledge graph for efficient storage, retrieval and utilization of the domain knowledge. This framework could provide intuitive and efficient decision-making guidance for railway operators. Specifically, fault case records from Chinese railway Cab Integrated Radio communication (CIR) equipment are selected as the data source. KEFD first mines the standardized entities and relations from CIR diagnosis documents, transforming the textual data into a fault event-oriented diagnosis knowledge graph. Subsequently, KEFD utilizes a fault event detection method incorporating the extended structural embedding and distant supervision to accurately extract diagnosis information and add to the graph. Finally, the diagnosis task is transformed into a matching problem between nodes of fault phenomena and maintenance measures based on DistMult. Empirical validation on the CIR fault diagnosis dataset demonstrates that event extraction performance of KEFD reaches 54%, realizing 3% improvement over other baseline models. By fully leveraging rich structural and semantic information in the knowledge graph, KEFD significantly enhances the performance of fault event extraction and intelligent diagnosis, serving for the practical needs of railway departments.

Keywords: Fault diagnosis · Knowledge management · Text mining · Railway equipment

1 Introduction

With the rapid development of digital technologies and automation, the modern railway system has gradually transformed from manual operation to more intelligent and autonomous modes of control [1]. Meanwhile, as a close combination of mobile equipment, control systems, and infrastructure, the modern railway system has a dynamic yet intricately networked operating environment [2]. Due

X. Tang et al. (Eds.): KSS 2024, CCIS 2269, pp. 330–344, 2025.
https://doi.org/10.1007/978-981-96-0178-3_23

to this complex inter-connectivity, the risks and failure modes of railway system are always perform randomness with heterogeneous characteristics, which has brought rising demands for efficient railway equipment maintenance and fault diagnosis mechanisms [3]. Especially, the fault diagnosis of railway Cab Integrated Radio communication (CIR) equipment, which acts as the "central nervous system" for scheduling railway operations, has gradually attracted more and more attention due to its importance to the operation safety [4,5]. There is a rising demand for reliable fault diagnosis framework toward CIR equipment that can detect, locate and isolate system faults, and provide recovery actions to mitigate faults [6].

Fault diagnosis is an essential technique for ensuring the stability of complex equipment. Effective diagnosis methods provide timely and reliable decision supports to the operators about the process states [7]. Various approaches to fault diagnosis have been developed and they can be divided into three main categories: data-driven approaches [8], analytical model-based approaches [9], and knowledge-based approaches [10]. It is worth noting that operational data is necessary for all approaches mentioned above [11]. While various diagnosis methods have been proposed, challenges arise when applying them to railway CIR equipment. The first challenge is the scarcity of labeled fault samples. Traditionally, effective fault diagnosis techniques are implemented based on sufficient, high-quality operating data [12]. However, the sophistication of the modern railway system precludes direct and real-time monitoring through sensors, severely limiting the collection of digital fault samples from CIR equipment [13]. This hinders the establishment of supervised diagnosis models. The second challenge is that the results output by previous statistic or deep learning diagnosis models often lack interpretability. In critical railway operations, especially in high-speed railway situation, safety-related dispatching and driving decisions largely rely on human experience [14]. Therefore, it is necessary for operators to understand the decisions made by diagnosis model and judge their reliability from a mechanistic perspective. Besides, due to the heavy workload and experience deficiency, operators sometimes cannot effectively and rapidly process multi-source heterogeneous information output by complex diagnosis models, as well as have difficulty making real-time and accurate maintenance decisions [15]. To support human decisions integral to safe railway operating effectively, it is important to improve the interpretability and clearness of results output by diagnosis models.

To address these problems, this study combines data-driving with knowledge-based approach, proposing a novel knowledge-based railway equipment fault diagnosis framework (KEFD). Given the strengths of knowledge graph (KG) in semantic representation, complex relation modeling and knowledge inference [16], it serves as the carrier of the knowledge model in this study. Firstly, following standardized analysis process of equipment structures and fault mechanisms, the pattern layer of the CIR equipment diagnosis KG is determined and example data is filled to construct a preliminary diagnosis KG. Next, to alleviate the training data scarcity, distant supervision idea is utilized to generate automatic labeling dataset for fault event extraction. To make full use of the rich structural and semantic information contained in the preliminary diagnosis KG, an event extraction model incorporating the extended structural embedding is proposed and the extraction result would be added into the diagnosis KG. This complete CIR equipment diagnosis KG is then leveraged to realize intelligent diagnosis and provide maintenance recommendations in an interpretable and clear way according to operators' experience and operating conditions. This study constructs a comprehensive diagnosis KG with domain prior knowledge, facilitating the storage, retrieval and utilization of the diagnosis knowledge in this field.

2   Literature Review

2.1   Domain Knowledge Modeling Based on Knowledge Graph

Domain knowledge modeling studies based on ontology theory exist across industries. [17] collected and integrated multi-source power asset data to provide a unified knowledge base. Specifically, the Knowledge Graph (KG) has been widely used for industrial ontology modeling. [18] described the standardized process of product design and manufacturing with KG, which could largely promote efficient interdepartmental sharing, exchange and reuse of product knowledge. [19] developed a device atlas assisting robots with assembly through workflow information collaboration. [20] constructed and expanded a domain-specific KG using CRF, meanwhile built an engineering application integration system, knowIME, which leveraged the domain KG to support manufacturing equipment. Additionally, [21] semi-automatically built a power corpus and extracted entities and relations with BiLSTM-CRF. The constructed KG could effectively monitor and identify abnormal refueling behaviors. These efforts demonstrate applications of KG across various industrial domains. On the contrary, domain knowledge modeling research appears relatively scarce in railway safety area.

2.2   Fault Diagnosis for Railway Equipment

By analyzing the characteristics of fault maintenance logs, [22] proposed a combination of top-down and bottom-up methods to construct a fault KG of in-vehicle equipment. [23] established a fault ontology for electric vehicle motor systems, and performed qualitative analysis on fault causes by fusing abnormal fault tree models, forming a priority diagnosis program that improved efficiency. In railway CIR equipment fault diagnosis, a comprehensive fault diagnosis KG is still lacking, largely hindering knowledge storage, retrieval, management and sharing. Additionally, existing KG-based fault diagnosis approaches mainly rely on qualitative analysis and pattern matching, without effectively capturing the rich structural and semantic data hidden in maintenance records. The KEFD framework proposed in this study integrates qualitative with quantitative analysis methods, which could significantly improve the efficiency of fault diagnosis.

3   Research Framework

This study aims to develop an intelligent fault diagnosis framework combining with data-driven and knowledge-based approaches for railway CIR equipment. Figure 1 shows an abstract overview of this framework, which consists of fault knowledge modeling, fault event extraction, and fault diagnosis.

4   Methodology

4.1   Fault Knowledge Modeling

Knowledge modeling based on ontology theory has increasingly supported industrial research in recent years [24,25]. While existing studies effectively proposed ontology frameworks for knowledge management, their construction processes primarily relied on expert experience [26] and sensor data [27] without fully integrating domain prior knowledge and unstructured fault records. This resulted in issues such as unreasonable ontology structures and limited domain coverage. To address these issues, this study proposes a normative knowledge modeling process for railway CIR equipment fault diagnosis.

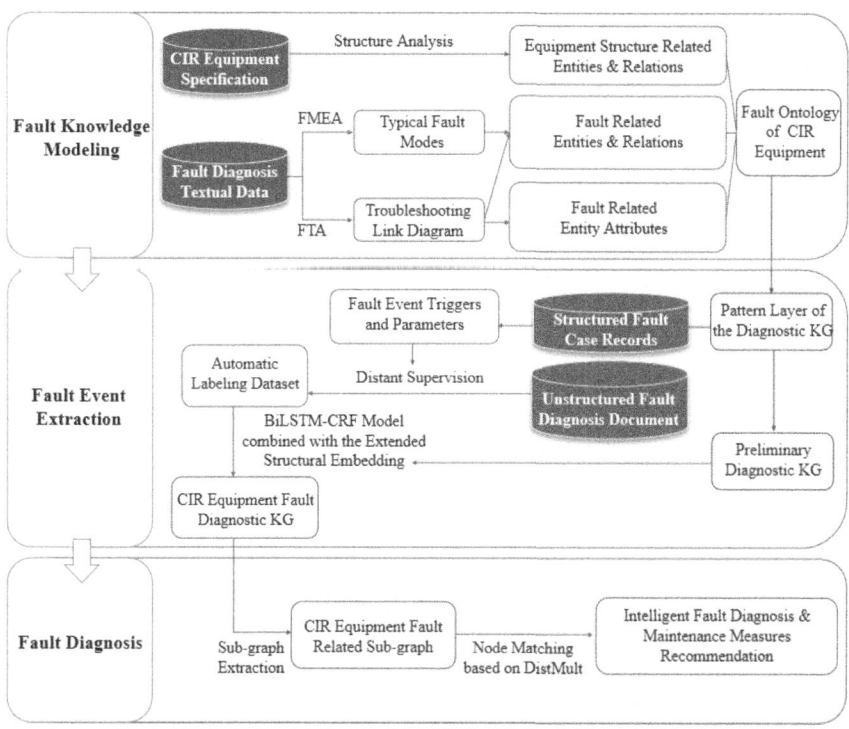

Fig. 1. An Abstract overview of the proposed framework

Equipment Structure Analysis. Railway CIR equipment has a complex mechanical structure with many decentralized yet centrally-controlled modules and components. Each functional unit performs dedicated duties and cooperates under host control, collectively realizing overall CIR functions.

According to the equipment specification, there are four main functional modules in CIR equipment: Host computer, Man-Machine Interface (MMI), Locomotive safety information comprehensive monitoring (TAX) device, and connection components. The detailed structures and functions of each module are then analysed. Drawing on these structural and functional analyses, a hierarchy diagram is presented in Fig. 2 to describe the physical structure of CIR equipment.

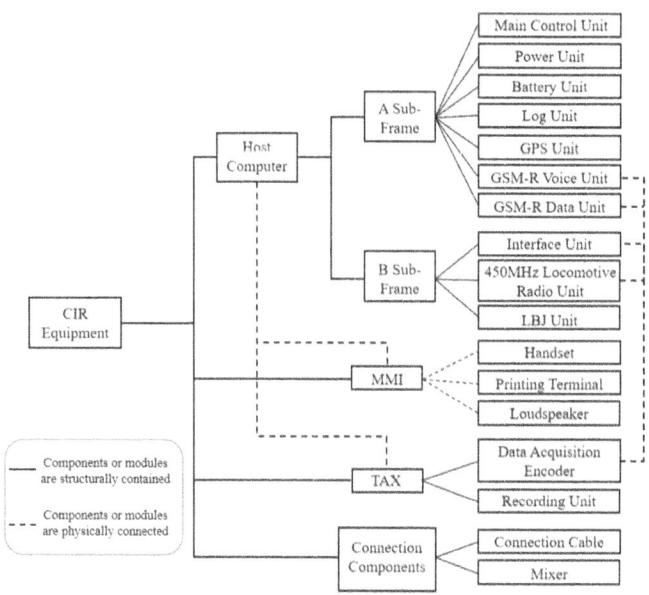

Fig. 2. CIR Equipment Structure Hierarchy Diagram

Equipment Fault Mechanism Analysis. According to the maintenance manual and fault case records, the fault mechanism of CIR equipment is obtained using Failure Mode and Effects Analysis (FMEA) and fault tree analysis (FTA).

FMEA includes failure mode analysis, failure cause analysis, failure impact analysis, failure detection method analysis and compensation measures analysis. Referring to operating instructions and manuals, five main fault types are identified: functional, display, transmission, parameter, and physical. After that, fault case examples are qualitatively analyzed to identify related modules, components, phenomena and possible causes. Table 1 shows an example of the FMEA process of the abnormal transmission fault.

After identifying the typical fault modes and corresponding causes of the equipment, this study also utilizes FTA to decompose the underlying cause of the fault. Figure 3 is an example of the troubleshooting measures and criteria for railway CIR equipment functional faults.

Table 1. FMEA process of Railway CIR Equipment Abnormal Transmission Fault

Module	Component	Phenomenon	FMEA Analysis		
			Human	Device	Environment
A Sub-frame	GSM-R Unit	GSM-R cannot initiate or receive a call	SIM card is not installed properly	1. Antenna failure 2. Communication abnormal between units	
B Sub-frame	450 MHz Locomotive Radio Unit	450 MHz Radio cannot call and receive		1. Communication abnormal between units 2. Failure of B sub-frame/450 MHz radio	No GSM-R Network

Entity, Relation and Attribute Definition. After conducting analyses of the equipment structure and fault mechanism, the pattern layer of the diagnosis KG could be defined. Here the event KG rather than general KG was chosen to store the fault knowledge for the following two reasons. Firstly, it is difficult to model the event as the first-class entity in general KG because of the rich element information included. While in the event KG, events could be easily modeled as first-class entities. Secondly, causal and dependent relationships between fault events cannot be directly inferred from a general KG structure as the edge only indicates associative connections. However, the event KG provide a natural and explicit representation of the causal relationships between fault events which are essential to equipment fault diagnosis.

According to the equipment structure analysis, the equipment structure-related entities in diagnosis KG could be divided into two levels: module and component. The module entities are A sub-frame, B sub-frame, MMI, TAX, and connecting components. The component entities are sub-class of corresponding modules. According to the fault mechanism analysis, fault-related entities could be divided into four parts: phenomenon, cause, measure and maintenance personnel.

After defining entities, relational and data attributes are introduced to describe associations between entities. Regarding the equipment structure and fault mechanism analysis, the relational attribute includes "be parts of", "occur in", "lead to", and "be solved by". The data attribute includes "judgment criteria", "equipment parameter", and "equipment status".

Based on the above definition, the pattern layer of the diagnosis KG is constructed. The results of the structure and fault mechanism analysis can be

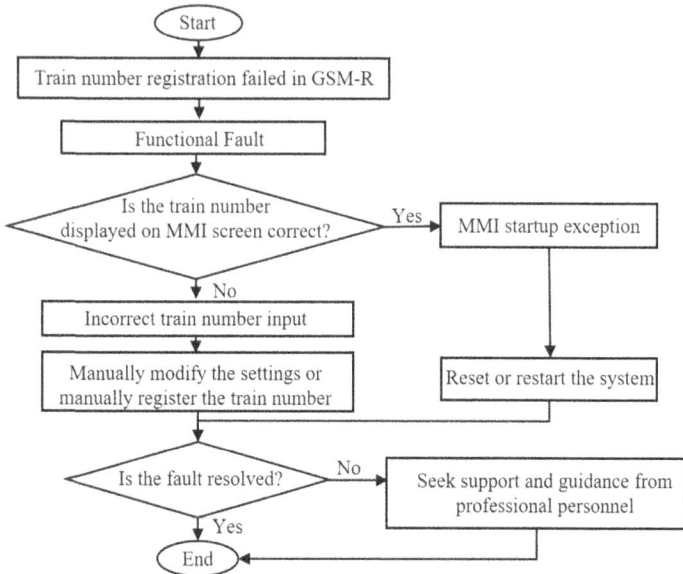

Fig. 3. Troubleshooting Link Diagram of Functional Fault

directly mapped to the data layer. After that, a preliminary diagnosis KG is finished.

4.2 Fault Event Extraction

Currently, railway CIR equipment lacks sensor data and structured fault case records - with most information available as unstructured text without manual annotations. To address these shortcomings, distant supervision (DS) method could be utilized by aligning a given structured knowledge base with a big text corpus to extract fault event [28]. This study extends the framework proposed in [29] for financial event extraction to the equipment fault diagnosis domain.

Automatic Data Labeling Technique. First, an automatic labeling technique is developed by constructing a dictionary of event triggers and key parameters. Distant supervision approach was built on the assumption that given an existing triple fact r(e1, e2) in KB, where r is the relation type between entity pair (e1, e2), all sentences containing e1 and e2 will be annotated as the relation type r. This idea can be extended to the event extraction. The dictionary of event triggers and key parameters are defined for each type of fault-related entity (phenomena, causes, measures and corresponding sub-class). It is worth noting that the dictionary is continuously updated and improved in collaboration with CIR equipment experts. Table 2 shows part of the dictionary.

According to the dictionary, triggers and key parameters are automatically labeled from large textual corpus (maintenance manual). After that, structured

Table 2. Typical Event Triggers and Key Parameters

Event Type	Event sub-class	Triggers	Key Parameters
Phenomenon	Functional Fault	Invalid, inoperable	Current equipment status, fault object, fault component, fault module and discoverer
	Display Fault	Blinking, white screen	
	Transmission Fault	Unable to connect to	
	Parameter Fault	Too high/low	
	Physical Fault	Appearance damaged	
Cause	Connection Abnormal	Abnormal connection	Fault object, fault component, fault module and fault personnel
	Component Abnormal	Out of control	
	Physical Damage	Aging, not durable	
	System Factor	Crash, system failure	
	Human Factor	Improper operation	
	Environment Factor	Network is not covered	
Measure	Replace Component	Replace, replace	Test benchmark, test parameters, maintenance objects, maintenance parts, maintenance modules
	Adjust Setting	Modify, adjust	
	Repeat	Retry, reinstall	
	Restart	Restart, reset	
	Factory Maintenance	Return to factory	

fault records can be mapped to unstructured diagnosis data, identifying event references and identifying corresponding triggers and parameters to generate sentence level labeled data. To validate the quality of the automatic labeling dataset, automatic labels are compared to manual labels with Fleiss' Kappa coefficients. As a result, automatic labels show high consistency with manual labels as Fleiss' Kappa exceeds 0.61 in phenomena, causes and measures - indicating reliability of this dataset.

An Event Extraction Model Incorporating the Extended Structural Embedding. After constructing the automatic labeling dataset, the next task is proposing an event extraction model to intelligently extract fault events, which could identify fault information contained in unstructured documents and supplement them to the preliminary fault diagnosis KG. KG could provide a concise and intuitive abstraction for various domains, from where latent semantic and structural information can be captured effectively [30]. However, previous studies did not fully utilize this advantage. This study proposes an event extraction model incorporating the extended structural embedding.

Structural Node Embeddings on Knowledge Graph (KGstruc2vec). Given an undirected network, struc2vec can automatically learn latent representations for nodes based on structural similarities [31]. Let G=(V,E) denote the network consisting of vertex set V and edge set E. Let $f_k(u, v)$ represents the structural distance between u and v when considering k-hop neighborhoods, where $k = 1, 2, ...k^*$ and $k^* \leq \delta$. In particular, the $f_k(u, v)$ is defined as:

$$f_k(u, v) = f_{k-1}(u, v) + g(R_k(u), R_k(v)) \tag{1}$$

where $k \geq 0, f_1 = 0$, $R_k(u)$ and $R_k(v)$ denote the degree sequences of the nodes at distance k from node v and u respectively. $g(D_1, D_2)$ measures the distance between the ordered degree sequence D1 and D2, which is calculated with Dynamic Time Warping (DTW) approach [32]. After getting the structural distance, the struc2vec construct a multilayer weighted graph to encode the structural similarity between nodes. Each layer consists of a weighted undirected complete graph with node set V and C_n^2 edges. Then random walk algorithm is applied to generate context for nodes. Finally, the struc2vec applies the random walk and skip-gram model to learn the representations of nodes.

However, the struc2vec algorithm still has limitations. First, struc2vec is unable to handle directed graphs [33]. Second, when applying this algorithm to generate structural embeddings of nodes in KGs, it fails to account for the semantic differences between edges. KGstruc2vec was proposed to address these shortcomings.

We applied a multi-dimensional variant of DTW to represent the extended structural embeddings. Considering a KG which includes n types of relation, the modified distance function is:

$$g(R_k(u), R_k(v)) = g_{(}R_{r1,in,k}(u), R_{r1,in,k}(v)) + g_{(}R_{r1,out,k}(u), R_{r1,out,k}(v))$$
$$+... + g_{(}R_{r1,in,k}(u), R_{r1,in,k}(v)) + g_{(}R_{rn,out,k}(u), R_{rn,out,k}(v)) \tag{2}$$

After getting the distances, we constructed a multi-layer weighted graph to encode the structural similarity between nodes and generated the node embeddings with the random walk and skip-gram language model. We have also improved the efficiency of the algorithm by calculating distances between a limited set of node pairs and constraining the depth of the random walk process.

BiLSTM-CRF Event Extraction Model Incorporating the Extended Structural Embedding. As BiLSTM-CRF performs well on natural language processing (NLP) tasks, this study applied it to extract fault events from automatic labeling dataset. The extended structural embedding obtained from the preliminary diagnosis KG would be introduced into the event extraction model. Additionally, in order to reduce ambiguity from domain knowledge in word segmentation, Word Level Features (WLF) would also be introduced into the BiLSTM-CRF sequence labeling model. The word embedding used in this study are shown in Fig. 4, where CE, WE, SE mean character, word and extended structural embedding respectively.

Relying on the fault event extraction results, this study identified fault events mentioned in unstructured maintenance documents, and combined the related fault phenomena, causes and maintenance measures in output. After that, the

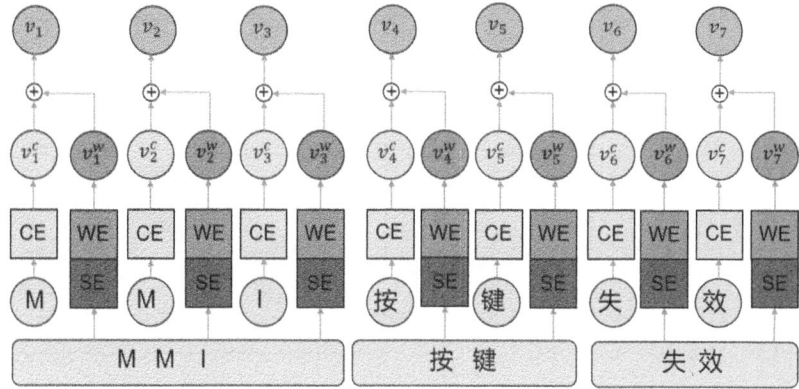

Fig. 4. Representation Combined with Word-level Features

results would be supplemented to the diagnosis KG referring to the pattern layer definition. Then the complete diagnosis KG is finished.

4.3 Fault Diagnosis

Based on the complete CIR Equipment Fault diagnosis KG (Fig. 5), this study transforms the diagnosis task into a matching problem between the phenomenon and measure node solved by DistMult. When inputting a fault phenomenon description, all possibly related judgment criteria are extracted and presented to operators to support the troubleshooting. The framework would then provide decision support according to operators' experience level and operating conditions. For experienced operators with adequate maintenance capabilities, it provides diagnosis criteria, collects feedback to accurately locate the phenomenon, and recommends detailed maintenance measures; While for operators without the adequate capability or sufficient condition to troubleshoot the fault, it extracts fault-related sub-graphs from the diagnosis KG to help operators understand the cause and provide all possible advice for them to try. In a word, this framework could provide the diagnosis result in an interpretable and clear way to support decision.

5 Experiment

5.1 Data Description

This study referred to the fault maintenance ledger data (including 2,797 maintenance records in the two years from 2016 to 2018) obtained from the railway CIR equipment maintenance corporations. Considering the verification needs, 80% ledger data (2238 cases) were selected for the construction of diagnosis KG, and the remaining 20% ledger data were used for the subsequent empirical test to prove the effectiveness of the framework.

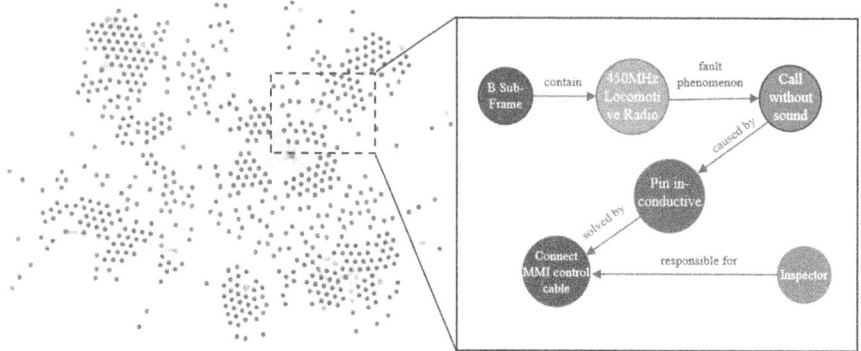

Fig. 5. Railway CIR Equipment Fault diagnosis KG

5.2 Experiment Result

In this study, the proposed BiLSTM-CRF-WLF event extraction model with the extended structural embedding, namely BiLSTM-CRF-WLF-SE, was applied to the automatic fault labeling dataset of railway CIR equipment. For comparison, widely used sequence labeling models were also applied. They are BiLSTM, BiLSTM-WLF, BERT and BERT-CRF. In ablation experiments, BiLSTM-CRF and BiLSTM-CRF-WLF models without the extended structural embedding were applied to verify the effectiveness of word-level features and structural embeddings. The precision, recall, and F1 values are reported in Table 3.

Table 3. Event Extraction Comparative Experimental Results

	CIR Automatic Labling Dataset		
	Precision	Recall	F1 score
BERT	48.21	50.44	49.30
BERT-CRF	63.06	43.07	51.19
BiLSTM	57.74	36.98	45.09
BiLSTM-CRF	53.52	44.12	48.37
BiLSTM-CRF-WLF	53.35	49.15	51.16
BiLSTM-CRF-WLF-SE	52.95	55.24	54.07

When applying the BiLSTM-CRF-WLF-SE model to the fault dataset, the results showed that the BiLSTM-CRF model incorporating the extended structural embedding and word-level feature achieved the best performance. The statistical analysis was then conducted to prove the significance of the improvement. The results of the paired t-test show that the improvement in classification performance achieved by BiLSTM-CRF-WLF-SE compared to the best baseline approach (BiLSTM-CRF-WLF) is statistically significant with a p-value < 0.05.

A robustness test was conducted to evaluate the model's application on other public datasets. The plane crash dataset proposed in [34] was used. This English corpus from Wikipedia covers 193 news texts about plane crashes after 1987. It is labeled by entity-level tags generated from distant supervision, focusing on extracting passenger numbers, aircraft types, and other crash facts. The BiLSTM-CRF-WLF-SE event extraction model and other baseline models were applied to this dataset.

Table 4. Event Extraction Comparative Experimental Results

	Plane Crash Dataset		
	Precision	Recall	F1 score
BERT	57.85	55.48	56.64
BERT-CRF	70.67	49.38	58.14
BiLSTM	65.29	43.68	52.34
BiLSTM-CRF(-WLF)	60.02	55.01	57.41
BiLSTM-CRF-WLF-SE	59.54	60.76	60.14

The results in Table 4 show that the BiLSTM-CRF model, which combines the word-level feature and extended structural embedding, still achieves the best classification results on the replacement dataset.

6   Fault Diagnosis Application

An empirical test was carried out on the remaining 20% of historical fault records (totally 559 cases) based on KEFD. With the input of inspection feedback according to diagnosis criteria, the accuracy of recommending the optimal maintenance measure reached 71.91%. Even for operators without the adequate capability or sufficient condition to troubleshoot the fault, the probability of providing the sub-graph contained the optimal solution reached 91.59%, which proves the effectiveness of the proposed framework.

7   Conclusion

This study presents a knowledge-based framework for intelligent CIR equipment fault diagnosis (KEFD), which realizes a transformation from small fault samples and unstructured data to a well-structured knowledge model. To evaluate the framework, we applied KEFD to real-world maintenance case records from railway CIR equipment. KEFD first defined diagnosis entities, relations, and attributes through the equipment structure and fault mechanism analysis, transforming textual data into an event-oriented KG served for fault diagnosis. Then KEFD leveraged a novel distant supervision method incorporating the extended structural embedding with domain knowledge to automatically extract fault events and finished the diagnosis KG. Finally, KEFD transformed the diagnosis task into a matching problem between fault phenomenon and recommended measure nodes based on DistMult method. This framework demonstrates the potential of diagnosis KG in solving equipment fault diagnosis tasks.

This study has both theoretical and practical implications. Theoretically, existing industrial ontology construction research did not sufficiently integrate external equipment knowledge and unstructured fault records, which would lead to inadequate ontology coverage and missing of important equipment diagnosis knowledge. This study combined ontology theory with FMEA/FTA methods, constructing knowledge models for railway CIR equipment fault diagnosis systems and extracting related fault events based on distant supervision idea. Meanwhile, the knowledge modeling process proposed in this study has a strong cross-domain applicability. Practically, KEFD proposed in this study is able to provide operators with interpretable and clear decision support when diagnosing faults. It could also realize efficient knowledge storage, retrieval and utilization to support practical accident risk identification and rectification, serving for the realistic needs of railway industry.

This research has several limitations that need improvement in the future work. Firstly, as CIR equipment lacks sensor data, failure influence analysis was not conducted in FMEA process. More historical maintenance records should be collected in the future to obtain important FMEA indicators like failure severity, frequency, and risk sequence degree to improve diagnosis accuracy. Secondly, the event extraction process in this study requires KG-based data, increasing application costs as most fault datasets are unstructured text. Syntax analysis could be applied in the future to help capture node relationships from unstructured data, which would further improve the applicability of the framework across diagnosis scenarios.

Acknowledgement. This study was funded by National Natural Science Foundation "Research on Recommendation System Based on Knowledge Map and Link Prediction and Its Application in Equipment Health Management" (grant number: 72071194).

Disclosure of Interests. The authors have no competing interests to declare that are relevant to the content of this article.

References

1. Athavale, J., Baldovin, A., Paulitsch, M.: Trends and functional safety certification strategies for advanced railway automation systems. In: 2020 IEEE International Reliability Physics Symposium (IRPS), pp. 1–7. IEEE (2020)
2. Barnatt, N., Jack, A.: Safety analysis in a modern railway setting. Saf. Sci. **110**, 177–182 (2018)
3. Guo, Q., Li, Y., Song, Y., et al.: Intelligent fault diagnosis method based on full 1-D convolutional generative adversarial network. IEEE Trans. Ind. Inf. **16**(3), 2044–2053 (2019)

4. Fan, J., Fan, J., Liu, F.: Research of K-MEANS analysis model on high-speed railway CIR device maintenance. In: 2017 IEEE International Conference on Prognostics and Health Management (ICPHM), pp. 199–204. IEEE (2017)
5. Qu, J., Liu, F., Meng, H.: A method for CIR fault diagnosis based on improved tri-training in big data environment. In: 2018 IEEE 3rd International Conference on Cloud Computing and Big Data Analysis (ICCCBDA), pp. 213–218. IEEE (2018)
6. Chi, Y., Dong, Y., Wang, Z.J., et al.: Knowledge-based fault diagnosis in industrial Internet of Things: a survey. IEEE Internet Things J. **9**(15), 12886–12900 (2022)
7. Feng, L., Zhao, C.: Fault description based attribute transfer for zero-sample industrial fault diagnosis. IEEE Trans. Ind. Inf. **17**(3), 1852–1862 (2020)
8. Mattos, W.J., Araújo, R.E.: Fault diagnosis in DC-DC power converters based on parity equations. In: 2020 International Young Engineers Forum (YEF-ECE), pp. 13–18. IEEE (2020)
9. Jalayer, M., Orsenigo, C., Vercellis, C.: Fault detection and diagnosis for rotating machinery: a model based on convolutional LSTM, Fast Fourier and continuous wavelet transforms. Comput. Ind. **125**, 103378 (2021)
10. Liu, B., Wu, J., Yao, L., et al.: Ontology-based fault diagnosis: a decade in review. In: Proceedings of the 11th International Conference on Computer Modeling and Simulation, pp. 112–116 (2019)
11. Li, W., Li, H., Gu, S., et al.: Process fault diagnosis with model-and knowledge-based approaches: advances and opportunities. Control. Eng. Pract. **105**, 104637 (2020)
12. Xu, K., Kong, X., Wang, Q., et al.: A bearing fault diagnosis method without fault data in new working condition combined dynamic model with deep learning. Adv. Eng. Inform. **54**, 101795 (2022)
13. Gao, Y., Gao, L., Li, X., et al.: A hierarchical training-convolutional neural network for imbalanced fault diagnosis in complex equipment. IEEE Trans. Ind. Inf. **18**(11), 8138–8145 (2022)
14. Wu, G.: Design on fault diagnosis expert system for railway signal equipment. In: 2018 6th International Conference on Machinery, Materials and Computing Technology (ICMMCT 2018), pp. 36–41. Atlantis Press (2018)
15. Liu, C., Yang, S.: Using text mining to establish knowledge graph from accident/incident reports in risk assessment. Expert Syst. Appl. **207**, 117991 (2022)
16. Reinanda, R., Meij, E., de Rijke, M.: Knowledge graphs: an information retrieval perspective. Found. Trends Inf. Retr. **14**(4), 289–444 (2020)
17. Yang, Y., Chen, Z., Yan, J.: Multi-source heterogeneous information fusion of power assets based on knowledge graph. In: IEEE International Conference on Service Operations and Logistics, and Informatics (SOLI), pp. 213–218. IEEE, New York (2019)
18. Chhim, P., Chinnam, R.B., Sadawi, N.: Product design and manufacturing process based ontology for manufacturing knowledge reuse. J. Intell. Manuf. **30**(2), 905–916 (2019)
19. Ding, Y., Xu, W., Liu, Z., et al.: Robotic task oriented knowledge graph for human-robot collaboration in disassembly. Procedia CIRP **83**, 105–110 (2019)
20. Yan, H., Yang, J., Wan, J.: KnowIME: a system to construct a knowledge graph for intelligent manufacturing equipment. IEEE Access **8**, 41805–41813 (2020)
21. Fan, S., Liu, X., Chen, Y., et al.: How to construct a power knowledge graph with dispatching data? Sci. Program. **2020**, 1–10 (2020)
22. Xue, L., Yao, X., Zheng, Q., et al.: Study on the construction method of fault knowledge map of on-board equipment for high-speed train control. J. Railway Sci. Eng. **20**(1) (2023)

23. Wang, X., Guo, Z., Wang, H., et al.: On-line fault diagnosis of motor based on multi-level fault tree analysis. Mod. Manuf. Eng. (11), 137–143, 39 (2023)
24. Bischof, S., Schenner, G.: Rail topology ontology: a rail infrastructure base ontology. In: Hotho, A., et al. (eds.) ISWC 2021. LNCS, vol. 12922, pp. 597–612. Springer, Cham (2021). https://doi.org/10.1007/978-3-030-88361-4_35
25. Yang, C., Zheng, Y., Tu, X., et al.: Ontology-based knowledge representation of industrial production workflow. Adv. Eng. Inform. **58**, 102185 (2023)
26. Zangeneh, P., McCabe, B.: Ontology-based knowledge representation for industrial megaprojects analytics using linked data and the semantic web. Adv. Eng. Inform. **46**, 101164 (2020)
27. Li, R., Mo, T., Yang, J., et al.: Ontologies-based domain knowledge modeling and heterogeneous sensor data integration for bridge health monitoring systems. IEEE Trans. Ind. Inf. **17**(1), 321–332 (2020)
28. Mintz, M., Bills, S., Snow, R., et al.: Distant supervision for relation extraction without labeled data. In: Proceedings of the Joint Conference of the 47th Annual Meeting of the ACL and the 4th International Joint Conference on Natural Language Processing of the AFNLP, pp. 1003–1011 (2009)
29. Yang, H., Chen, Y., Liu, K., et al.: DCFEE: a document-level Chinese financial event extraction system based on automatically labeled training data. In: Proceedings of ACL System Demonstrations, pp. 50–55 (2018)
30. Hogan, A., Blomqvist, E., Cochez, M., et al.: Knowledge graphs. ACM Comput. Surv. (CSUR) **54**(4), 1–37 (2021)
31. Ribeiro, L.F.R., Saverese, P.H.P., Figueiredo, D.R.: struc2vec: learning node representations from structural identity. In: Proceedings of the 23rd ACM SIGKDD International Conference on Knowledge Discovery and Data Mining, pp. 385–394 (2017)
32. Deriso, D., Boyd, S.: A general optimization framework for dynamic time warping. Optim. Eng. **24**(2), 1411–1432 (2023)
33. Steenfatt, N., Nikolentzos, G., Vazirgiannis, M., Zhao, Q.: Learning structural node representations on directed graphs. In: Aiello, L.M., Cherifi, C., Cherifi, H., Lambiotte, R., Lió, P., Rocha, L.M. (eds.) COMPLEX NETWORKS 2018. SCI, vol. 813, pp. 132–144. Springer, Cham (2019). https://doi.org/10.1007/978-3-030-05414-4_11
34. Reschke, K., Jankowiak, M., Surdeanu, M., et al.: Event extraction using distant supervision In: Language Resources and Evaluation Conference (LREC 2014), pp. 4527–4531. Association for Computational Linguistics, Louisiana (2014)

Author Index

X. Tang et al. (Eds.): KSS 2024, CCIS 2269, pp. 345–346, 2025.
https://doi.org/10.1007/978-981-96-0178-3

GPSR Compliance

The European Union's (EU) General Product Safety Regulation (GPSR) is a set of rules that requires consumer products to be safe and our obligations to ensure this.

If you have any concerns about our products, you can contact us on ProductSafety@springernature.com

In case Publisher is established outside the EU, the EU authorized representative is:

Springer Nature Customer Service Center GmbH
Europaplatz 3
69115 Heidelberg, Germany

The manufacturer's authorised representative in the EU is Springer
Nature Customer Service Centre GmbH, Europaplatz 3, 69115 Heidelberg,
Germany. If you have any concerns regarding our products, please
contact ProductSafety@springernature.com

Printed and bound by CPI Group (UK) Ltd, Croydon, CR0 4YY

05/05/2026

02103581-0003